Praise for *Making War to Keep Peace*

"Those who assume that neoconservatism is a platform with all its planks shared by those to whom the label is affixed, not to say those who knew Kirkpatrick only by reputation, will find surprises in store in the pages of *Making War to Keep Peace*, her tour d'horizon of American policy since the 1980s."
—*New York Times Book Review*

"*Making War to Keep Peace* discloses to the public that one of America's most insightful geopolitical thinkers opposed the war in Iraq. While fully supporting the invasion of Afghanistan, she concluded Iraq lacked the preconditions for a successful transformation. Liberal societies evolve. The process can be helped but not imposed. The situation in Iraq may not be hopeless, but the U.S. would see with greater clarity if more individuals such as Kirkpatrick inhabited the inner circle."
—*Richmond Times-Dispatch*

"In her posthumously published book, *Making War to Keep Peace*, Reagan's UN ambassador, Jeane Kirkpatrick, clearly draws the distinction between the Reagan Doctrine—which called for a thorough, prudential analysis of the costs and benefits to U.S. interests in determining when to use means other than direct U.S. military intervention to support friendly forces within unfriendly regimes—and the Bush Doctrine, which used U.S. forces to directly invade Iraq without having taken a methodical analysis of the potential pitfalls of the invasion's aftermath, in which U.S. troops would be charged with the task of trying to rebuild a nation in a culture they know little about and that has no tradition of representative government."
—**TERRY JEFFREY**, nationalreview.com

"As one of the most thoughtful of Ronald Reagan's foreign policy advisors, Jeane Kirkpatrick recognized that it is not only important to know our history, but crucial to understand the lessons of that history in order to formulate a wise foreign policy course. Leading the reader through U.S. incursions in Kuwait, Somalia, Haiti, the Balkans, Kosovo, and finally Afghanistan and Iraq, Kirkpatrick explains how and when the use of military force is justified and when it is not. *Making War to Keep Peace* is a must-read for those who not only want to know what those incursions were all about, but also who want to understand the lessons learned—and not learned."
—**ALFRED S. REGNERY**, publisher, *The American Spectator*

"Jeane Kirkpatrick's legacy as scholar and diplomat is further enhanced by this vital new study, *Making War to Keep Peace*. It is an honest and penetrating analysis of the huge penalties nations will pay when the courage to do what must be done is lacking and when flawed strategies are not revised. She cautions us to 'unflinchingly acknowledge our mistakes as well as celebrate our successes' and 'learn from both.' Trenchant and fearless in her analysis, Jeane Kirkpatrick does not hesitate to assign blame for policy failures. It was this fundamental honesty that made her such a valuable advisor to presidents."
—RICHARD V. ALLEN, former national security advisor

"In her unflinching search for truth, Jeane Kirkpatrick can never be considered a traditional 'conservative Republican' or even a 'liberal Democrat' as we use the terms today. Kirkpatrick's willingness to speak the truth in the world of power was her signature—until the last days of her life, and even now in this book. In the 1970s she published in *Commentary* her groundbreaking article 'Dictatorship and Double Standards,' severing ties with her Democratic background at profound personal loss. This single article alone has had historic impact on the world for decades. . . . Armed with her words and straight talk, Kirkpatrick's work at the UN earned her the accolade from Mikhail Gorbachev as 'the most dangerous woman in the world.' Here, in *Making War to Keep Peace*, Kirkpatrick challenges leaders to be willing to make war—but to remember the price of keeping peace before and after war is waged. She instructs how to exercise might—and restraint—in perilous and often-misguided efforts in peacekeeping and nation building. Kirkpatrick helped Ronald Reagan exercise the moral authority that brought down the Berlin Wall and Iron Curtain without firing a single shot. Now she has delivered us another historic work with a worldview which, as it was through her life, will be revered by many, even if its advice is considered too costly to be dared by more than a few."
—JACK KEMP, former U.S. representative and
secretary of Housing and Urban Development

About the Author

JEANE J. KIRKPATRICK was the U.S. Ambassador to the United Nations from 1981 to 1985 and a member of the National Security Council during the Reagan administration. She was also a senior fellow at the American Enterprise Institute, the founder of Empower America, and a professor of government at Georgetown University. She died in December 2006.

MAKING
WAR
TO
KEEP
PEACE

MAKING
WAR
TO
KEEP
PEACE

Trials and Errors in American
Foreign Policy from Kuwait to Baghdad

Jeane J. Kirkpatrick

HARPER ● PERENNIAL

NEW YORK ● LONDON ● TORONTO ● SYDNEY ● NEW DELHI ● AUCKLAND

HARPER ● PERENNIAL

A hardcover edition of this book was published in 2007 by HC, an imprint of Harper-Collins Publishers.

HarperCollins books may be purchased for educational, business, or sales promotional use. For information please write: Special Markets Department, HarperCollins Publishers, 10 East 53rd Street, New York, NY 10022.

FIRST HARPER PERENNIAL EDITION PUBLISHED 2008.

Designed by Kris Tobiassen

Library of Congress Cataloging-in-Publication Data is available upon request.

ISBN 978-0-06-137365-7

08 09 10 11 12 ID/RRD 10 9 8 7 6 5 4 3 2 1

DEDICATED TO MY GRANDCHILDREN,
LAURA, JONATHAN, LILY,
MICHAEL, KATHERINE, AND RYAN:

*Freedom and democracy are now
yours to honor and protect.*

CONTENTS

MAKING
WAR
TO
KEEP
PEACE

INTRODUCTION

The end of the cold war, which began with the dramatic fall of the Berlin wall in 1989, freed the United States and Western Europe from a major military threat for the first time since Hitler marched into Poland, Austria, and Czechoslovakia before the outbreak of World War II. It was, as former Soviet president Mikhail Gorbachev described, "our common victory."[1] Freedom and democracy swept Eastern Europe, enabling Western Europe to concentrate on the construction of a united Europe, on other new alliances, and on the creation of new patterns of international politics, and that is what they have done.

For a brief period, Americans found themselves deeply involved in the world at a time of relative peace, where no imminent catastrophe and no powerful enemy threatened our civilization. Or so we thought. This new era for Americans did not include a need for new foreign policy goals.

At a fundamental level, however, U.S. foreign policy goals remained unaffected by the end of the cold war. Our commitment had always been to the spread of democracy in the world. We had been promoting development, peace, and prosperity worldwide all along. Contrary to popular impression, we had never really been consumed by the sole task of "containing" the former Soviet Union.

Even during the cold war, in accordance with our fundamental goals and values, we had worked to foster peace between Israel and her Arab neighbors; sought to prevent dominance of the Persian Gulf by a hostile power; and supported the survival of Taiwan, the freedom of Afghanistan, and the end of apartheid in South Africa. We had concerned ourselves with rising living standards, expanding exports, promoting world trade, and discouraging human rights violations in China, Burma, North Korea, and elsewhere. We had labored to prevent aggression and proliferation of nuclear technology, to resolve regional conflicts, and to end wars. All of these were important goals of U.S. policy throughout the cold war.

Dealing with the Soviet Union and communism had been our most urgent task because only the Soviet Union had the power and the will to dominate significant numbers of other nations. Communism had been an expanding imperial tyranny, and Marxism, the most powerful intellectual paradigm of the century, provided its justification. The United States and its NATO allies were seen by the Soviet Union as an obstacle to the achievement of Soviet goals. The urgent task of containing this national power was the focus of much attention, and most American goals in the post–cold war world are those we pursued during the cold war. But we never sought to conquer the world or the USSR itself. We never threatened to "bury" them. There never were two superpowers locked liked scorpions in a bottle in a twilight struggle to a bitter end.

During and after the cold war, three goals and values have provided the foundation for U.S. policy. The first goal was, is, and, I think, should be to preserve our own freedom, independence, and well-being. This requires preserving the integrity of our institutions and maintaining a capacity to defend the United States itself from potential adversaries and the weapons of the era. That is why we badly need an effective defense against missiles.

The second goal of U.S. policy was, is, and should be to help (in ways consistent with our resources) to preserve and to expand the number and vitality of democratic governments in the world, because they share our civilization, respect the rights of their own citizens and neighbors, and contribute to the sum of peace and well-being in the world.

The third goal of American foreign policy was, is, and should be to prevent (or help to prevent) violent expansionist leaders from gaining control of the governments of major states.

While geographic, economic, demographic, and historical factors influence the policies of governments, ultimately a nation's foreign policy flows from the character of its regime, its culture, and the purposes of its political elite. Those purposes reflect the principles, beliefs, habits, values, and goals of its political class.

We have had ample opportunity in this century of wars and revolutions to observe what happens when violent elites, who espouse coercive ideologies, gain access to the resources of states. They start with murder and denial of freedom in their own states, and move on to war, which may spill over to their neighbors and sometimes to genocide.

Under Joseph Stalin's dictatorship, the deadly purge period in the Soviet Union was directed first at his citizens; then his tyranny turned outward and spread to the Baltic and Eastern European states. The stateless Osama bin Laden's al Qaeda and its allied groups continue to kill and wreak havoc in the United States, Europe, and the Middle East.

It is important to recall that each of the violent antidemocratic movements of the century came as a surprise: Bolshevism, Fascism, Nazism, Chinese Communism (especially its Cultural Revolution), the "killing fields" of Cambodia, and the fanatical ayatollahs. Each shocked an unsuspecting, disbelieving world with its explosive violence and aggressive policies. Each was led by "new men" unknown to the then dominant political class. No one knew where or when the next violent leader or movement might appear. Could the Federal Republic of Germany or any other country in Europe breed a group as violent, obsessive, and skilled in the achievement of power as the Nazis? I don't think so. Could a transitional Russia prove the birthing room for a movement as violent and as skilled in uses of power as the Bolsheviks who abruptly terminated Russia's transitional democracy of 1917? I don't think so. Could an aggressive, expansionist, violent leader in Serbia unleash mass murder once again in the very heart of Europe? It has already happened. However, in this century, not all the surprises have been bloody or hopeless. Just as no one anticipated any of the previous disasters, no one anticipated the sudden abandonment of Marxism or turn to democracy and free markets in Central and Eastern Europe and elsewhere. As President Ronald Reagan articulated, "It is the march of freedom and democracy which will leave Marxism-Leninism on the ash heap of history as it has left other tyrannies which stifle the freedom and muzzle the self-expression of the people."[2] Seven years after this statement, his words came true.

We cannot protect ourselves or others against the resurgence of aggressive powers or the reoccurrence of evil unless we face the fact that tyranny and war have the same source—*in persons who use force to expand their control of others*. These individuals use force to gain power inside and outside their own countries. When would-be dictators manage by one means or another to get to the top of government, they seek to bend that government's resources to their own purposes to malevolently maximize their own power. They destabilize existing institutions—often ruthlessly—and create wars. The evidence against these interlopers has been plentiful in this century. We have watched expansionist dictatorships make war in Europe and Asia. We have seen the ambitions of Saddam Hussein exposed in his wars against his people and his neighbors. We have the legacy of Slobodan Milošević's ceaseless efforts to extend his own power by expanding Serb control over the whole of the former Yugoslavia. We continue to confront the Taliban in Afghanistan.

Such men and policies, such wars and revolutions, unsettle the foundation of national stability, which relies on principles of legitimacy and patterns of politics. Ruptures of regimes and changing principles of legitimacy advance the ambitions of these violent elites, helping them impose themselves by force. This happened in the periods after World War I and after World War II, and it happened with the end of the cold war.

After the cold war, the urge to conquest quickly reasserted itself across the world. New and legitimate political institutions were weakly rooted and often unable to cope with the resurgent ambition of former elites who remain rivals for power. In Africa, war and famine resulted. In Europe, Slobodan Milošević's determined aggression pressed forward at the expense of smaller states on the new frontiers of Kosovo, Croatia, Slovenia, and Bosnia-Herzegovina. Ethnic rivalries fed new fires on other borders in Ukraine, Moldova, Georgia, Azerbaijan, Armenia, and Russia. Sectarian violence continues to devastate post–Saddam Hussein's Iraq, challenging the future of Prime Minister Nouri al-Maliki's burgeoning government while the expanding reach of terrorist groups, such as Hezbollah, and the states that sponsor them, continues to threaten the national sovereignty of Lebanon, its neighbors, and elsewhere.

NEW WORLD ORDERS

The end of the cold war marked the third time in a century that the United States was confronted with the need and the opportunity to try to control violence, aggression, and war to create peace in Europe and the world. And for the third time in a century, the U.S. government—starting with the administration of George Herbert Walker Bush—and those who helped shape its foreign policy chose to treat the situation as a fundamentally international challenge, one in which international organizations and alliances should be primary arbiters of policy. The first Bush administration envisioned what Bush himself called a "new world order," in which an unprecedented number of problems were conceived as transcending national boundaries, their solutions requiring collective action and administration through the United Nations.

This was the third time in the twentieth century that America had faced such a fundamentally new playing field in international relations, and the third time that our solution involved trying to control the outbreak of war through the use of contracts and peacekeepers—that is, to bring about a world without power. Influential Americans and Europeans imagined they might control or even eliminate war through international action and organization. As we shall see, none of these efforts has been successful.

One main reason these initiatives failed is that they relied on the idea that war could be effectively restrained by juridical means. That idea is peculiar to our time, in part because collective international efforts have proved unable or unwilling to take the decisive action required to constrain the power of violent, aggressive, expansionist leaders and regimes. In 1907, the Hague Convention first imposed limitations on the right to wage war. In 1919, the Covenant of the League of Nations obligated members of the League to submit "any dispute likely to lead to a rupture to arbitration or to inquiry by the Council [and] in no case to resort to war until three months after the award by the arbitrators or the report of the Council." In 1928, the Kellogg-Briand Pact stated that "the High contracting parties solemnly declare, in the name of their respective peoples that they condemn recourse to war for the solution of international controversies, and renounce it as an instrument of national policy in their relations with one another."

The creation of the League of Nations at the time had stimulated high hopes among persons of idealistic and internationalist bent around the world. Soon it proved necessary to face the truth that this organization was incapable of maintaining international peace. Treaties were violated and aggression went unpunished.

The failure of the League of Nations has often been explained for reasons other than a lack of will to meet aggression with force. However, when Italy's Benito Mussolini's troops charged into Ethiopia in 1935 as part of his plan to expand Italy's influence in Africa, neighboring League members sat idle while tens of thousands of defenseless Ethiopians perished. Despite watersheds in history such as this, when the failure of will among nations led to tragedies of epic proportion, the failure of the League of Nations is often blamed instead on the harsh terms of the Versailles Peace Treaty, which was said to have engendered a bitterness and desire for revenge, on the inadequate structure and powers of the League itself, and on the failure of great powers, especially the United States, to try hard enough to make the institutions work. But I believe the League failed when major European powers proved unable and/or unwilling to take the decisive action required to constrain the power of violent, aggressive, expansionist leaders and regimes in Russia, Italy, and Germany between 1917 and 1936.

The crowning achievement of the effort to constrain war by juridical means was, of course, the establishment of the United Nations and the promulgation of its Charter. In its preamble, the Charter declares its determination "to save succeeding generations from the scourge of war" by requiring in Article Two that members act in accordance with the principles of respect for the sovereign equality of all states, the peaceful resolution of disputes, the nonuse or threat of force, and nonintervention in internal affairs of others.

Yet the limits of such measures were demonstrated again in 1945, when the United Nations failed to protect the territorial integrity of Eastern Europe after Stalin demonstrated a will to conquest wholly incompatible with the provisions of the UN Charter.

Lessons from History

This book considers the United States' third try at a new world order—at constructing a world without power. Woodrow Wilson had proposed such a world after World War I. Franklin Roosevelt after World War II.

Roosevelt assured a joint session of the Congress on March 1, 1945, that the Yalta Conference spelled "the end of the system of unilateral action, exclusive alliances and spheres of influence, and balances of power and all the other expedients which have been tried for centuries and have always failed. We propose to substitute for all of these a universal organization in which all peace-loving nations will finally have a chance to join." Soon after, Roosevelt's high hopes and earnest plans were interrupted by his death and betrayed by Stalin's conquest of Eastern Europe.

But, after forty-five years of cold war, there came to power in the United States other presidents and administrations ready to act on Roosevelt's assumption and agenda. The administrations of George H. W. Bush and Bill Clinton also believed it should be possible to "end the system of unilateral action, exclusive alliances, and spheres of influence and rely on a universal organization to settle disputes among nations." They achieved office at a time, moreover, when the international environment was more favorable to international cooperation and collective action than at any time since 1946. The world seemed ready to try again to build a new world order, and America, as we had earlier in this century, seemed ready and willing to take the lead.

These two presidents' approaches to international strategy were different. George H. W. Bush, who was constantly briefed and kept abreast of foreign affairs, seemed to govern more from his gut than from methodical and analytical decision making—a trait that appears to have resurfaced in the administration of his son, George W. Bush. Bill Clinton was less interested in foreign affairs than his predecessor or successor, yet he did analyze the situations with which he was presented—so much so that some critics charged his administration with inaction on matters of foreign affairs. We shall see how our former and current presidents' different priorities, some noble, others perhaps naïve, have set the course for U.S. foreign policy and, inadvertently, guided us into some of the difficult challenges we face today.

When President George H. W. Bush spoke eloquently of a "new world order," he sought to demonstrate how collective security could produce peace. Toward the end of the Gulf War, he explained why he believed it had become realistic to think of such a new order and what it might look like. Its principal characteristic would be a long period of peace. Conflicts would be few, and they would be managed by peacekeepers operating on the basis of collective security and multinational effort.

No one expected that the end of the cold war would be a preface to something so far from this century-long dream. Americans expected global peace, but what did follow the fall of the Berlin Wall were multiple, small wars closely resembling the wars of the past, with the United States being drawn into military conflicts, sometimes unilaterally and other times in tandem with the United Nations, as it pursued its ends of peace again and again. This book reviews the process by which the U.S. government found itself embroiled in one conflict after another, confronting escalating costs in economic and human terms while examining our critical mistakes, our successes, and how they directly affect our future and our accountability to the world community.

Twelve years after the Wall tumbled down, on September 11, 2001, a threat long simmering in the margins of global events violently thrust its fury on America's soil and into the forefront of foreign policy. Again Americans were shaken out of the false sense that war as foreign policy had been retired because all strongmen had been defeated. Now we were confronted by yet another coercive ideology wearing yet another face. Today, more than ever, understanding the lessons learned *and not learned* from the past is crucial if we are to chart a wise foreign policy course for the future of our nation.

As President George W. Bush has said, "Freedom is once again assaulted by enemies determined to roll back generations of democratic progress. Some call this evil Islamic radicalism; others, militant Jihadism; still others, Islamo-fascism.[3] This global threat confronting us today tests and weakens the fragile foundation of the idyllic and hopeful dream that peace can be kept without making war."

1.

IRAQ INVADES KUWAIT

On August 2, 1990, Saddam Hussein's invasion of Kuwait shattered the peace and optimism of the summer. This was the first clear act of international aggression after the dramatic changes in the Soviet Union and Eastern Europe had ended the cold war. The United States, the Gulf States, and their allies were not ready for the invasion. To some Americans, it recalled Hitler's swift moves across Europe at the start of World War II and the consequences of appeasing an aggressor. To others, it recalled the Soviet Union's surprise invasion of Afghanistan in 1979 and the long, terrible war that followed.

The story of how President George H.W. Bush and the United States responded to this foreign policy challenge is a chronicle of the birth of the new world order. It raises the fundamental questions of how the United States decides what is in its national interest, when it should use military force, the nature of our relationship with the United Nations and the world, and how long our responsibilities to a nation or people persist after military intervention. The crisis also offers an object lesson on the danger of waiting for international consensus when time is of the essence.

President Bush's initial response resembled that of Harry Truman when North Korean forces attacked South Korea in 1950. "By God," Truman said, "I'm going to let them have it!" Later, more introspectively,

Truman wrote: "If the Communists were permitted to force their way into the Republic of Korea without opposition from the free world, no small nation would have the courage to resist threats and aggression by stronger Communist neighbors. If this was allowed to go unchallenged, it would mean a third world war, just as similar incidents had brought on the Second World War." [1]

Saddam Hussein's invasion was not part of a global contest between two superpowers. But it was a clear-cut case of aggression, and Bush had to act.

In fact, Bush was an activist. He hated bullies and was prepared to use American power unilaterally to bring them to order. When Manuel Noriega stole the elections in Panama in 1989, the Organization of American States (OAS) did nothing. [2] When Noriega "declared war" on the United States and murdered a U.S. serviceman in the Canal Zone, Bush moved quickly. He was not inhibited by the lack of an OAS resolution of approval and seemed little concerned when Democratic congressmen complained about the use of force without multinational sanctions.

Iraq's invasion of Kuwait was a clear act of aggression across an international border, but in this case its meaning for the United States was less clear. Panama was in our own neighborhood, and the two countries had special ties. The United States had played a role in the creation of Panama, and the canal was built in large measure by U.S. citizens and with American money. Iraq was on the other side of the globe—not part of an historic American sphere of interest.

THE FIRST POST–COLD WAR CONFLICT

Desert Storm was not the war the United States had planned for. Before 1989, strategic thinkers had assumed a continuing political-military competition with an expansionist Soviet Union that was ready to exploit any weakness and profit from any crisis. Containing Soviet expansion and regional violence had been the principal goal of U.S. policy for decades. The Soviet Union had been our main adversary in a global struggle, but Iraq's invasion of Kuwait was not caused by Soviet expansionism.

Iraq was neither an historic enemy of the United States nor a global power, but Saddam Hussein nevertheless posed a serious challenge. He was a ruthless ruler with a boundless appetite for power and an unlimited capacity for violence, a man who needed war like fire needs oxygen. He had made war against his own countrymen; for eight years he made war against Iran; now he made war against Kuwait. Saddam made no distinction between legal and illegal weapons; military and civilian targets; children and adults; men and women; Persians, Kurds, Jews, and Arabs. All within his reach were potential targets. His powerful army threatened the Gulf oil that was vitally important to Europe and Asia; his attack on Kuwait dramatized the vulnerability of traditional regimes in the area.

Bush himself had a long-standing interest in the Persian Gulf, especially Kuwait. He had spent time in the region and knew the rulers and the oil companies. It was clear to him that something had to be done to undo the aggression. But this would be much more difficult than any previous use of force in his administration—it would be a global confrontation, with global implications.

Until then, the United States had preserved normal relations with this repressive government, though Bush and most other Americans disapproved of the Iraqi regime for its autocratic character, brutal practices, use of chemical weapons against Iran, and repeated threats of violence against Israel. Bush had signed a Presidential Directive on October 2, 1989, that found normal relations with Iraq to be in the U.S. interest.[3] The Bush team believed that political realism sometimes required the United States to deal with unsavory regimes in the interest of strategic goals. Bush saw his policy as essentially a continuation of the Reagan administration's; the Bush team and the State Department had hoped that a policy of cooperation would moderate the policies of the government of Iraq.[4] America's principal allies had followed suit, tilting toward Iraq during the Iran-Iraq war.

But neither the conciliatory policy of the Western nations, nor the offer of new credits using enhanced trade relations and economic incentives, induced better behavior from Iraq. Instead, Saddam grew bolder and more threatening. On April 2, 1990, he publicly confirmed that Iraq possessed chemical weapons, which he had already used against Iran and against Iraqi dissidents. He also threatened that, if attacked, "We will make the fire eat up half of Israel."[5] Still, U.S. policy toward Iraq continued to emphasize American flexibility.

Saddam decided on an easier target than Israel. By mid-July 1990, evidence had begun to accumulate of his hostile intentions toward Saudi Arabia, as well as a buildup of Iraqi troops and weapons along the border with Kuwait. He began by emphasizing grievances and demands: Kuwait must stop stealing Iraqi oil from the Ramali oil field and must repay the $2.4 billion it had "stolen." The good intentions and official assurances that the United States desired "to improve relations with Iraq,"[6] issued by U.S. ambassador April Glaspie and echoed by other western powers, failed to deter Saddam, however, and the Gulf states were reluctant to make defensive moves. Only the United Arab Emirates (UAE) asked the United States for a demonstration of support and cooperated in a joint exercise with U.S. naval forces.

Kuwait—small, vulnerable, and essentially defenseless—was an easy mark. On August 2, 1990, after the appeals of other governments had failed, Iraqi troops swept into Kuwait on the flimsy pretext of restoring the historic borders of greater Iraq. Kuwait immediately called on the United States and other friendly nations for help, invoking the self-defense provision of Article 51 of the United Nations Charter:

> Nothing in the present Charter shall impair the inherent right of individual or collective self-defence if an armed attack occurs against a Member of the United Nations, until the Security Council has taken measures necessary to maintain international peace and security.[7]

That same day, by a 10 to 0 vote, the UN Security Council passed Resolution 660, condemning Iraq's aggression and calling for an immediate, unconditional withdrawal of Iraqi troops. The resolution contained strong language but made no threat of force to counter force. This attack on a nonaggressive Arab brother added to a growing list of indications that the Iraqis intended to move soon on Saudi Arabia. On August 5, Bush said flatly, "This will not stand, this aggression against Kuwait."[8] The next day, the Security Council passed Resolution 661 by a vote of 13 to 0, reaffirming the right of individual or collective self-defense in response to the armed attack by Iraq against Kuwait. Acting under Chapter

VII of the UN Charter (the chapter under which forceful measures are authorized), the Security Council imposed mandatory economic sanctions on Iraq, including a trade and financial embargo on all but medical and humanitarian goods and payments.[9] Cuba and Yemen abstained, but the lopsided majority, including all five permanent members, testified to the unusual degree of unity in the Security Council.

On August 8, Bush announced the deployment of U.S. troops to Saudi Arabia in Operation Desert Shield to deter an attack on that country and block Iraq's advance while the embargo took its toll. In his memoir, Bush's secretary of state, James Baker, described his concern about the "formidable political realities" of the crisis. Baker said he reminded Bush, "[T]his has all the ingredients that brought down three of the last five presidents: a hostage crisis, body bags, and a full-fledged economic recession caused by forty-dollar oil."[10] Bush knew the risks but was determined to take whatever action was necessary.

Baker describes the Bush team's strategy: "We would begin with diplomatic pressure, then add economic pressure, to a great degree organized through the United Nations, and finally move toward military pressure by gradually increasing American troop strength in the Gulf." This approach would "lead a global political alliance aimed at isolating Iraq."[11] Eventually, in the military phase, five hundred thousand U.S. troops and several hundred thousand soldiers from twenty-five other countries were assembled in the Saudi desert.

Bush's strong words and determination surprised even some of his own aides. He later said, "I had decided . . . in the first hours that the Iraqi aggression could not be tolerated . . . I came to the conclusion that some public comment was needed to make clear my determination that the United States must do whatever might be necessary to reverse the Iraqi aggression."[12]

Bush's bold statement that Saddam's invasion "will not stand"— made without consulting Congress or the Security Council—was a giant step toward a major U.S. military commitment in the Gulf, and toward a new world order in which such outrages would not go unpunished.

Bush Saw Vital U.S. Interests in Kuwait

The stakes were high for Bush. At issue were the independence of the Gulf states and the control of their oil; the Arab governments' ability to act in their own self-defense; Bush's capacity for effective leadership in confronting a major military and diplomatic challenge; the Europeans' capacity to act in their own interest; the ability of all to mount a collective action through the United Nations; and the future of Saddam Hussein, his dangerous government, his powerful army, and his capacity to establish hegemony in the Gulf. Bush repeatedly stated that this action would serve as a precedent for dealing with future aggression. "We will succeed in the Gulf," he said in his State of the Union speech in late January 1991. "And when we do, the world community will have sent an enduring warning to any dictator or despot, present and future, who contemplates outlaw aggression. The world can therefore seize this opportunity to fulfill the long-held promise of a new world order—where brutality will go unrewarded, and aggression will meet collective resistance." [13]

So George Bush assumed leadership of a campaign to counter Saddam Hussein's invasion of a small country on the other side of the world in which the United States had important, but not vital, interests. Why? Because through years of service in government he had grown accustomed to the idea of the United States as the world leader; because U.S. presidents since Eisenhower had defined the Persian Gulf as an area of vital interest; because Bush had personal experience in the area and hated bullies; and because he supported the idea of a collective response and believed he could carry out a collective action through the UN.

Many Americans and others believed that Bush acted because the United States needed to preserve its access to the Gulf region's oil, but it was the *world* that needed that access. At that time, the United States got only 13 percent of its oil from the Middle East. [14] Japan, in contrast, received 70 percent of its oil from the region, and other U.S. allies were heavily dependent on Middle East oil. [15] The interests that made the president act were broader than oil and less tangible. He believed that the United States had a national interest in world order, peace, and American leadership. [16] The conquest of Kuwait would give Iraq control of 20 percent of the world's known oil reserves, and Iraq clearly had designs on Saudi Arabia's oil as well.

Still, taking the United States into war was an awesome responsibility. Disputes among Americans over the use of force had been bitter since the Vietnam War sparked the antiwar movement that led a demoralized Lyndon Johnson to retire rather than seek reelection in 1968. These disputes spoiled the Democratic Convention in 1968, split the Democratic Party, and prepared the way for Richard Nixon's electoral victories in 1968 and 1972. They inspired university riots and embittered U.S. politics at every level, destroying the habitual civility that characterized American political life. The result was widespread reluctance among politicians to use force—a reluctance that came to be called the "Vietnam syndrome" and that affected the attitudes of both the Carter and Reagan administrations toward the use of force.

Reagan's secretary of defense, Caspar Weinberger, articulated the following principles—the Weinberger Doctrine—to govern the use of U.S. military forces through the eight years of Reagan's presidency:

1. The engagement must be deemed vital to our national interest or that of our allies.

2. We should put forces in combat only if we did so wholeheartedly and with a clear intention of winning.

3. We should have clearly defined objectives.

4. We should reassess the relation between our objectives and forces and maintain a clear preponderance of force.

5. We must have the support of the American people and Congress.

6. The commitment of forces should be a last resort.[17]

As Reagan's vice president for eight years, Bush was very familiar with these principles, as was Weinberger's chief of staff, General Colin Powell. Marginally altered, the principles came to be called the Powell Doctrine, and they guided Bush as he led the country into war. Bush made two significant additions: He sought a mandate from the UN Security Council specifically authorizing military action, and he assembled a multinational force (for symbolic as well as military reasons). Bush deferred action until all alternatives to war had been exhausted, but it was clear from the beginning that he was determined to secure the withdrawal of Iraqi forces.

There were differences in the Bush administration over whether Iraq's aggression against Kuwait was an adequate cause for the United States to become involved in war. Powell saw containing Saddam in Kuwait and protecting Saudi Arabia as the objectives, and favored delaying or avoiding military action. Baker generally agreed with Powell.[18] Bush, however, thought Iraq's forces *must* withdraw from Kuwait.

Authorization to Use Force

The most pressing question was whether the United States and its allies should undertake military action under Article 51 immediately or wait until they had explicit Security Council authorization for the use of force. Even though Article 51 recognizes the right of a state to self-defense and collective defense against aggression, some lawyers argued that forceful action requires explicit permission from the Security Council. Bush was committed to collective action through the UN, as was Baker.[19]

Baker wrote, "There was no doubt in my mind the President would authorize force if necessary—and we were very careful all along to preserve our options under Article 51 of the UN Charter."[20] Baker wanted the United States to act as part of a coalition that would share the burden. "There was simply no percentage" in going it alone,[21] he wrote, although his opposition to unilateral action was a matter of practical considerations rather than principle. Without a broad multilateral coalition, Baker argued, "we would never have achieved the sort of solidarity from the Arab nations that was crucial to isolating Saddam."[22] He worried that Arab anger would be turned against the United States, and he pointed out to Saudi Arabia's ambassador, Prince Bandar Bin Sultan, that once the war was under way, Americans would be killing Arabs in Kuwait. "No problem," Bandar replied.[23]

"From the very beginning," Baker wrote, "the President emphasized the importance of having the express approval of the international community if at all possible."[24] Bush saw the advantages in securing the acquiescence of the Security Council, building a large coalition of allies and regional powers, and ensuring the approval of the Soviet Union and the U.S. Congress before any military action. Baker wanted a specific Security Council mandate.[25]

Britain's then prime minister, Margaret Thatcher, thought otherwise. Thatcher believed Iraq's invasion of Kuwait was a clear violation of international law and that Article 51, the Security Council resolution, and the call for Iraq's withdrawal were sufficient authorization for collective military action. She did not think it was necessary to go back to the UN for approval of new action, and she worried that Bush might "go wobbly" rather than acting with decision.[26]

As Baker recalled, Thatcher "wasn't the least bit shy in expressing her serious misgivings about our preference for pursing a multilateral course."[27] Bush's national security adviser, Brent Scowcroft, supported Thatcher's view. Where Baker and Powell favored maximum consultation, maximum UN authorization, and maximum delay, Thatcher thought Bush and Baker went too far in involving the United Nations in each step. In her memoir, *The Downing Street Years*, she recalled the "almost interminable argument between the Americans—particularly Jim Baker—and me about whether and in what form United Nations authority was needed for measures against Saddam Hussein. I felt that the Security Council resolution which had already been passed, combined with our ability to invoke Article 51 of the UN Charter on self-defense, was sufficient."[28]

Thatcher was reluctant to involve the UN in details of the operation because "there was no certainty that the final wording of a resolution, which was always open to amendment, would be found satisfactory. If not, it might tie our hands unacceptably."[29] She added:

[A]lthough I am a strong believer in international law, I did not like unnecessary resort to the UN because it suggested that sovereign states lacked the moral authority to act on their own behalf. If it became accepted that force could only be used—even in self-defense—when the United Nations approved, neither Britain's interests nor those of international justice and order would be served. The UN was for me a useful—for some matters vital—forum. But it was hardly the nucleus of a new world order. And there was still no substitute for the leadership of the United States.[30]

Later developments proved the prescience of Thatcher's observations.

The problem with seeking detailed authorization from the United Nations, Thatcher said, was that the UN was always engaged in a search for consensus, which she described as "the process of abandoning all beliefs, principles, values, and policies in search of something in which no one believes, but to which no one objects; the process of avoiding the very issues that have to be solved, merely because you cannot get agreement on the way ahead. What great cause would have been fought and won under the banner, 'I stand for consensus'?"[31]

Thatcher points up the basic problems in multinational decision-making and action. Governments have distinct interests and perspectives that make consensus difficult to achieve, and the need for consensus makes multilateral processes slow, cumbersome, indecisive, and inconclusive. Thatcher had already encountered this problem in negotiations on the European Union (EU), where she invoked the moral authority of the nation against the Brussels bureaucrats—and was frequently outvoted.

Instead of acting under Article 51 to assemble forces immediately and act in defense of Kuwait, as Thatcher urged, Bush chose to seek specific authorization from the Security Council for each new step and to painstakingly build support through Baker's extensive personal conversations around the world.

In the end, discussions in the Security Council and in various national capitals produced the consensus Bush and Baker desired, a broad alliance, and the Security Council resolutions, including an ultimatum. Yet four months elapsed between the first resolution condemning the invasion on August 2 and the ultimatum on November 29, proclaiming that Iraq must withdraw its troops from Kuwait by January 15, 1991, or be driven out by force.

During those critical months between the invasion and the beginning of the military operation, Kuwait was devastated. Its people were murdered, raped, tortured, and dispersed, its resources plundered and destroyed, the nation sacked. Much of this devastation could have been avoided by more rapid action. Though some time was required to assemble the necessary forces, earlier air attacks would have slowed the progress of Saddam's forces. Resolution 660 on August 2 had already demanded that Iraq withdraw from Kuwait. Resolution 678 added a deadline and authorized all necessary means to enforce the ultimatum.

As the United States and the coalition went again and again to the Security Council for specific authorization, Margaret Thatcher went along, but she remained unconvinced to the end. She would write later that the slow beginning and unsatisfactory outcome of the Gulf War (by which she meant the U.S. decision to end the war and to leave Saddam's military base intact) were direct consequences of the earlier decision to seek detailed authorization from the Security Council and to rely on the consensus of a diverse group of allies.[32] Had the consensus road not been taken, she thought, the war might have been carried to a more definitive end. But Bush and his administration had their own views about legitimate force in the post–cold war world.

Bush assumed the task of mobilizing U.S. forces, rallying a broad coalition in the Security Council, rallying support in Congress for military action if that proved necessary, and leading the military forces to victory.[33]

Building Consensus in the UN and the United States

The process of negotiating resolutions, constructing a coalition, and building majority support in the Security Council for the use of force was a landmark in the evolution of post–cold war conduct. The decision to defer action to build a consensus was unprecedented, as was giving priority to multilateral processes and collective action over a rapid response. Military action, when it came, had wide and explicit approval from the international community and the American government.

Secretary Baker visited the heads of state in the major countries in Europe and the Middle East to explain the situation, argue the case for collective action, and secure their support and financial contributions. He later wrote that he and Bush saw this first international crisis of the post–cold war era as a critical opportunity to establish precedents.

Bush and Baker hoped to persuade Saddam to abandon his occupation of Kuwait before the United States began a war, but he refused. Though strong and shrewd, Saddam repeatedly underestimated his opponents. He probably believed that the Gulf monarchies were too decadent to resist his attack on Kuwait, and that the rest of the Arab world was too divided to unite against him. Perhaps he believed he was already the leader of the united "Arab nation" he imagined. His speeches made clear that he believed other Arab governments would rally around him. His writing made clear that he saw himself as ferocious and unconquerable. He wrote:

> I've always preferred to make my decisions without the involvement of others. My decisions are hard, harsh, just like my desert. I've always related my behavior to the desert. Usually it looks so quiet and kind, but suddenly it erupts with rage, mightily fighting the gusts of storms and gales. And this outburst of the desert's rage gave me the feeling that I was on the brink of the end of time.[34]

Saddam gave many of the governments and people within his reach the feeling that they too were on the brink of the end of time.

In an effort to make Saddam understand the opposition he faced, Bush advertised the strength and sophistication of the U.S. military power being assembled, and reiterated his personal commitment to the enforcement of the UN resolution. As Iraqi forces began to round up Westerners at gunpoint, Bush prepared for action. He provided leadership in the UN, where the United States was represented by Thomas Pickering, a skillful, politically savvy career ambassador.

Bush and Baker worked hard to coax a diverse group of nations to participate without regard for the political quality of the allies assembled. When French president François Mitterrand made the point that Kuwait "is not the kind of government I would like to send French soldiers to be killed for,"[35] Bush and Baker made the case for realpolitik. They believed that world politics sometimes drives governments to make alliances based on necessity as well as shared moral principles.[36] Both understood that international politics may sometimes require alliances with unpalatable leaders or regimes. If opposing aggression, constructing new collective security arrangements, helping maintain independence and order in the world, and protecting world access to Middle East oil required collaborating with some governments whose principles and practices did not meet American standards, so be it.

Bush was experienced in dealing with the governments of the Gulf and the world, and he responded eagerly to the challenge. "I enjoy trying to put the coalition together and keep it together," he told reporters at a press conference in October. Dealing with the Gulf was more fun than dealing with Congress on taxes.[37] But there were many pitfalls on the way to collective action.

Peace offensives aimed at delaying U.S. action were launched, resisted, and defused. Bush believed a negotiated settlement was possible, but only on the condition that it did not leave Iraq the hegemonic power in the Gulf. Although he wanted to avoid war, he opposed appeasement. He had learned well the lessons of Munich and Vietnam. Still, he sent Baker to Baghdad after November 30 for a conversation with Iraq's foreign minister Tariq Aziz, in one last effort to secure Iraq's withdrawal from Kuwait.

Confronted with unsatisfactory options, as well as Arab ambiguities, American impatience, and Saddam's ambition, Bush chose his course. He believed that if Saddam should succeed in this act of brute force, his image and power in the Arab world would be dangerously enhanced; the governments of Saudi Arabia, Egypt, the Emirates, and Oman would be in continuing danger; America's reputation in the world would suffer; American influence would shrink; and the opportunity to establish a new world order would be greatly reduced.

For Americans, the challenge was to show Saddam that he could not succeed in the effort to annex Kuwait, and that there would be devastating consequences for trying. The United States had to communicate to Saddam that it would use the large military forces assembled in the Middle East. Any wavering in purpose, any seeming uncertainty, would undermine the effort. The United States was convinced that if Saddam Hussein could be made to believe he had only two options—withdraw from Kuwait or face total war—he would withdraw.

UN SECURITY COUNCIL RESOLUTIONS
CONCERNING IRAQ'S INVASION OF KUWAIT

RESOLUTION 660, AUGUST 2, 1990: The Security Council condemned the invasion of Kuwait; demanded Iraq's unconditional, immediate withdrawal; and called on both countries to begin negotiations. (Adopted 14-0; Yemen did not participate)

RESOLUTION 661, AUGUST 6: The Security Council imposed a trade and financial embargo on Iraq and occupied Kuwait; established a special sanctions committee to implement the resolution; and called upon UN members to protect the assets of Kuwait around the world. (Adopted 13-0; Cuba and Yemen abstained)

RESOLUTION 662, AUGUST 9: The Security Council declared Iraq's annexation of Kuwait null and void. (Adopted by unanimous vote)

RESOLUTION 664, AUGUST 18: The Security Council demanded the immediate release of foreigners from Iraq and Kuwait and the right of diplomats to visit their nationals, and insisted that Iraq rescind its order closing diplomatic and consular missions in Kuwait. (Adopted by unanimous vote)

RESOLUTION 665, AUGUST 25: The Security Council called on UN members with ships in the region to enforce sanctions by inspecting and verifying cargoes and destinations. (Adopted 13-0; Cuba and Yemen abstained)

RESOLUTION 666, SEPTEMBER 13: The Security Council reaffirmed that Iraq was responsible for the safety and well-being of foreign nationals, and provided guidelines for the delivery of food and medical supplies. (Adopted 13-2; Cuba and Yemen against)

RESOLUTION 667, SEPTEMBER 16: The Security Council condemned Iraqi aggression against diplomats and diplomatic compounds in Kuwait and demanded the immediate release of foreign nationals. (Adopted by unanimous vote)

RESOLUTION 669, SEPTEMBER 24: The Security Council emphasized that only the Special Sanctions Committee had the power to permit food, medicine, or other humanitarian aid shipments to Iraq or occupied Kuwait. (Adopted by unanimous vote)

RESOLUTION 670, SEPTEMBER 25: The Security Council expanded the economic embargo to include air, except for humanitarian aid authorized by the Special Sanctions Committee, and called on UN member nations to detain Iraqi ships. (Adopted 14-1; Cuba against)

RESOLUTION 674, OCTOBER 29: The Security Council demanded that Iraq stop mistreating Kuwaiti and other foreign nationals and reminded Iraq of its liability for damages to foreigners or their property resulting from the invasion and occupation of Kuwait. (Adopted 13-0; Cuba and Yemen abstained)

RESOLUTION 677, NOVEMBER 28: The Security Council condemned Iraq's attempts to change Kuwait's demographic composition and condemned Iraq's destruction of Kuwaiti civil records. (Adopted by unanimous vote)

RESOLUTION 678, NOVEMBER 29: The Security Council emphasized Iraq's failure to withdraw from Kuwait as requested in Resolution 660 and subsequent resolutions; authorized use of "all necessary means" to implement 660 if Iraq continued in this behavior; and set January 15, 1991, as the date by which Iraq must withdraw forces or be driven out. (Adopted 12-2-1 Cuba and Yemen against; China abstains)

The New United Nations

That it was possible to forge a consensus among the five permanent members of the Security Council (the United States, the United Kingdom, France, China, and the Soviet Union[38]) and others regarding Iraq's invasion of Kuwait was a dramatic demonstration of how much the world had changed in a short time and what those changes meant for the UN. For the first time since the organization's founding, a series of Security Council resolutions condemning aggression could be passed without cold war vetoes. A large majority of the member states condemned Iraq's invasion and annexation of Kuwait, imposed sanctions, and authorized their enforcement by all necessary means. As George Bush told the UN General Assembly in September 1991:

> The United Nations, in one of its finest moments, constructed a measured, principled, deliberate, and courageous response to Saddam Hussein. It stood up to an outlaw who invaded Kuwait, who threatened many states within the region, who sought to set a menacing precept for the post–cold war world. The coalition effort established a model for the collective settlement of disputes.[39]

This response was possible because of a new pattern of interaction in the UN. If Iraq, bound to the Soviet Union by a friendship treaty, had invaded Kuwait two years earlier, the Security Council could not have passed a resolution condemning the invasion. The Soviet Union would have called Iraq's aggression a "liberation" that fulfilled age-old aspirations for the unity of two peoples wrested apart by colonial powers.[40] (This was Iraq's own rationale for the invasion.) Iraq could have counted on 100 percent support from the Soviet bloc and wide support from the nonaligned nations, among which the Soviets had broad influence. Internal divisions would have neutralized the Arab bloc and the Islamic Conference, and a long procession of speakers would have declared the condemnation of the invasion to be simply another machination of imperialist and Zionist powers, and Kuwait a corrupt remnant of the colonial era.

For decades, aggression by the Soviet Union and its allies had been defined as liberation. For example, the Soviet invasions of Czechoslovakia and Afghanistan were described as acts of international fraternal solidarity. But that era had passed. Instead of defending Iraq as a treaty ally, Dr. Nikolay Shishlin, political adviser to the Soviet leadership, pointed out that Iraq broke its 1972 friendship treaty with the Soviet Union when it invaded Kuwait. Shishlin said, "The people regard [the invasion of Kuwait] as a crime and a criminal act." Soviet president Mikhail Gorbachev's closest adviser, Aleksandr Yakovlev, called the invasion a violation of "a moral law."[41]

For the first time, Soviet officials read the UN Charter as it was meant to be read, and the Soviet government joined in the series of resolutions. China, which usually avoids casting a veto alone, also joined in. An unprecedented consensus was born among the permanent members that held throughout the Gulf War.

Securing the Support of Congress

Some legal scholars have argued that a mandate from the UN Security Council is all the authorization required to legitimize U.S. military action. But that argument assumes that the Security Council's decision overrides the requirement of the U.S. Constitution that Congress must declare war and tacitly transfers that authority in the U.S. government from Congress to the president, who decides how the United States will vote in the Security Council. President Bush saw his obligations differently. Just as he sought explicit approval from the Security Council for the use of force against Saddam Hussein, he also sought the support of Congress in committing U.S. troops to the conflict.

Bush remembered the bitter debates when Lyndon Johnson failed to secure congressional support before plunging U.S. forces more deeply into the war in Vietnam. He wanted clear authorization from Congress, not a Gulf of Tonkin Resolution. UN Security Council Resolution 678, which set a deadline for Saddam's withdrawal, was expected to help with Congress, but the Democratic leadership in both the House and the Senate—including House Speaker Tom Foley (WA), Senate Majority Leader George Mitchell (ME), and Armed Services Committee chairman Sam Nunn (GA)—opposed the resolution requested by Bush.

Most Democrats argued, along with Nunn, that rather than risking thousands of lives on a war on the other side of the globe, the United States should allow more time for the economic sanctions to produce the desired results.[42] Most Democratic leaders in the Senate argued that Saddam could not indefinitely withstand the economic pressure of the sanctions. Mitchell charged that Bush was about to make the decision "prematurely."[43] Senator Claiborne Pell (D-RI) thought Iraq's military strength was already being eroded.[44]

Many arguments were heard for and against granting Bush the power to send American forces into combat. Some questioned the efficacy of economic sanctions in general and of these particular sanctions against this particular adversary. Others wondered whether denying Bush the authority to use force, or making him wait for such authority, would undermine U.S. credibility and render the threat in UN Security Council Resolution 678 (and subsequent UN resolutions) hollow.

Senator Joseph Lieberman (D-CT), an articulate supporter of Bush's request, emphasized practical issues: the effect of any delay on the coalition against Saddam, on the uneasy allies in the Gulf, and on U.S. and other troops deployed in the area. Lieberman also raised the prospect that waiting might give Saddam time to perfect weapons of mass destruction.

The debate considered whether Bush had exhausted all options and whether congressional authorization would encourage him to turn to war. It considered whether protecting the Gulf was a *vital* U.S. interest or merely an important one, and why countries that were more dependent on Gulf oil were not playing a much larger role in its protection. It was a searching debate, resolved by a narrow, largely Republican majority in the Senate (52 to 47) in favor of granting the authorization Bush wanted. The House approved the joint resolution by a wider margin (250 to 183).

In the time it took to debate these issues in the administration, in Congress, and among the allies, great harm had been done to the people and the country of Kuwait, and Israel and Saudi Arabia had been subjected to great danger. Yet the result was that no reasonable charge of precipitous, ill-considered, or unauthorized action could be made against the Bush administration. The legitimacy of his action had been firmly established, and this had unquestionable value for the U.S. polity, even as it raised doubts that aggression could be quickly and effectively countered through the laborious processes of the UN Security Council (and Congress).

Building Support Among the American Public

Bush needed support not only in the UN, in the Gulf, and in Congress, but from the public as well. Vice President Dan Quayle made numerous speeches explaining the nature of the adversary and emphasizing Saddam's desire to be the leader of a new Arab superpower. Quayle said:

> To that end, he spent some fifty billion dollars on arms imports during the 1980s alone. He has launched two wars of aggression during this period . . . at a cost of some one million lives thus far. He has built the sixth largest military force in the world. He has acquired a sizable stockpile of both chemical and biological weapons . . . and he has launched a massive program to acquire nuclear weapons.[45]

Former president Richard Nixon entered the discussion with his own evaluation of the U.S. stake in the Gulf and of why the United States should act: "[because] Saddam Hussein has unlimited ambitions to dominate one of the most important strategic areas in the world. . . . Because he has oil, he has the means to acquire the weapons he needs for aggression against his neighbors, eventually including nuclear weapons."[46]

Like Bush, Nixon characterized the world's response to this aggression as a precedent:

> We cannot be sure . . . that we are entering into a new, post–cold war era where armed aggression will no longer be an instrument of national policy. But we can be sure that if Saddam Hussein profits from aggression, other potential aggressors in the world will be tempted to wage war against their neighbors.
>
> If we succeed in getting Mr. Hussein out of Kuwait in accordance with the UN resolution . . . we will have the credibility to deter aggression elsewhere without sending American forces. The world will take seriously U.S. warnings against aggression.[47]

Saddam's Threats

The Gulf War reflected not only one man's ambitions, but also that man's misunderstanding of his relative power position. In spite of the sustained efforts of Bush, Mitterrand, and UN secretary-general Javier Perez de Cuellar to make Saddam understand the strength and determination of the forces assembled against him, the Iraqi leader continued to underestimate the power of his adversaries. His speeches, and the commentaries in the Iraqi press in the days before the expiration of the UN deadline, made plain that Saddam seemed oblivious to Bush's promise that he faced an imminent choice between withdrawal and destruction.

Saddam sometimes insisted that the issue was not Kuwait at all but the liberation of Palestine. He said he was engaged in a crusade to eliminate the terrible injustice of Israeli occupation that had been inflicted on the Arab world. The Iraqi army, he told his people, would "achieve several aims in one battle," eliminating injustice, poverty, and foreign hegemony.[48] As Saddam saw it, liberating Kuwait from Iraq's aggression was but a pretext, a smokescreen to obscure the real U.S. goal, which was to establish its hegemony over the Gulf and its oil and to dominate the world. According to *Al Qadisiya*, a Baghdad newspaper, "America wants to control oil resources in such a way that will make the oil resources needed by the rest of the world come under its hegemony. Thus, America will regain its lost influence by governing all other countries; it will give oil to anyone it wants and deprive anyone it wants."[49] The choice, as Saddam described it, was between American dominance of Gulf oil, on one hand, and the elimination of global injustice, poverty, and occupation on the other. It would not be a local or regional war, Baghdad radio insisted. "In one way or another, it will spread all over the world, where more than a billion Muslims from Indonesia to West Africa will view this battle as a war against colonialism."[50]

Meanwhile, Saddam's state-controlled press told him that "[a] lot of Arabs consider President Saddam Hussein the only Arab leader who dares to challenge Israeli occupation of Arab territories and believe he might rescue them."[51] It seemed clear from Baghdad that, because America's objectives were so ignoble, God would be on the side of Iraq, as promised in the Koran: "To those against whom war is made, permission

is given to fight, because they are wronged, and verily, God is most powerful for their aid." [52]

Saddam and his lieutenants threatened blood and destruction. They
promised to burn their enemies with "a great fire that does not go out," to
"drown them in rivers of blood," and to destroy Israel. Saddam saw himself as the fearless champion of the Arab nation who would rally the
faithful to the ultimate jihad. He saw the American president as facing a
defeat "terrible and total." [53] But Bush had a different plan.

Saddam understood that he was no match for the United States militarily. In October he opened a new front, this time against the Americans'
will. He understood that it was Bush's determination to confront Iraq
that was primarily responsible for the forces assembling against him. He
sought to frighten his adversaries with talk of jihad, threats of terrorism,
and predictions of heavy casualties. He tried to split the heterogeneous
anti-Iraq coalition—accusing the Saudis of defiling Muslim holy places,
charging Morocco with being a Zionist agent, and seeking tirelessly to inflame the Palestinian issue.

Then, in mid-October, came new hints of an interest in peace—just
two days after the Iraqi minister of information had said there was no
room for any compromise and one day after the *New York Times* published a series of interviews in which Jordan's king Hussein warned that
war would be catastrophic for the region. [54] Iraqi officials began to encourage hopes for a diplomatic solution, hinting that Iraq might withdraw from Kuwait and retain only the strategic island of Bubiyan, an oil
field at the Iraq-Kuwait border, and a few other special privileges over
Kuwaiti territory.

Presumably, Iraqi officials understood that there was no better way
to prevent the United States and its allies from using their superior force
than to hold out the prospect of a diplomatic solution based on a compromise. In such situations, the mirage of a peaceful alternative to war
breeds false hope and diminishes the will to fight, though the "solution"
may be only the first step on the road to defeat. The classic textbook
example is Neville Chamberlain's "peace in our time" compromise at
Munich.

If Saddam Hussein did not understand how vulnerable the West is to
appeals to peace, his good friend Yasir Arafat did. Words like "negotiated

settlement," "peaceful solution," and "compromise" are the political equivalents of the rubber hammers with which physicians test our reflexes. But any compromise that gave Iraq a piece of Kuwait would have rewarded Saddam's aggression and left him stronger than ever and emboldened to target other governments in the region. On behalf of the coalition, Bush and Baker rejected the siren song of appeasement. "It's our position that he should not in any way be rewarded for his aggression," Baker said.[55]

As Carl von Clausewitz observed, as long as an aggressive man remains armed, he can be persuaded to abandon his aggression by "one single motive alone, which is that he waits for a more favorable moment for action. . . . If the one has an interest in acting, then the other must have an interest in waiting."[56]

Preparations for war went forward. At the end of October 1990, 200,000 more U.S. troops were ordered to the Gulf, doubling the total in the region. Saddam had made not one move to end the devastation and plunder of Kuwait or defend its suffering people, and now the U.S. and allied troops and materiel required to restore Kuwait's independence were being moved into position. Saddam Hussein had been given ample notice of the seriousness with which the United States and its allies regarded his aggression, though he may not have understood that unless he withdrew his forces, they would be driven out.

"If we desire to defeat the enemy," wrote von Clausewitz, "we must proportion our efforts to his powers of resistance. This is expressed by the product of two factors which cannot be separated; namely, the sum of available means and the strength of the will."[57] Bush had assembled the necessary means. Now he needed to demonstrate a will to use them equal to Saddam's will to resist.

As Eliot Cohen, Professor of Strategic Studies at Johns Hopkins University, wrote at the time: "The longer he [Saddam] has to fortify Kuwait, to prepare his armies and people for war and to lay the groundwork for a campaign of terror and subversion overseas, the harder he will make it for us."[58]

Saddam had been given plenty of time. It was time to begin the liberation of Kuwait.

Dreaming of a New World Order

Bush was determined not only to turn back Saddam Hussein's aggression, but to create a new system of international security that would deter or defeat future aggression. This new world order would be his legacy. "The civilized world is now in the process of fashioning the rules that will govern the new world order beginning to emerge in the aftermath of the cold war," he told *Newsweek* in November 1990. "When we succeed, we will have shown that aggression will not be tolerated. We will have invigorated a United Nations that contributes as its founders dreamed. We will have established principles for acceptable international conduct and the means to enforce them." [59]

George Bush, a man who had once publicly proclaimed to be devoid of "the vision thing," had a clear vision of America's role in the post–cold war world. He shared that vision with Congress and with the American people in a series of speeches that explained who we were, what we must do in the Gulf, and why. In his January 1991 State of the Union Address, he spelled out his version of American exceptionalism and explained why American forces were halfway around the world:

> We know why we are there: We are Americans, part of something larger than ourselves. For two centuries, we've done the hard work of freedom. And tonight, we lead the world in facing down a threat to decency and humanity.
>
> What is at stake is more than one small country; it is a big idea: a new world order, where diverse nations are drawn together in common cause to achieve the universal aspirations of mankind—peace and security, freedom, and the rule of law.
>
> For generations, America has led the struggle to preserve and extend the blessings of liberty. And today, in a rapidly changing world, American leadership is indispensable. Americans know that leadership brings burdens and sacrifices. But we also know why the hopes of humanity turn to us. We are Americans; we have a unique responsibility to do the hard work of freedom. And when we do, freedom works. [60]

Like Wilson, FDR, and Truman before him, Bush sought a more peaceful world community based on law, and he believed the United States had a special obligation and calling to build that community.

Bush acted not because the United States had formal or close ties to Kuwait (we did not) or because the United States was dependent on Gulf oil (we were not), but because he saw a broader national interest in preserving the independence of Kuwait and the Gulf states—all of which he believed to be threatened by Saddam's appetite—and in preventing Saddam, or any other dictator, from gaining control of the Gulf's vast resources.

Like most of his generation, Bush believed that it was essential that aggression not be permitted to succeed. He knew that the American response in this crisis would establish a precedent that would influence how the United States and the UN dealt with other crises, other dictators, and other acts of aggression. He knew that the League of Nations had never recovered from its inaction in the face of Mussolini's conquest of Abyssinia in the 1930s. This second, broader objective of setting a precedent was the reason Bush rejected a "Libyan solution" of simply acting unilaterally to turn back Saddam and sought authorization for the use of force from the UN Security Council.

Instead, Bush and Baker wanted to liberate Kuwait *and* strengthen the UN. "The credibility of the United Nations is at stake," Baker told the *Washington Post*. "It's very important that when the United Nations . . . passes resolutions and takes actions . . . that those resolutions and actions be implemented."[61] Moreover, they did not want to appear trigger-happy, so they made a point of exhausting all other options before they turned to force. In his address to the nation on January 16, 1991, announcing that military action had begun, Bush spoke as if for the world, emphasizing the deliberate, orderly, lawful course that had been taken, explaining that "sanctions were tried for well over five months" and that the United States and the UN had "exhausted every means" to achieve a peaceful end to the crisis.[62]

THE COMPLEXITY OF COLLECTIVE ACTION

The process of dealing with Saddam through the United Nations, and establishing new principles of international conduct, demonstrated the serious obstacles to an effective system of global collective security and military action.

Even in this singular moment in history—with cold war divisions overcome and the blocs partially neutralized, and with the United States armed and ready—it was still difficult to deal with a clear-cut case of international aggression. No major power had a stake in prolonging or exacerbating the conflict, and regional solidarity had been shattered by the aggression of one Arab nation against another. Yet the process of consultation among allies and the building and preserving of a consensus was cumbersome, time consuming, and, in a fundamental sense, irrational. Even governments that were in basic agreement disagreed, equivocated, and lost time. To expect a fifteen-member Security Council representing all regions and cultures to plan a war policy—or a peace policy—is a tall order. The perspectives and interests of the five permanent members differed, and their views were not easily reconciled.

The ten nonpermanent members of the council also had widely varying interests in the issue. As a neighbor in the region, North Yemen had a direct stake. But Colombia, Cuba, Ethiopia, Finland, the Ivory Coast, Malaysia, Romania, and Zaire had the power to decide the outcome, though they had no special knowledge of the countries or regions and no direct national interest in the conflict.

No one knew how stable the Security Council consensus would be or what role the council would play in managing the conflict once it was under way. This diverse body seemed perpetually on the verge of serious disagreement that threatened to rend the fragile coalition. Baker believed it was necessary to shuttle around the globe and spend hours negotiating with Security Council members, even though a consensus had been expressed in the first, unanimous resolution condemning the invasion.[63] Anyone with experience in multinational diplomacy knows that it is much more difficult to preserve a consensus for action than to express condemnation. Again and again, it was necessary to overcome both straightforward and devious resistance to sustain the consensus.

Bush and Baker were especially anxious about the role Israel might play in the conflict. They feared that Israel's entry into the war would rupture the diplomatic and military cooperation with Arab governments, but it was clear that Saddam intended to target the Jewish state.

When Yasir Arafat and Jordan's king Hussein rallied to Saddam's side, the threat to Israel increased dramatically. Keeping Israel's forces out of the war became a priority of the Bush administration. Two good friends of Israel—Larry Eagleburger, the deputy secretary of state, and Paul Wolfowitz, the assistant secretary of defense for international security policy—were dispatched to Jerusalem to persuade the Israelis to let the United States defend them, rather than responding to attacks themselves. The United States had no effective defenses against the Iraqi Scud missiles that were falling on Israel; it could only appeal to the Israelis to do what they had always refused to consider: delegate their self-defense to another government.

According to a study published in the British journal *Nature* in January 1993, of thirty-eight Scud missiles launched at Israel by Iraq during the Gulf War, ten hit Israeli cities, meaning that more than one in four penetrated the Patriot missile defense system. The system was only effective because it was intercepting slow, rather primitive high-explosive-armed Scuds. As a retired army lieutenant general and former director of the Defense Intelligence Agency noted:

> Had the Scuds been armed with a fusing system for chemical warheads, the Patriot would have been useless. The reason? In many cases, the Patriot intercepted incoming Scuds only eight thousand feet or less from the ground. Had the Scuds been chemically armed, poison chemicals would have rained down on their targets. . . . Of course, with nuclear warheads, intercepts close above the targets would be to no avail.[64]

Though many Americans did not realize it at the time, the performance of the Patriots demonstrated that the United States had no missile defense.

Allied Complications

And then there were the French and the Germans.

The twelve foreign ministers of the European Community (EC) had initially promised to coordinate their Gulf policies with the United States, but as the January 15 deadline for Iraq's withdrawal from Kuwait approached, the French and German foreign ministers, Roland Dumas and Hans Dietrich Genscher, called the EC into session. U.S. officials speculated that by convening in the one international arena that excluded Americans, the Europeans meant to take the decision for war or peace out of the hands of the Bush administration. The EC had no tradition of common action on foreign affairs, so why did Germany—which had accepted no significant responsibility with regard to the Gulf crisis— suddenly seek to play a major role?

The Bush administration reacted cautiously, fearing that Saddam would perceive the European initiative—which called for Iraq's withdrawal from Kuwait in exchange for assurances that the United States forces would not launch attacks on Iraq—as evidence of disunity in the coalition. In fact, there were potentially important differences between the United States and Europe with regard to the Gulf—differences that would not become clear until the question of ending the war arose.

Such divergences of policy had existed between the United States and its Franco-German allies from the outset of the crisis. In part these differences were a matter of style, but they stemmed from the fact that the Europeans were simply less indignant about the destruction of Kuwait. They were less concerned about the danger Saddam posed to stability in the Gulf, less committed to Israel, and less interested in engendering a new world order. These factors led some European governments to resist making a major commitment of money and people to the effort; Washington, in turn, found its confidence in the solidarity of some of its allies wavering.

François Mitterrand sought to draw Saddam into negotiations on Kuwait with promises of a conference on Arab-Israeli settlements, which Israel and the United States opposed and most Arab states supported. Mitterrand pledged that France would be willing to discuss all Middle Eastern problems at one or more international conferences once Iraq had withdrawn from Kuwait. "With Kuwait under occupation," he said, "nothing is possible. With Kuwait evacuated, everything is possible."[65] But neither the United States nor Israel would acquiesce in a deal that linked the Gulf War's end to regional initiatives unrelated to Iraq's invasion. The United States had long opposed an international conference on the Arab-Israeli conflict, but supported direct negotiations between

Israel and her neighbors, as called for in Security Council Resolutions 242 and 338.

Efforts to find a negotiated settlement to the Kuwait occupation intensified as the January 15 deadline approached. Some urged the United States to strike a deal that would permit Saddam to save face. The U.S. government took the position that this was exactly what we should not do. There should be a price for invading, occupying, and devastating a neighboring country—and it should include losing face.

Americans argued that Saddam and Iraq, having trashed Kuwait, disrupted the region, cost the United States more than $30 billion, and cost other members of the coalition perhaps $30 billion more, should not be permitted to walk away without a substantial penalty. At the very least, Saddam would have to withdraw unconditionally from Kuwait and compensate his victims and their allies for the economic costs of his violence. He could not undo the human misery and death, but he could provide the financial compensation called for by Resolution 674.[66]

France and Germany agreed that Iraq must honor the UN demand to withdraw, but Mitterrand told the General Assembly that after Saddam released foreign hostages and announced his intention to withdraw, negotiations between Iraq and the coalition could begin on the withdrawal's timing and related details. The United States—and the UN resolution—said there could be negotiations only after "complete and unconditional" withdrawal of Iraqi forces and the restoration of the legitimate government of Kuwait.

Meanwhile, Bush focused on winning the war. His understanding of the size, strength, and character of the enemy helped his team make accurate estimates of what would be needed to defeat Saddam. And no asset was more important than his personal clarity about U.S. goals in the Gulf. He was determined to avoid the damaging and demoralizing incrementalism of the Vietnam War, with its endless wrangling in and out of Congress.

In the Vietnam War, the search for a political solution had become an excuse for deferring needed military replenishments and not seeking victory. In fact, the very idea of winning had been discouraged in many quarters as evidence of an insufficient interest in peace or a parochial concern with strictly national perspectives.

In the Gulf conflict, Bush emphasized from the outset that if Saddam refused to withdraw completely and unconditionally from Kuwait, it would be necessary to defeat him definitively. Bush rejected any suggestion that avoiding the use of force and finding a political solution should be the goal of negotiations. He declined to become involved in endless negotiations, even when they were promoted by Mikhail Gorbachev, with whom he was working to develop relations. His refusal must have been frustrating to Saddam, who was accustomed to the tactics of endless delay common in the diplomacy of his region. A French diplomat of my acquaintance observed, "Saddam didn't understand that he was dealing with Anglo-Saxons. He thought he was in a game that could go on forever."

Bush signaled that a full force would be assembled, equipped, and utilized from the outset. He stated clearly that there would be no sanctuary. Although various UN diplomats and coalition members urged that attacks be limited to Kuwait, which would have ensured that Iraq suffered no damage, Bush refused to discuss such restrictions. If war came, the coalition would strike at the heart of Saddam's power—inside Iraq.

At every stage—at home and abroad—Bush encountered questions, debate, and opposition; one by one, he overcame them.

THE WAR BEGINS—FOR AMERICANS

On January 16, 1991, Bush announced that allied air forces, as he called them, had begun attacks in Iraq and Kuwait. "The twenty-eight countries with forces in the Gulf area have exhausted all reasonable efforts to reach a peaceful resolution," he told the nation, "and have no choice but to drive Saddam from Kuwait by force."[67] He promised that Saddam's nuclear and chemical potential would be destroyed and that Iraqi forces would leave Kuwait. "The legitimate government of Kuwait will be restored to its rightful place and Kuwait will once again be free."[68] Noting the five-month delay, he acknowledged flatly that "while the world waited, Saddam Hussein systematically raped, pillaged, and plundered a tiny nation no threat to his own." He quoted a U.S. Army lieutenant: "If we let him get away with this, who knows what's going to be next."[69]

Bush did not micromanage the war; he operated on the principle that military decisions should be left to the military. Secretary of Defense Richard Cheney, Joint Chiefs of Staff Chairman Colin Powell, and the force commanders made the strategic and tactical military decisions. Bush concentrated on dealing with the heads of state in the coalition, with Congress, and with the American people.

Throughout, Bush proved himself a man of great steadiness who sat at his own center of gravity and communicated confidence in the goals he set for the nation and the coalition. His confidence was contagious and he turned in a remarkable performance, supported by the dazzling power and precision of America's high-tech weaponry, its highly professional military forces, and effective leadership in the Pentagon. Together, they buried the ghost of Vietnam in the desert of Kuwait.

THE WAR ENDS—FOR THE COALITION

In approximately one hundred hours, coalition forces captured 73,700 square kilometers of territory and cut the Iraqi army to pieces—leaving only seven of forty-three divisions capable of operating effectively. Casualties for the coalition were few: for the United States, 148 dead and 458 wounded; for other coalition members, 92 dead and 318 wounded.[70]

Then, after the collapse of Iraq's army, Bush decided to end the war—leaving Saddam Hussein in power and his Republican Guard nearly intact. Charles Lane wrote in the *New Republic*, "In the final hours of the ground war . . . Colin Powell counseled the President to call off the fighting. It was the only 'option' he proposed."[71]

In an address to the nation on February 27, 1991, Bush announced the suspension of hostilities in decidedly international terms:

> Kuwait is liberated. Iraq's army is defeated. Our military objectives are met. Kuwait is once more in the hands of Kuwaitis in control of their own destiny. . . .
>
> No one country can claim this victory as its own. It was not only a victory for Kuwait but a victory for all the coalition partners. This is a victory for the United Nations, for all mankind, for the rule of law, and for what is right.

After describing the terms of the cease-fire, Bush added:

> . . . I have asked Secretary of State Baker to request that the UN Security Council meet to formulate the necessary arrangements for this war to be ended.[72]

Thus, Bush placed the fate of Saddam Hussein not in the hands of the coalition that had defeated him, but in the hands of the UN Security Council, some of whose members, like Russia, had merely acquiesced in the Gulf War and would remain profoundly ambivalent, even hostile, to the use of force against a recalcitrant Saddam Hussein in the aftermath of this military action.

Bush's defense of Kuwait against Iraq's invasion was instinctively understood and broadly supported by a large majority of the American people, but many of these same Americans did not understand why Saddam's forces were spared or why his subsequent massacre of Iraqi civilians was permitted. Nor did they understand why Israel, the smallest state in the region and the only democracy, had been pressured to remain passive under attack. A sudden end of a war that spares the aggressor and gives him the chance to fight again and to repress minorities again requires much explaining. Over the years, Congress has pressed many inquiries about the end of Desert Storm.

The quick military success made clear to all that the U.S. military had learned important lessons from Vietnam. Saddam had also analyzed the Vietnam War, but he learned the wrong lessons. He had concluded that Americans lacked the weapons and the will necessary to defeat Iraq's well-equipped and experienced army. He underestimated the American people and George Bush, and he continued to do so after his forces had withdrawn from Kuwait. Saddam did not stop making war; he merely found another target.

THE WAR ENDS

Elie Kedourie, one of the true experts on the Middle East, described as a great mystery the Bush administration's decision to abruptly end Operation Desert Storm, allowing "the tanks and the helicopters and the soldiers to escape and be used again by a ruler whom no law can contain and no popular suffrage can dismiss from office."[73] The result of this policy was a massacre of civilians unequaled in the region since Syria's Haffez Assad slaughtered some thirty thousand Sunni Arabs at Hama in early 1982.

Kedourie, who died in 1992, believed that no one ever truly understood why the Bush administration permitted Saddam's forces to survive. At a June 1991 hearing of the House Foreign Affairs Subcommittee on Europe and the Middle East, when assistant secretary of state John Kelly was asked whether the allies should have continued on to remove Saddam from power, he responded:

> It might well have [created a more stable Middle East], but when the president made his decision to terminate hostilities against Iraq, it was in a view that the liberation of Kuwait had been assured and the Security Council resolutions had been implemented. To have continued to drive into Iraq would have, in our belief, exceeded the authorization of the Security Council resolutions and would have resulted in great numbers of additional casualties, American and other. And it was not the belief of the administration that the pursuit deep into Iraq and the additional casualties were either authorized or called for by either the Security Council resolution or the measure that the American Congress had voted.[74]

Some speculated that the administration's decision to stop the fighting was motivated by a desire to preserve Iraq's territorial integrity from Iran. But, as Kedourie rightly observed, "Why the integrity of a state with such a continuous record of violence and malfeasance should have been thought worth preserving is quite mysterious."[75]

Bush himself took pains to explain his decision, which was grounded in his belief that Saddam Hussein could not long remain in control of his country. In this he was challenged even by his own coalition commander. When General Norman Schwarzkopf appeared on the *MacNeil/Lehrer NewsHour* on March 27, 1991, he claimed that he had not wanted to end the war when Bush did but wanted to finish the job. "Frankly, my recommendation had been . . . to continue to march," Schwarzkopf insisted.

"We could have completely closed the door and made, in fact, a battle of annihilation."[76] But Bush's position was firm: "We are not targeting Saddam and we have no claim on Iraqi territory," he told a press conference shortly after the end of combat operations.[77]

It seemed clear to many observers that Bush wanted to see Saddam out of power, but he wanted the Iraqis to remove him. On February 28, Bush wrote in his diary: "He's got to go, and I hope those two airplanes that reported to the Baghdad airport carry him away. Obviously when the troops straggle home with no armor, beaten up, fifty thousand . . . and maybe more dead, the people of Iraq will know. Their brothers and their sons will be missing never to return."[78]

A month later, Bush told a group of reporters:

> With this much turmoil, it seems to me unlikely that he can survive. People are fed up with him. They see him for the brutal dictator he is. They see him as one who has tortured his own people and . . . took his country into a war that's devastating for them.[79]

But Bush underestimated the strength of Saddam's grip on Iraq and the lengths to which he would go to stay in power.

Among the reasons offered by observers to explain Bush's decision to end the war were pressures from Saudi Arabia and Egypt, pressures from Moscow, and concern about becoming bogged down in an occupation of Iraq. Whatever the contributing factors, he seems to have had misgivings even as the moment of decision was upon him. On February 25, two days before announcing the end of the war, he contemplated the possible ramifications of the decision in his diary. "It seems to me that we may get to a place where we have to choose between solidarity at the UN and ending this thing definitively," he wrote. "I am in favor of the latter because our credibility is at stake. We don't want to have another draw, another Vietnam. . . . We're not going to permit a sloppy ending where this guy emerges saving face."[80]

Bush later commented that "I was not about to let Saddam slip out of Kuwait without any accountability for what he had done, nor did I want to see an Iraqi 'victory' by default, or even a draw. Either he gave in completely and publicly, which would be tantamount to a surrender, or we would still have an opportunity to reduce any future threat by grinding his army down further."[81] And yet the concerns remained. On February 28—the day after the war ended—he wrote in his diary: "Still, no feeling of euphoria. . . . After my speech last night, Baghdad radio started broadcasting that we've been forced to capitulate. I see on the television that public opinion in Jordan and in the streets of Baghdad is that they have won. . . . It hasn't been a clear end—there is no battleship *Missouri* surrender. This is what's missing to make this akin to WWII, to separate Kuwait from Korea and Vietnam."[82]

In *A World Transformed*, written seven years after the Gulf War, Bush gives perhaps his fullest explanation of why he chose not to lead the coalition on to Baghdad to remove Saddam from power:

> The end of effective Iraqi resistance came with a rapidity which surprised us all, and we were perhaps psychologically unprepared for the sudden transition from fighting to peacemaking. . . . We soon discovered that more of the Republican Guard survived the war than we had believed or anticipated. . . . We were soon disappointed that Saddam's defeat did not break his hold on power, as many of our Arab allies had predicted and we had come to expect. . . .
>
> While we hoped that a popular revolt or coup would topple Saddam, neither the United States nor the countries of the region wished to see the breakup of the Iraqi state. We were concerned about the long-term balance of power at the head of the Gulf. Breaking up the Iraqi state would pose its own destabilizing problems. . . .
>
> Trying to eliminate Saddam, extending the ground war into an occupation of Iraq, would have violated our guideline about not changing objectives in midstream, engaging in "mission creep," and would have incurred incalculable human and political costs. Apprehending him was probably impossible. . . . We would have been forced to occupy Baghdad and, in effect, rule Iraq. The coalition would instantly have collapsed, the Arabs deserting it in anger and other allies pulling out as well. . . . Furthermore, we had been self-consciously trying to set a pattern for handling aggression in the post–cold war world. Going in and occupying Iraq, thus unilaterally exceeding the United Nations

mandate, would have destroyed the precedent of international response to aggression that we hoped to establish.[83]

The way the war ended provoked swift criticism from, for example, Colonel David H. Hackworth, who reported based on extensive conversations with soldiers who fought in Operation Desert Storm:

> They all felt we left the job undone, even though we had been on the verge of total victory. . . . Early on, Gen. Colin L. Powell said Iraq was a snake and we were going to chop off its head and kill it. But that's not what we did. What we really did was chop off a piece of its tail. We completely missed the head. The snake, with its tail bobbed, slithered off to heal and fight another day.[84]

But the war was largely viewed as a phenomenal diplomatic and military success; the allied forces had achieved their goal of removing the Iraqis from Kuwait with few allied casualties. Strengthened by his belief that Saddam had been mortally weakened, President Bush thought it best to declare victory and bring the U.S. troops home. Another February 28 diary entry suggests that Colin Powell's beliefs mirrored his own: "Colin last night put it in perspective—this is historic and there's been nothing like this in history," Bush wrote. "Bob Gates told me this morning, one thing historic is, we stopped. We crushed their forty-three divisions, but we stopped—we didn't just want to kill, and history will look on that kindly."[85]

But Saddam was not finished. He wasn't even chastened. Soon he would turn his wrath, and the remains of his forces, on the Iraqis themselves.

SADDAM'S NEXT WAR

The survival of Saddam Hussein and his military forces left Iraq's civilians, especially its minorities, painfully vulnerable. Shortly after he capitulated to coalition forces, he began a violent, intensive war against Iraq's minorities, focusing his wrath on all those who might oppose him internally. In the months that followed, the Iraqi government's merciless treatment of its Kurdish and Shiite populations created terrible human suffering for more than 2.5 million refugees and precipitated a confrontation between two basic principles of international order. This conflict—between the human rights of refugees and the principle of noninterference in the internal affairs of states—initially delayed effective protection of Kurdish and Shiite ethnic groups.

A special report to the UN on the status of human rights in Iraq, written by Dutch diplomat and special rapporteur Max Van der Stoel and released in early 1992, described in gruesome detail a record of baseless arrests; unspeakable torture, including electric shock, burning, beating, rape, and the extraction of teeth and nails; and arbitrary executions of individuals, families, and whole villages. The report catalogues the torture and murder of children; the sudden, unexplained disappearances of Iraqi citizens; and a litany of capricious sentences before Saddam's despotic courts.

"It is clear," Van der Stoel reported, "that deliberate actions of the Iraqi government have caused refugee flows, forced urbanization and internal deportation."[86] At least two million people fled to the Kurdish hills in the spring of 1991. "Detailed reports allege the destruction of some four thousand villages affecting well over a million people," Van der Stoel wrote. Kurdish property was stolen and farmland mined, and people were gassed and denied food, fuel, and medicine through an internal blockade—a kind of "siege within a siege."[87] Iraq's one million Assyrians also suffered massacres, forced relocation, and the systematic destruction of their villages, churches, and schools. The Shia of southern Iraq, too, were special targets of Saddam's wrath—thousands were arrested, imprisoned, tortured, and executed.

Van der Stoel concluded that the Iraqi actions amounted to no less than genocide. After carefully documenting the charge, meticulously reviewing both detailed testimony from the victims and the Baghdad government's own explanations, he concluded that "[t]he Government of Iraq has systematically violated and continues to violate its international human rights obligations. . . . The number of victims suffering from these violations is certainly in the hundreds of thousands, if not much higher."[88]

Humanitarian Intervention

In the freezing spring of 1991, Saddam's forces drove fleeing Kurds from their homes to the borders of Iran and Turkey, creating great human misery and threatening to destabilize Turkey and the always-tense relations in this area. In response, the Bush administration proposed passage of Security Council Resolution 688, which defined these massive human rights violations as a threat to international peace and security and providing the Security Council with a first-ever justification for the use of force in the internal affairs of a nation. A few months later, when warring Somali warlords and acute food shortages threatened hundreds of thousands of Somalis with starvation, Bush turned again to the Security Council to secure authorization for the use of force in the internal affairs of a member state. These authorizations to use force to provide humanitarian aid in a militarily risky setting differed from the authorization to drive out Iraqi forces. They were enacted after the heady victory in the Gulf War, when the idea that military force could and should be used for purposes beyond the protection of a nation's vital interest, was gaining acceptance.

A major innovation in the international law of human rights occurred on April 5, 1991, when the UN Security Council—led by the United States—passed a resolution condemning Iraq's repression of its civilian population as "a threat to international peace and security" and, thus, that it was the proper concern of the Security Council. Resolution 688 affirmed these two principles by a 10 to 3 vote (Cuba, Yemen, and Zimbabwe voted no, and China and India abstained). The two principles are consistent with the UN Charter, but both were major departures from conventional UN doctrine.[89]

In the past, the most brutal repression by a government of its own population was treated as an internal matter that was beyond the jurisdiction of the Security Council—even if it created a million refugees and put destabilizing pressure on neighboring states. Until the adoption of Resolution 688, Article 2(4)'s noninterference principle was accorded a position of paramount importance in most UN discussions of international law.[90]

Adherence to the principle of noninterference had prevented action to stop even the most terrible human rights violations. When Idi Amin killed tens of thousands of Ugandans in the 1970s, and Pol Pot starved, beat, and worked to death approximately two million Cambodians from 1975 to 1979, the Security Council took no action. When Ethiopia's Mengistu Haile Mariam created a massive famine with his forced "villagefication" policy, the Security Council took no action. These humanitarian catastrophes were regarded as internal matters, as were China's Great Leap Forward of 1958 and the decade-long Cultural Revolution that began in 1966, each of which slaughtered untold millions. Only South Africa (which because of its racist government structure was considered an illegitimate state) was regularly scrutinized and condemned by the UN Security Council for its treatment of its own population.

Resolution 688 could only be passed because the cold war had ended, the Soviet bloc had collapsed, and the major governments were changing their views about the proper business of the Security Council. China, which could have blocked the passage of the resolution (as it could have blocked the passage of the Gulf War resolutions), did not, probably because of its powerful distaste for standing alone.[91]

In early April 1991, the Security Council considered a second measure to save the Kurdish refugee population. British prime minister John Major argued for the creation of secure enclaves inside Iraq, protected by the United Nations, that would be large enough to include population centers. The European Community supported Major's position, but the United States, the Soviet Union, and China expressed reservations about the violation of Iraq's territorial integrity. UN support for the safe haven concept would reappear in the Bosnian conflict, again giving respect for human rights priority over respect for territorial integrity.[92]

These proposals were a clear indication that the new world order would have substantially higher standards of conduct than the old and would give a greater priority to the rights of people compared with the rights of whatever government is in power.

George H. W. Bush's Vision

These emerging views were congruent with George H. W. Bush's vision. He was convinced that Americans had a special mission and he frequently spoke of it.

He shared with Wilson, FDR, and Truman the twentieth-century American dream of a world of law and peace preserved through collective action—a world order based on "peaceful settlement of disputes, solidarity against aggression, reduced and controlled arsenals, and just treatment of all peoples." [93] Bush described his dream in a speech to the UN General Assembly on September 21, 1992. His adult life, he said, had been marked by successive conflicts between tyranny and freedom, and deep divisions between totalitarianism and democracy. Now, with the end of the cold war, he dreamed of transcending these divisions:

> I believe we have a unique opportunity to go beyond artificial divisions of a first, second, and third world to forge, instead, a genuine global community of free and sovereign nations—a community built on respect for principle, of peaceful settlement of disputes, fundamental human rights, and the pillars of freedom, democracy, and free markets. [94]

The three dominant challenges of this new world would be to keep the peace, prevent proliferation of weapons of mass destruction, and promote prosperity for all in an open economic order. Bush believed the United Nations would have a special role in meeting these challenges: to provide and coordinate peacekeeping, enforce nonproliferation, and eliminate the walls that divide people and prohibit trade.

Bush saw the defining characteristics of the new world order as a global perspective, a proclivity for multilateral engagement and collective action, a lesser reluctance to use force, and a greater deference to and broader reliance on the UN for pursuit of American foreign policy objectives. He gave high priority to the institutionalization of the Security Council's role in actions under Article 51, emphasizing the council's role in each phase of the response to the Iraq invasion.

After Iraq withdrew its forces from Kuwait, Bush sponsored a major expansion of the Security Council's jurisdiction to include humanitarian intervention into the internal affairs of states. In April 1991, his efforts resulted in Resolution 688 and the creation of a UN mission to enforce surveillance. Within the year, Bush sponsored a resolution calling for the use of force, if necessary, under Chapter VII to deliver humanitarian relief to starving Somalis. Each of these actions substantially expanded the jurisdiction of the Security Council into areas from which it had previously been excluded.

Through his repeated moves into multilateral arenas, his more frequent use of force, his reliance on collective action and UN Security Council permission to act, and his repeated expansion of UN jurisdiction, Bush significantly altered U.S. policy and expectations concerning the use of force and the role of multilateral institutions.

In Kuwait, Bush wanted more than an extra layer of legality for his decision to turn back Saddam Hussein's invasion. He wanted to establish and strengthen a precedent for collective response to aggression through the UN. In *A World Transformed*, he describes how he sought to use the Gulf War to reinvigorate the Security Council:

> Building an international response led us immediately to the United Nations, which could provide a cloak of acceptability to our efforts and mobilize world opinion behind the principles we wished to project. Soviet support against Iraq provided us the opportunity to invigorate the powers of the Security Council and test how well it could contribute. . . .
>
> It was important to reach out to the rest of the world, but even more important to keep the strings of control tightly in our hands. In our operations during the war itself, we were . . . attempting to establish a pattern and precedent for the future. . . .
>
> The Gulf War became . . . the bridge between the cold war and post–cold war eras. . . . Superpower cooperation opened vistas of a world where, unlike the previous four decades, the permanent members of the UN Security Council could move to deal with aggression in the manner intended by its framers. . . . [W]e emerged from the Gulf conflict into a very different world.[95]

Desert Storm was a collective action, taken through the United Nations, in which a number of countries joined together to defend a member state against international aggression with authorization of the Security Council. This is the one use of force clearly foreseen and accepted in the UN Charter.

Operating under a clear mandate and with Bush's leadership, the United States organized and led a predominately American multinational force in a massive, successful effort that quickly achieved its stated goal: the withdrawal of Iraqi forces from Kuwait. As Bush explained later, the war's beginning and end were guided by Security Council resolutions. U.S. forces, he said, did not pursue and destroy Saddam's forces because the authorizing resolution limited the scope of military action, calling only for Iraq's withdrawal from Kuwait.

But what kind of precedent for what kind of new world order did the Gulf War set?

From 1948, when the United Nations deployed 259 peacekeepers to oversee the armistice between Israel and the Arab states, until approximately the end of the cold war, UN peacekeeping was carried out according to the principles articulated by Secretary-General Dag Hammarskjöld in the Suez Crisis in 1956. The Hammarskjöld model assumed a conflict between states, a cease-fire, or between the parties to the conflict, the consent of the conflicting parties to the peacekeeping mission, the neutrality of the peacekeepers, and minimum use of force by peacekeepers.[96] The model postulated a multinational military action authorized by the Security Council under Chapter VI of the UN Charter. But the Gulf War was assuredly not a peace operation—it was a war. The forces dispatched to enforce the resolution did not have Saddam's consent. They did not rely on peacekeeping rules of engagement or on the principle of minimum force. Instead, U.S. leaders, applying the lessons of Vietnam and the Weinberger-Powell Doctrine, operated on the principle of overwhelming force, congressional and popular support, decisive action, and victory as a goal.

Desert Storm was carried out not under the command and control of the UN secretary-general, but under U.S. commanders collaborating with those of more than two dozen other countries, several of which were principal U.S. allies. It was a coalition of the willing under American leadership. Javier Perez de Cuellar, the UN's secretary-general at the time, interposed no obstacles; in fact, he helped as he was able. The Security Council passed the resolutions that authorized the war's foundational policies. The Secretariat assisted with coordination. The Gulf War was successful and efficient in achieving its limited objectives, though its slow start gave Saddam a prolonged opportunity to inflict damage on the people and resources of Kuwait. The war's early end left the Republican Guard intact and Saddam in power and strong enough to impose murder and mayhem on the Kurds and Shiites in Iraq.

Despite the dazzling demonstration of American military power and the professionalism of U.S. forces, the Gulf War displayed some of the characteristics of later, unsuccessful multinational operations. James Baker's five-month-long effort to secure and preserve consensus and to elicit financial commitment came at a very high price, especially considering that a consensus existed in the Security Council for condemning Saddam's invasion from the day of the invasion, and that most of the money to wage the war was contributed by a mere five nations: the United States, Saudi Arabia, Kuwait, France, and Great Britain.

Still, Desert Storm was a clear example of collective action against international aggression. By his actions and words, Bush committed the United States to the principles of the UN Charter and the resolutions of the Security Council. The Gulf War was successful, in spite of the disadvantages of war by committee and the difficulties of recruiting a coalition and maintaining a consensus in the Security Council.

Although as a military operation Desert Storm had been a great success, it quickly became clear that the threat posed by Saddam to the region or to Iraq's minorities had not been eliminated. By mid-March through early April 1991, Saddam's forces drove fleeing Kurds from their homes toward the borders of Iran and Turkey, creating great human misery and threatening the always-tense relations in the area.

After the Iraqi withdrawal, the Bush administration had turned its efforts to the delivery of humanitarian assistance to Kurds, Shiites, and Somalis, and the enforcement of the terms of the armistice when the Bush team broke new ground again. Bush returned to the UN, undertaking more intrusive activities through the Security Council, and proposed the passage of Resolution 688 that defined massive human rights violations by their government as a threat to international peace and security. Resolution 688, which passed on April 5, signaled a new era in which UN member states could use force not just to respond to aggression but also for humanitarian purposes, and it provided the Security Council with the justification for the first time ever to use force when engaging the internal affairs of a nation.[97]

By providing the Security Council with more power and a wider reach as part of Bush's New World Order vision, 688 may inadvertently have set the stage for future expansionist secretary-generals' fastidious attempts to increase their role and the UN role within the world community. In December 1992, when warring clan leaders, food shortages, and natural disasters threatened hundreds of thousands of Somalis with starvation, Bush again turned to the Security Council, securing authorization under Chapter VII to help create a secure environment in which humanitarian assistance could be delivered to the suffering Somalis. Resolution 794 permitted the U.S. the use of "all necessary means" to accomplish the mission.[98]

THE NEW PEACEKEEPING

Soon after the Gulf War ended, the United Nations chose a new secretary-general, the French-educated Egyptian Copt Boutros Boutros-Ghali (with George Bush providing the necessary U.S. vote); the Americans chose a new American president, William Jefferson Clinton; and there followed a veritable explosion of UN activities involving the use of force. For the new secretary-general and many governments, including ours, UN peacekeeping became the method of choice for dealing with conflict. The number, variety, and scope of peacekeeping operations grew and expanded. These operations involved the United States and others in unprecedented interventions in the internal affairs of member states, often undertaken in haste, and under new doctrines whose implications had barely been explored.

Some peacekeeping operations after the end of the cold war fit the conventional Hammarskjöld model, but most did not. "Peacekeeping" operations were undertaken in conflicts within states as well as between them, in situations where there was no armistice or cease-fire, and in those in which there was only shaky consent to the mission on behalf of the conflicting parties. Some operations involved new activities: monitoring human rights practices, observing or overseeing elections, and repatriating refugees. So diverse have the concept and practices of *peace-keeping* become that the term may refer to any activity—diplomatic, military, humanitarian, political, or economic—whose purpose is to contribute to the peace, security, and well-being of a group or people, and which is carried out by a multinational force under UN auspices.

The expansion of peacekeeping operations took place rapidly and haphazardly, in response to pressing, often unanticipated, problems and new, often unexamined, ideas about multinational action. Delivering humanitarian assistance to civilian populations of Kurds, Shiites, Somalis, Croatians, and Bosnians caught in bitter conflicts within or among nations became a principal occupation of UN peacekeepers. The instabilities of the post–cold war period, the Clinton administration's enthusiasm for multinational activities, and Boutros-Ghali's expansive bureaucratic appetite and elastic doctrine of peacekeeping encouraged a dramatic expansion of UN jurisdiction based on new views about the functions appropriate to states, regional organizations, and the UN.

POSTSCRIPT: THE ARMISTICE

As Iraqi spokesmen reiterated tirelessly, the Gulf War victory did not defeat Saddam Hussein. It only drove him out of Kuwait and forced inspections on plants where his regime was believed to be developing nuclear and other weapons. Under extreme duress, Saddam had agreed to the sanctions imposed as a condition of the cease-fire.[99] But soon it was clear to all that he was violating a number of the terms. It was also clear that nothing much happened when he did. On February 19, 1992, the Security Council noted that Iraq's "failure to provide full, final and complete disclosure of its weapons capabilities" was a breach of Resolution 687. Nothing happened. A similar Security Council notice on March 11 produced a similar result.[100]

Iraq violated the cease-fire agreement repeatedly. It refused to participate in the work of the Boundary Commission demarcating the boundary between Iraq and Kuwait, refused to permit delivery of food and medicine to Kurds in northern Iraq, and defiantly used fixed-wing aircraft (specifically forbidden by the cease-fire and the Safwan Accords of 1992) to attack Shiite villages in southern Iraq. And, of course, Saddam repeatedly refused to provide access to multiple sites for inspection. In the summer of 1992, a confrontation over access to documents took place outside Baghdad's Department of Agriculture between Iraqi government officials and UN inspectors. Some inspectors—but no American, French, or British personnel—were permitted to enter the building.

These acts of noncompliance were accompanied by Iraqi demands that the sanctions be lifted. They were met by American, British, and French threats, but these were rendered hollow when the chief of the UN team decided to accept Saddam's conditions for inspection. Saddam's offer of limited access to the Department of Agriculture building, and his demand for a veto over membership on the inspection team, were treated as acceptable, even though the Iraqis had already had ample opportunities to remove any materials they desired to protect from UN eyes.[101]

Saddam's conditions may have seemed less offensive to the UN team and its leader, Ambassador Rolf Ekeus, than they did to most observers because they were similar to the conditions under which the parent body of the inspection team—the International Atomic Energy Agency (IAEA)—regularly operates. Many people think the IAEA is authorized to police violations of safeguards and of the Nonproliferation Treaty (NPT). But the IAEA inspection system is not designed to be efficient in catching cheaters; the agency negotiates with member states the conditions of access to their facilities, much as Ekeus negotiated with Saddam.

In regular inspections, conducted under regular rules, the IAEA inspects only facilities *declared* by member states, and these only when nuclear material is present. The agency is not authorized to search for undeclared weapons or facilities. The composition of an inspection team must be approved by the nation being investigated. Members may veto classes of people—for example, Americans—if it judges them unacceptable as inspectors. The number of inspectors and their access to the country can also be controlled. The requirement (imposed by some, but not all, countries) that a UN inspection team procure visas provides early warning of any intention to inspect.[102] Saddam sought an escape from the special requirements of the cease-fire—penalties for his aggression in invading Kuwait—and the restoration of Iraq's previous rights as a UN member, and he largely achieved his goal.

The IAEA does not provide much protection against governments that lie and cheat. The agency is governed by its member states, some of which are themselves actively engaged in cheating on the principle of nonproliferation and lying about it. Iraq, Iran, North Korea, Libya, and Syria are all members of the IAEA and sit on its governing board. So are other states that have developed or are working hard to develop a nuclear capacity outside the NPT regime.[103] Iraq is a signatory to the NPT and had long served on the governing board of the IAEA—during the same period when it was working surreptitiously to develop a capacity to produce nuclear, chemical, and biological weapons of mass destruction.

The IAEA suffers problems typical of UN and quasi-UN agencies. It is politicized and tends to be dominated by a third world agenda. Its priorities frequently reflect those of developing countries. Technical competence is only one of the criteria for employment in the IAEA.

Try as it will (and it frequently tries very hard), the IAEA can only monitor countries that are willing to be monitored. It does valuable but limited work in monitoring safeguards for the peaceful use of nuclear energy, but its capacity to deal with problems like those posed by Iraq, or the broader problems of nuclear weapons proliferation, is limited. Today the agency is unable to monitor, much less to police, the activities of Syria, Iran, Pakistan, and India, among other states. What the IAEA can do in the field of nuclear energy is much like what UN peacekeeping forces can do with regard to peace: it can be helpful when the parties are willing.

By provoking a series of confrontations throughout the 1990s, Saddam sought to undermine and discredit the terms of the Gulf War cease-fire, divide the Security Council, and undermine its resolve. Just as the United States and the United Kingdom had assumed responsibility for forcing Saddam out of Kuwait, they tried to force him to abide by the terms of the cease-fire and abandon his pursuit of weapons of mass destruction. The United States repeatedly used force to uphold the Security Council resolutions, which were violated or ignored by Iraq, often with the tacit acceptance of coalition allies and UN officials.

American lives were put at risk and billions of dollars were spent in these confrontations with Iraq. At a hearing of the Senate Appropriations Committee concerning supplemental appropriations for continuing operations in Bosnia and Iraq in March 1998, Senator Pete Domenici (R-NM) estimated the extra cost of the continuing Iraqi headache at $5.3 billion through 1998, over and above the ordinary costs of defense in the Persian Gulf. Each confrontation cost from tens to hundreds of millions of dollars. A single cruise missile costs more than a million dollars; in one December 1998 strike, hundreds of the missiles were fired at Iraqi targets by U.S. and U.K. forces. And although these expenses were incurred by the United States to enforce Security Council resolutions, the expenses were not credited to the United States under its UN assessments.

During the Clinton years, Saddam's repeated provocations took on a familiar pattern. When he believed the timing was right to extract a concession, he imposed limitations on the ability of the United Nations Special Commission (UNSCOM) to carry out its mandate. The Clinton administration would reject Iraq's demand and order a military buildup, and Saddam would check international opinion and tailor his actions accordingly. If he sensed lukewarm resistance, he would make demands that either weakened the sanctions regime or diminished the ability of the inspectors to do their job. Negotiations would then take place in which the U.S. administration, acting through a divided Security Council, would split the difference by granting a concession—for example, allowing an increase in Iraqi oil exports. Clinton would declare that diplomacy, yet again, had triumphed over force. From time to time there were variations—the personal intervention of the UN secretary-general or the good offices of a nation sympathetic to Iraq, such as Russia—but the outcome was usually the same: the undermining of the sanctions regime and the inspectors, and another victory for Saddam.

Again and again, Saddam sought to build support for lifting the sanctions while simultaneously hiding weapons of mass destruction. In each confrontation, the Security Council split over the appropriate response. Often, the United States could not even convince the council to support the use of force to enforce its own resolutions, which demonstrated the council's impotence in the role of sheriff of the new world order. And as the crises continued to occur, the allies, who had shown such unity during the Gulf War, chose to pursue their own national interests at the expense of multilateralism.

CHANGING PURPOSES, CHANGING PROBLEMS, AND CHANGING PARADIGMS

The policy elites positioned to take charge of American foreign policy at the end of the cold war had many ideas about the nature of the world to come—ideas that were quite different from those that had dominated American politics during the cold war.

In the past, the strategies that had dominated discussion in the United Nations and the liberal international law community emphasized the inviolability of state sovereignty, the prohibition against intervention in the internal affairs of states, and the illegitimacy of force in international relations. Each of these principles was questioned and revised or abandoned in the first years after the cold war ended, giving way to new arguments in favor of collective action through the United Nations. Actually, a state's rights to sovereign control and inviolability of its territory and people, the principle of nonintervention in a state's internal affairs, and the illegitimacy of using force in international affairs had never been accepted as binding by the Soviet Union, the United States, or most other governments. For the Soviet Union, a socialist revolution had priority over the sovereign inviolability of another state's territory.

These principles were further subordinated when Bush mobilized force against Saddam, and when the United States and other governments responded to Saddam's massive violations of human rights in Iraq and to the imminent threat of starvation in Somalia. And they would be called into question again by the Clinton administration's decision to use force to restore Jean-Bertrand Aristide's government in Haiti. The prohibition against the use of force, which had seemed compelling to many American liberals a few years earlier, no longer seemed absolute; they were more likely to see the use of force as a variable, depending on the decisions of the Security Council and who occupied the White House.

As the cold war wound down, the view had been expressed again and again—in publications and speeches, in public policy circles, and in and out of the Bush administration—that future problems (hunger, chaos, anarchy) would transcend national boundaries and would require resolution by multinational groups acting collectively through global institutions in multilateral arenas. For the first time, experts and analysts argued that, whenever possible, the United States should act collectively through global organizations rather than nationally or unilaterally. More and more frequently, the most inclusive and complex multilateral institution in the world, the United Nations, was cited as the preferred arena and instrument for action—on aggression, famine, starvation, and nation building; for containing civil war in the Soviet Union and "restoring democracy" in Haiti; for whatever problem was at hand.

The Bush and Clinton administrations saw almost no limits to U.S. involvement; they conceived the United States as potentially engaged everywhere in the world, as needed. Both presidents sought to deal with international crises through multilateral action coordinated through the UN.

George Bush was the first to take steps toward a system of multilateral military operations authorized by the Security Council. "What is at stake," Bush said, "is more than one small country. It is a big idea, a world where brutality will go unrewarded and aggression will meet collective resistance." [104] It was Bush who set the precedent of greater reliance on and cooperation through the UN. He broke new ground again and again, expanding the Security Council's jurisdiction by seeking to have Saddam Hussein's violation of Kurdish human rights declared a threat to international peace and security, and sending armed forces to deliver humanitarian relief in Somalia. He stretched the UN's jurisdiction and expanded U.S. involvement with the UN.

Through his repeated moves to multilateral arenas, his resort to the use of force (in Panama, Kuwait, and Somalia), and his break with the Weinberger-Powell Doctrine, Bush took the first long steps toward the new world order and set new limits. The Gulf War had involved a matter of genuine national interest, had enjoyed widely popular support, had a clearly articulated goal, and could be won by mobilizing overwhelming force—yet Bush returned repeatedly to the Security Council for approval. Just as he sought the approval of the U.S. Congress because he wanted to demonstrate his respect for the American Constitution, he demonstrated respect for the rule of international law by engaging the UN.

Since Bush already had a strong legal and moral foundation on which to base U.S. military assistance to the Kuwaitis, why did he invest so much time and effort courting a broad group of marginally interested nations in multilateral fora? The answer lay in Bush's dream of establishing the new world order as a reality. As Laurence Martin, director of the Royal Institute of International Affairs, observed:

> George Bush's proclamation of an impending New World Order in the midst of the Persian Gulf crisis was more than a celebration of a rebirth for the United Nations system, now released from its cold war freeze by the emerging consensus on the Security Council. It was the third international organization, collective security based utopia to be announced to the world by an American president in this century.[105]

After Bush's departure from office, Bill Clinton quickly transformed the operation in Somalia from modest, conventional peacekeeping to a more expansive nation-building role. Clinton's team had arrived in office with high hopes for peacekeeping, yet soon after the inauguration the Clinton administration joined an accelerating worldwide trend toward using force more frequently in the form of multinational operations under UN auspices and command. Clinton upgraded U.S. troop commitments to a UN peacekeeping force in Somalia and promised forces to help implement any peace agreement achieved in Bosnia. His policies quickly engaged the United States in more new conflicts than ever before.

The long-term ramifications of decisions made during the Gulf War would not crystallize quickly. After the success of Desert Storm, American troops left Iraq and stayed out for over a decade as the policy to contain Saddam prevailed. No one predicted, then, that the carefully crafted international alliance to keep the peace with Saddam in 1991 would be torn asunder in 2003, casting the United Nations and the United States far apart on the world stage. Nor could conventional thinking have predicted the other, more ominous, unintended consequence: Bush's decision to send U.S. troops into Saudi Arabia to keep peace and offer protection from Saddam would later provide the pretext for Osama bin Laden's making war on the United States.

2.

SAVING SOMALIA

The disastrous U.S. intervention in Somalia probably did more
to undermine worldwide perceptions of the efficacy of U.S. military power than any event in recent memory. . . . American actions in Somalia influenced the calculations of leaders in Haiti
and Bosnia as they confronted U.S. threats simultaneously.[1]

BARRY M. BLECHMAN AND TAMARA COFMEN WITTES,
*Defining Moment: The Threat and Use of Force in American
Foreign Policy Since 1989*

The end of the cold war left the United States stronger than ever—
stronger than any other nation in the world—and the ensuing Gulf War
demonstrated the power and skill of American military forces. In Washington, Paris, and other world capitals, and at UN headquarters, foreign
service professionals and foreign policy strategists discussed how to use
these resources at a moment when our military capacity outstripped the
dangers, and the collapse of the Soviet Union created opportunities for
new relations among nations. And what better place to start building the
new world order than Somalia—one of the poorest countries of East
Africa, beset by hunger and near anarchy?

The independent state of Somalia was born in 1960 out of the rem-
nants of colonial empires and indigenous clans. Somalia had the characteristics of many new African nations: weak borders, a weak sense of

national identity, a weak central government, and strong subnational clans.[2] Although economically and technologically undeveloped, with a dismally low gross national product and literacy rate, Somalia normally produced enough food for its population. A decade of attempts at democratic self-government ended in 1969 with a coup that brought General Siad Barre to power. Somalia's location on the Horn of Africa gave it strategic interest for major powers during the cold war, but that interest faded with the demise of superpower competition. Barre ruled Somalia by force until shortly before its problems burst onto the world's television screens.

By 1990, Somalia had become a good example of what was becoming known as a "failed state"—a people without a government strong enough to govern the country or represent it in international organizations; a country whose poverty, disorganization, refugee flows, political instability, and random warfare had the potential to spread across borders and threaten the stability of other states and the peace of the region.[3] At the end of the cold war there were several such failed states in Africa, any one of which could theoretically have been considered "a threat to international peace and security"[4] and thus an appropriate object of concern of the UN Security Council and a potential candidate for international peacemaking or peacekeeping.

To Somalia's complicated problems, the American foreign policy establishment brought an optimistic perspective and very good intentions. The United States had no significant national interest, economic or strategic, in Somalia and no history of significant involvement. But in late 1991, American officials were moved by the Somalis' urgent need for food, medicine, and order. Ultimately, the response to that need involved the United Nations, the Organization of African Unity (OAU), the International Committee of the Red Cross (ICRC), the United States, and roughly twenty-nine other nations.

At first, this involvement took the form of a judicious, humanitarian peacekeeping mission, begun during the administration of George H. W. Bush, which avoided the temptation to overreach. During the Clinton administration, however, the Somalia mission took a more ambitious—and arguably irresponsible—direction, toward what became known as "assertive multilateralism." It was in Somalia that the United States first ventured onto the slippery slope between peacekeeping and nation building through the use of force, and learned a lesson about relying on inefficient and insufficient UN forces to do dirty work that we are not equipped or assigned to do.

SOMALIA'S DISINTEGRATING GOVERNMENT

By the time it attracted international attention, Somalia was in terrible shape. Barre had governed for twenty years as a typical African strongman, but he had found it more and more difficult to maintain control of the country. At the end of the 1980s, disorder intensified and violence and repression spread. In May 1990, Barre arrested his leading opponents, and his personal guards fired into a crowd at a soccer match, killing sixty-five people. He promised elections the following February, but before they could be held, rebel forces drove his government from power. On January 27, 1991, Barre fled Mogadishu. Almost immediately, the opposition split into multiple factions. The United Somali Congress (USC) named Ali Mahdi Mohamed as interim president, while another group named Umar Arteh Ghalib as interim prime minister.[5] Both were rejected by Mohammed Farah Aideed, the leader of a third faction.

Fighting broke out among the factions, refugees multiplied, and famine developed. In December 1991, UN secretary-general Javier Perez de Cuellar consulted with the OAU, the League of Arab States (LAS), and the Organization of the Islamic Conference (OIC) in an effort to find a peaceful solution to the conflict. Perez de Cuellar dispatched his undersecretary-general for political affairs, James O. C. Jonah, to the area, but no peaceful solution was forthcoming. War and anarchy spread, interrupting the normal cycles of planting and harvesting, thus causing famine. By January 1992, the International Red Cross was reporting a widespread danger of starvation in Somalia.

On January 23, 1992, in Resolution 733, the Security Council called for a cease-fire, an arms embargo, political reconciliation, and increased humanitarian assistance for the nation. Pictures of starving Somalis appeared on television screens in Europe and America, and humanitarian organizations and nongovernmental organizations (NGOs) stepped up efforts to deliver food. UN observers were dispatched. They and the

NGOs reported the proliferation of armed profiteers, warring clans, and blocked ports, which made it difficult and dangerous to get food through to the hungry.

Boutros Boutros-Ghali's appointment as secretary-general in January 1992 put a French-educated Egyptian, with decades of interest and experience in Africa in general and Somalia in particular, at the helm of the United Nations.[6] Boutros-Ghali's appointment was strongly backed by Egyptian president Hosni Mubarak and French president François Mitterrand and was supported by U.S. secretary of state Lawrence Eagleburger. Boutros-Ghali also had the support of the Nonaligned Movement (NAM), the Socialist International, and the OAU. He believed that Africa deserved more attention, and made it clear that Africa would be a high priority for the UN.

Some Security Council members regarded the spreading war in the former Yugoslavia as a more urgent humanitarian and strategic problem than Somalia, but that view was not shared by either the new secretary-general or the Bush administration. Two key members of the Bush team—Eagleburger and Bush's national security advisor, Brent Scowcroft—were determined that the United States not become involved in the former Yugoslavia, a country they knew well and in which they had long personal service. The two crises competed for attention throughout 1992. In both countries, spreading military conflicts were causing widespread human misery and death, although the problems were very different.

In February 1992, representatives of the OAU, the OIC, and the LAS persuaded two of the warring clan leaders—Ali Mahdi and Aideed—to sign a cease-fire agreement that promised security for humanitarian assistance and deployment of military observers from each of the principal factions. Boutros-Ghali named Mohamed Sahnoun, a skilled Algerian diplomat, as his special representative for Somalia.

Sahnoun's Assessment

By the time Sahnoun arrived in Mogadishu in March 1992, 250,000 to 300,000 Somalis had died of hunger and malnutrition. Most of the nation's livestock had been lost, and half a million Somalis had taken refuge in neighboring countries—mainly Ethiopia, Kenya, and Djibouti.

Women and children were dying at the rate of roughly three thousand a day, and four and a half million people were in urgent need of food. The ICRC estimated that two million people were at risk of death from starvation.[7] But the rivalry between Aideed and Ali Mahdi intensified, and the cease-fire did not hold. In June, Sahnoun reported, "Somalia is today a country without central, regional, or local administration and without services: no electricity, no communication, no transport, no school, no health services."[8]

In the south, clan warfare and starvation were widespread, agricultural cycles were disrupted, and basic services had been devastated by bombardment and war. In the north, major cities were without electricity or running water, and violence was ubiquitous. The tools and equipment necessary to live were missing, broken, disrupted. Displaced people needed food, medical assistance, and security. UN personnel, NGOs, journalists, and others on the ground documented the breakdown of authority and order and the resulting anarchy, in which gangs engaged in extortion, profiteering, and intimidation. Faction leaders were unable to control armed youth and fighters. Chaos reigned.

Sahnoun summarized the situation: "Lawlessness, banditry, and looting have taken the place of major fighting and open factional hostilities. Marauding armed groups, loyal to no particular warlord but only to themselves, pose a grave threat to the safety of international personnel as well as the local population, and hinder the effective delivery and distribution of humanitarian supplies."[9]

On April 24, 1992, acting under Security Council Resolution 751, the UN launched an emergency humanitarian assistance program dubbed the United Nations Operation in Somalia (UNOSOM I), to expedite the delivery of food and deploy fifty unarmed military observers to monitor the cease-fire.[10] But this first effort fell far short of the mark; lawlessness and disorder only intensified through the month of May, making it nearly impossible for help to reach the Somalis. In early July, Aideed said he would permit the UN's fifty unarmed observers to monitor the cease-fire and speed the delivery of food.[11] But UN flights to Mogadishu were suspended that month, and Boutros-Ghali announced plans to send five hundred UN military personnel instead.

On July 27, the Security Council passed Resolution 767 (under

Chapter VI of the UN Charter, which does not authorize the use of force), requesting the secretary-general to make use of "all available means," including an airlift, to deliver humanitarian supplies. In August, at Boutros-Ghali's request, the Security Council passed Resolution 775, increasing UN forces in Somalia by three thousand (although these forces were never deployed).[12] The United States was also taking direct aim at the problem: In mid-August, President Bush announced that he was sending unarmed aircraft loaded with food to Somalia, and he followed up a month later with four U.S. ships. Eventually, Aideed agreed to allow the deployment of the additional troops to protect food supplies and to permit deliveries of food, but the promise was kept only briefly.

Mohamed Sahnoun saw Somalia's problem as fundamentally political, and he believed that the restoration of indigenous leadership and authority was essential.[13] He consulted widely, relying on personal contacts with local leaders and groups to organize a conference attended by representatives of all the regions. But the situation got worse. In mid-September, after an American plane carrying food was fired on, the United States suspended its airlift. A UN food warehouse in Mogadishu was looted. Siad Barre's forces, trying for a comeback, gained control of Bardera, the site of a large relief camp, and forced UN relief workers to close the airport that supplied it. In Mogadishu, the airport was shut down by armed gangs that demanded high fees for the privilege of landing. Five hundred Pakistani soldiers were pinned down, unable either to guard the airport or to defend the relief convoys. By this time, Robert Kaplan wrote in the *New Republic*, Somalia's political culture was "in such crisis that it won't even let you feed the inhabitants unless you send in the Marines."[14]

Sahnoun remained convinced that the breakdown of order and political consensus was the problem, and that political reconciliation— not adding forces—was the answer. In fact, he believed that the UN management itself was creating problems. In a memo to Boutros-Ghali, he reported:

- that most UN agencies were unable to complete their assigned tasks or do what was required to mount a massive relief effort;

- that some UN agency personnel were hardly leaving Mogadishu;

- that most UN agencies were reluctant to coordinate their activities with UNOSOM;

- that much of the food shipments were stolen and marketed by guards or looters;

- that troops from member states were too slow in arriving; and

- that the Russians had violated the international arms embargo with UN connivance, outraging some Somali factions. (UN personnel arrived on the same plane as Russians and Russian equipment for Ali Mahdi.) [15]

By this time, it was clear that General Aideed's faction was the one to be reckoned with in Somalia. But Sahnoun explained to the secretary-general that trying to disarm Aideed's clan alone, rather than targeting all clans at once, was a recipe for continuous civil war. Sahnoun thought he saw signs of progress in getting the local leaders to find common ground. "We have tried to move quickly to reinforce these positive trends," he wrote.[16] But Boutros-Ghali had different ideas. So instead of receiving help from the Secretariat, Sahnoun received two critical messages: one questioning his presence in the Seychelles, where he had organized a conference of political leaders to promote political reconciliation, and a second ordering him to refrain from criticizing UN agencies or personnel.

On October 26, 1992, Sahnoun resigned under pressure, attributing his departure to "bitter experiences with the UN bureaucracy."[17] A few months later, he reflected that "those who foolishly have been pushing for a greater buildup of forces without any kind of strategy bear a heavy responsibility in the tragic events of recent weeks in Somalia."[18] Sahnoun believed that animosities against the UN would spread quickly in Somalia. American journalist Michael Maren later confirmed Sahnoun's account of what went wrong, including the incompatibility of Boutros-Ghali's emphasis on a centralized military solution through the UN with Sahnoun's approach, which relied on local leaders. According to Maren, "UN headquarters in New York wanted something bigger and more sensational than one Algerian diplomat talking peace with Somalia's clan leaders. . . . Boutros-Ghali wanted his massive intervention; Sahnoun stood in the way. And Sahnoun's early successes in getting the factions to talk became a threat to the secretary general's plans."[19]

Boutros-Ghali did show a marked preference for a large multinational military operation, and he was eager to experiment with new kinds of collective military operations and new approaches to the problems of failed states of Africa.

The U.S. Decision to Intervene

It was clear from the beginning that the peacekeepers' mission in Somalia would be very different from any in which the United States or its military had previously participated—different from Operation Desert Storm, from other UN police actions, and from typical peacekeeping, in which the peacekeepers work with separate parties who have signed a cease-fire.[20]

In Desert Storm, the U.S.-led coalition was organized in response to Iraq's invasion of Kuwait—a classic case of international aggression. The Security Council had demanded the withdrawal of Iraq's forces, authorized member states to use all necessary means to effect Iraq's withdrawal, and called on them to provide the necessary armed force to enforce the demand.[21] The Kuwait crisis was conducted within existing legal and political structures, with clear and foreseen goals.

In Somalia, nothing was so clear. Here the problem was not international aggression but the breakdown of internal order and authority, aggravated by the presence of several armed factions and a lack of food. The original goal of the United States was to save the Somalis from starvation by delivering food.[22] But if the objective was simple, achieving it was not. The relief effort was hampered by war and anarchy, by the need to protect relief workers, by the lack of a government in Somalia, and by Boutros-Ghali's desire to limit U.S. independence in the use of force and establish a central role for himself in this and other such international operations.

The situation was further complicated because the use of force to resolve the internal problems of a nation is explicitly forbidden by the UN Charter,[23] which allows the Security Council to consider the use of armed force only where it is necessary "to maintain international peace and security."[24] Nonintervention in the internal affairs of states is a central tenet of the UN, as explicitly stated in Article 2.7: "Nothing contained in the present Charter shall authorize the United Nations to intervene in matters which are essentially within the domestic jurisdiction of any state or shall require the Members to submit such matters to settlement under the present Charter. . . ." Without a direct threat to international peace and security, the UN had no authority to intervene in Somalia's civil war. When confronted with a humanitarian disaster in the post–cold war world, however, concerns about this violation of the Charter were swept away.

It could not be claimed of Somalia (as it was claimed of the Kurds in northern Iraq) that a massive human rights violation had created a danger to international peace and security by driving refugees across borders.[25] For a long time, the crisis was contained in Somalia. There were other differences as well. In Kuwait, the goal was to return a government to power; in Somalia, there was no government.

If Somalia was no Kuwait, it was also no Lebanon, though it reminded some of October 1983, when 241 U.S. Marines, sent to Lebanon as part of a multinational peacekeeping force, were murdered by a truck bomber. But that multinational force was not a UN operation, and there was a big difference in the size of the operation: Where Ronald Reagan provided only a few hundred troops in Lebanon, George Bush would eventually send nearly twenty-eight thousand Americans to Somalia—heavily armed and authorized to use all necessary force. Nor was Somalia another Yugoslavia, where violence broke out among component states after the new Serbian president, Slobodan Milošević, sought to consolidate power, trampling on the Yugoslav Constitution and the historic powers of its states. The violence quickly spread to four states and threatened to spread farther. That situation—to which the United States had not committed troops—constituted a clear threat to international peace and security.

The most pressing problem in Somalia was widespread hunger—a terrible humanitarian problem, but one that did not threaten to ignite an international conflict. Yet Boutros-Ghali insisted that the real problem in Somalia was the nation's economic and political underdevelopment, and the violence, which was at least a theoretical threat to the region. He cited the emerging theory of failed states—and the links among famine, break-down of internal order, and international peace and security—to justify the use of force.

Whatever the underlying reasons for the breakdown in Somalia, President Bush and the United States Congress found the humanitarian crisis sufficient grounds for sending forces to deliver food and medicine.[26] Images of desperate, undernourished Somalis were appearing frequently on television screens by the fall of 1993, creating the phenomenon now known as the "CNN effect." The world became aware of the unfolding tragedy without knowing why the problem had suddenly arisen, how it had developed, or what to do about it. And, as the United States cast about for a strategy to address the problem, the hard facts of the situation—that the famine had both natural and man-made causes, and that war and anarchy on the ground would make solving it nearly impossible—were temporarily overlooked.

In August 1992, Congress passed a resolution calling on the president to seek UN action and deploy security guards to protect food shipments. Bush ordered a new round of food airlifts to Somalia on U.S. aircraft. On September 16, he ordered four U.S. Navy ships to the Somali coast. A few days later, however, a U.S. flight bearing food was fired on; Bush responded by suspending all U.S. flights to Somalia.

To protect further food shipments and personnel, Bush proposed to Boutros-Ghali that aid should be delivered by an American-led coalition outside the auspices of the UN. Under this plan, the United States would deploy up to thirty thousand troops to secure seaports, airports, and roads in central and southern Somalia. These troops would operate under American command and would remain for a limited time, with the UN taking control of the operation after three to four months. The objectives would be specific and minimal: to stabilize the military situation to the extent necessary to deliver relief supplies.[27]

Neither the United States nor the UN had ever undertaken an operation quite like this one, in a country with no government. But top officials in the Bush administration had experience with various kinds of military operations, and Bush himself was deeply interested in international operations, collective action, and promoting cooperation in a UN context. A number of Bush administration members shared the president's interest in trying a new approach to collective security and military operations. The United States had the food, the forces, and the transport capabilities to meet the Somalis' needs, and Desert Storm had left Americans with positive feelings about collective action, the role of the UN, and when it was permissible to use force.

In late November 1992, just two months before he left office, Bush decided to get involved in Somalia—on a large scale. After Secretary of State Larry Eagleburger assured the president that the Somalia operation was "doable,"[28] Bush resolved to move, and Eagleburger informally assured the UN that the United States was ready to take the lead in organizing and ensuring the delivery of food to Somalia. The result was the U.S.-led "peacekeeping" mission that began as "Operation Restore Hope" and became the Unified Task Force (UNITAF), which would take the United States and the UN alike into uncharted waters.

THE NEW PEACEKEEPING

The history of UN peacekeeping to this point had been brief and limited in its scope. It had begun in June 1948, when 259 peacekeepers were deployed to oversee an armistice between Israel and the Arab states. Subsequently, Secretary-General Dag Hammarskjöld drew up principles that shaped peacekeeping efforts in the Suez Crisis in 1956 and long afterward. UN peacekeepers would

1. oversee compliance with a cease-fire or armistice agreement that had already been negotiated . . . patrol a border or serve as a buffer between parties to a conflict;

2. be deployed with the consent of the parties to the conflict;

3. be neutral;

4. operate under the supervision of the secretary-general;

5. be lightly armed and use force only sparingly; and

6. not be drawn from the five permanent members of the Security Council.

Conventional peacekeeping under these principles was a useful tool for containing certain kinds of conflicts, though these missions did not solve problems and rarely, if ever, ended. Nearly every peacekeeping force dispatched in the postwar years is still serving to this day, fulfilling essentially the same function for which it was originally deployed. The forces deployed in 1948 to monitor the Arab-Israeli cease-fire were still there decades later; so were those deployed to the India-Pakistan border in 1949, to Cyprus in 1964, to the Golan Heights in 1974, and to Lebanon in 1978. Since very few troops were employed in UN peacekeeping, and none were American, the concept and practice of peacekeeping was not well understood in the United States, though support for UN-based collective, nonviolent use of force had begun to grow in American foreign policy circles by the end of the cold war.

By the early 1990s, such missions were multiplying thick and fast. As George H. W. Bush noted in a speech to the UN General Assembly in September 1991, "The United Nations has mounted more peacekeeping missions in the last thirty-six months than during the first forty-three years."[29] The increase had reached critical mass during Bush's tenure, when the number of UN peacekeepers quadrupled—from 11,000 to 44,000. By December 1994, the number had reached 80,800, and the U.S. share of peacekeeping costs was set at 31.7 percent of total UN spending on such missions.

In congressional hearings and in the press, Secretary of State James Baker defended the Bush administration's request for $460 million for peacekeeping in fiscal year 1993, up from $107 million in 1992—in addition to a supplementary request for $350 million for 1992. "Peacekeeping is a pretty good buy in my view," he said.[30] Baker defended the UN decision to assess the United States almost a third of the entire peacekeeping budget as consistent with the country's leadership role. "We have a preeminent and unique role in the United Nations as one of the Permanent Five."[31]

In calling for increased peacekeeping expenditures, Baker ventured an argument that his successors, Madeleine Albright and Warren Christopher, would echo later—that "common problems demand a collective security response from the international community" and that "multinational collective engagement is a bargain."[32] Peacekeeping,

Baker predicted, would prove an important tool in resolving post–cold war regional conflicts—and it would be less expensive than war. This argument incorporated two dubious presumptions: (1) that peacekeeping would prevent war, and (2) that conflicts would inevitably lead to war in the absence of peacekeeping.

Baker was not the only official who was more enthusiastic than cautious in his judgments about the potential utility of peacekeeping; many members of the foreign policy elite shared his optimism. On August 28, 1992, the United States joined in Security Council Resolution 775, which increased the number of UN forces in Somalia from five hundred to thirty-five hundred at the request of the secretary-general. Because of high hopes that the UN would play an important future role, the response was generally favorable, even though intervention in the internal affairs of a member state seemed to be a clear violation of the UN Charter and would take U.S. forces into a country where we had no vital interests.

The parameters of peacekeeping, too, were expanding. In *An Agenda for Peace*, a monograph he published shortly before taking office, Boutros Boutros-Ghali described a new vision of UN peacekeeping—one that included operations normally considered as war.[33] During Boutros-Ghali's tenure, peacekeeping came to refer to almost any activity in which conflict resolution was carried out by a multinational force under the auspices of the UN. This new conception dissolved the lines between humanitarian missions, peacekeeping, and military engagement. Some peace operations would involve no use of force, no danger, and no armed conflict; others would be coercive and dangerous. Boutros-Ghali even tried to absorb war itself into the category of peace operations, to change the goals from victory to accommodation, and to vest command and control in the secretary-general.

Somalia would be the first test of this new model of peacekeeping. Previously, peacekeeping operations had taken place under Chapter VI, which did not authorize the use of force. Security Council resolutions 688, which classified human rights within Iraq as a threat to international security, and 775, the resolution on Somalia, both authorized the use of force in interventions in the internal affairs of nations. Boutros-Ghali proposed to give the Secretariat jurisdiction over the conflict in

Somalia, which required some unprecedented concept-stretching to cover intervention in the internal affairs of a member state. The decision by the Security Council that "the magnitude of the human tragedy" constituted a threat to international peace and security (thus justifying the use of force under Chapter VII) was also new, though it had some precedent in Resolution 688.[34]

There were other differences, too. Boutros-Ghali believed that precedent, and Chapter VII, gave him the authority to recruit and organize forces, and to determine the rules of engagement under which they operated. Surprisingly, the Security Council accepted most of his claims, including his "right of oversight . . . in return for legitimizing the operation," and his contention that he was equipped to deal with a civil war.[35]

In *An Agenda for Peace* and a later essay, "Empowering the United Nations,"[36] Boutros-Ghali made a concerted effort to expand the jurisdiction of the secretary-general to include the resolution of disputes before they escalated into conflict. Under the rubric of *preventive diplomacy*, he grouped together the functions of *peacemaking* (as defined in Chapter VI); *peacekeeping* by military forces; *peace-building* actions that seek to prevent disputes; *preventive deployment* for a wide range of purposes, including facilitating the delivery of humanitarian assistance; *peace enforcement*; and broader functions of fact finding, intelligence, and analysis. More heavily armed missions to respond to aggression would be available on call, under the command of the secretary-general. Boutros-Ghali proposed a general shifting of authority, including financial authority, from the regional organizations to the secretary-general, in spite of the Charter's specific encouragement of regional arrangements to solve local disputes.[37] He also proposed offering "peacekeeping services" for the settlement of longstanding conflicts in regions including Angola, Cambodia, El Salvador, and Mozambique, as well as the ethnic conflict in the former Yugoslavia.

The same tendency to expand the authority and powers of the secretary-general was present in Boutros-Ghali's discussion of "intrusive" operations dealing with ethnic conflicts, failed states, new states, development, and sovereignty. The new definitions quickly led to new practices, which were then treated as established procedures. Yet this new conception was about to be tested as the United States moved toward action in Somalia—and problems soon followed.

Alternatives in Somalia: Boutros-Ghali's View

As the situation worsened, the secretary-general wrote to the Security Council on November 29, 1992, proposing five options. He obviously favored the fifth, which was to organize "a countrywide enforcement operation to be carried out under UN command and control . . . which would be consistent with the recent expansion of the Organization's role in the maintenance of international peace and security and which would strengthen its long-term evolution as an effective system of collective security."[38]

One aspect of this proposal raised red flags in Washington: it would give the secretary-general control of U.S. troops and the rules of engagement. Boutros-Ghali wanted the authority to determine when U.S. troops should be deployed to Somalia, and to decide when a secure environment for humanitarian relief had been established, clearing the way for U.S. troops to withdraw and UN troops to take over.[39]

But George Bush had no intention of relinquishing control over U.S. troops or over decisions about their weapons, rules of engagement, deployment, and withdrawal. He also clashed with Boutros-Ghali over how peacekeeping operations in Somalia should be conducted. At every stage in the development of UNITAF, Bush resisted efforts by Boutros-Ghali and others to expand the scope of the American mission. When Bush presented the original U.S. plan in late November, the secretary-general urged the United States to disarm the Somali factions before handing control of the operation over to the UN.[40] When American forces arrived in Mogadishu in December, Boutros-Ghali pressed the U.S.-led coalition to defuse all land mines, train a civilian police force, and create a civil administration. The Bush administration resisted the pressure, confining the discussion to the logistics of the upcoming transfer of the operation to UN control.[41]

In Boutros-Ghali's view, U.S. troops would have to disarm the country before a secure environment could be said to exist. Bush thought otherwise. "[T]he mission of the Coalition is limited and specific," he wrote to the secretary-general, "to create security conditions which will permit the feeding of the starving Somali people and allow the transfer of this security function to the UN peacekeeping force."[42]

Bush knew he had to set limits, because it was clear that the UN Secretariat saw Somalia as only the first in a series of potential military operations in African states threatened by social breakdown, civil war, and famine. Mozambique, Liberia, Sudan, and Angola were among the other failed states in perilous condition, and even before U.S. troops landed in Somalia, a State Department team was in Africa surveying problems far beyond Somalia's borders.

Despite political pressure to do something about Somalia, many in Congress were wary of the proposed military operation. Up to the day before the authorizing resolution was passed, serious doubts were expressed about the operation, whose first phase would be carried out under U.S. command and in whose second phase the United States would participate but not command. Senator Hank Brown (R-CO) spoke for many when he pressed questions about command of the troops and the rules of engagement, including the authority of U.S. leaders to decide when to withdraw.

> Do they fire back if they are fired upon? If we are attacked, do we call in air support? If troops are captured, do we pursue? . . . I think it is a mistake to commit military personnel to combat without a clearly defined mission and without the ability to protect themselves.[43]

The United States and the United Nations were entering new territory. Who would define the goals of a UN force? Who would decide when the job was completed? Who would determine the relationship between national and UN command? Peacekeeping in a war zone was a new idea with many unknowns.

Even the day before the Security Council passed Resolution 794, which authorized UNITAF, no agreement on the major questions had been reached. Boutros-Ghali continued to state his preference for a military operation under the control of the UN Secretariat and the secretary-general, but Bush knew that the Secretariat lacked the resources and experience to command a military operation of that size and complexity, and he refused. Boutros-Ghali acquiesced, but demanded that once UNITAF had established a secure environment for the delivery of humanitarian aid, the military command would be turned over to the United Nations. He emphasized that *countrywide* disarmament and enforcement was required, not just the establishment of order at ports, and that "member states would have to be ready to accept that the United Nations would command and control the operation."[44] Bush stated his doubts that the UN had the military capacity to carry out such an operation, but the secretary-general countered that China would veto any operation under U.S. command.

The Bush Phase in Somalia: Resolution 794

Even after Resolution 794 was passed on December 3, 1992, disagreement persisted on its wording and interpretation. The resolution featured a military dimension that called for the use of "all necessary means to establish as soon as possible a secure environment for humanitarian relief operations in Somalia."[45] The United States agreed to deploy approximately twenty-five thousand troops to satisfy this condition, alongside twelve thousand troops from other countries. The resolution called on member states to contribute troops and establish mechanisms for coordination between their military forces and the UN. By this point President Bush had lost his reelection bid to Bill Clinton, but the president-elect quickly endorsed the new resolution: "The United Nations has provided new hope for the millions of Somalis at risk of starvation. I commend President Bush for taking the lead in this important humanitarian effort."[46]

The next day, Bush took to the airwaves to announce Operation Restore Hope to the American people, announcing that the United States had decided to intervene in Somalia because 250,000 Somalis had died from famine and war and 1.5 million more were threatened, and because the humanitarian organizations couldn't handle the problem alone. Bush explained that the United States would lead an international coalition, providing it included clear Chapter VII authorization enabling U.S. forces to defend themselves effectively, and with the understanding that the UN would take over as soon as possible. He emphasized the limits of the undertaking, and its humanitarian purposes:

> Our mission has a limited objective—to open the supply routes, to get the food moving, and to prepare the way for a UN peacekeeping force to keep it moving. This operation is not open-ended. We will not stay one day longer than is absolutely necessary.[47]

On December 10, he wrote to the leaders of Congress:

> We do not intend that U.S. armed forces deployed to Somalia become involved in hostilities. Nonetheless, these forces are equipped and ready to take such measures as may be needed to accomplish a humanitarian mission and defend themselves if necessary; they will also have the support of any additional U.S. Armed Forces necessary to insure their safety.[48]

The Democratic Speaker of the House of Representatives, Thomas S. Foley, promised congressional backing for Bush's stance. "There is strong bipartisan support among the leadership for the action the president is taking," he told CNN. "I think that should be reemphasized. It should also be emphasized again that there is going to be international participation in a major way, including not only the present military operations but the carryover activity of the peacekeeping forces following our departure."[49]

Early signs were encouraging. The first U.S. Marines and Navy Seals entered Somalia on December 8; within a week, they had secured the airport and the port of Mogadishu without casualties. French, Turkish, Egyptian, and Algerian forces soon joined the U.S.-led coalition. Five hundred UN security personnel were assigned around Mogadishu. Eventually, there were twenty-five thousand American troops in Somalia and nine thousand troops from approximately twenty-eight other countries.

A CBS poll showed that 81 percent of the U.S. public supported the operation,[50] which the American people saw as a humanitarian relief mission, not a use of force or an act of war. Most Americans probably never noticed the difference between Bush's description of the mission and Boutros-Ghali's statement to the people of Somalia on December 8, in which he promised that "the unified military command . . . comes to feed the starving, protect the defenseless, and prepare the way for political, economic, and social reconstruction." [51] In a letter to Bush on the same day, Boutros-Ghali again sought to expand the U.S. commitment, emphasizing the need to disarm the lawless gangs and neutralize their heavy weapons to ensure security throughout Somalia. [52] He told the *New York Times* of "private commitments" made by the United States to pacify the country, implying that Bush was reneging on those commitments. [53]

But Bush had made it clear that he had always envisioned Operation Restore Hope as a brief operation with a clear goal: the rapid delivery of food to starving Somalis. His intention was to make the port, the capital, and the surrounding areas safe enough to deliver food. He was not willing to take on the more complicated and dangerous tasks of disarming the clans, undertaking political reconciliation, and creating a civil administration.

Bush believed that an effort to disarm Somalia—even if it were possible—would lead the United States deeply into the country's internal politics, and burden the United States and the UN with broad responsibilities for nation building. He stated as much when he first made the decision to send troops to Somalia. In his December 4 address, he said, "To the people of Somalia. . . . We come to your country for one reason only: to enable the starving to be fed." [54] He knew Congress would not support an open-ended commitment. As one senior official of the Bush team said, "If we go out and try to physically disarm people who don't want to be disarmed, we're talking about going to war against all the factions in Somalia. That isn't what we came over to do." [55]

In spite of Bush's efforts to be clear about the mission, however, misunderstandings multiplied between his administration and the UN Secretariat. And both had underestimated the difficulties of the operation and the amount of weapons in Somalia. What Eagleburger had called a "doable" operation seemed less doable every day. [56] Almost everything proved more difficult than expected. The extent of social, political, and economic collapse in Somalia came as a surprise to the Americans. The U.S. troops were more vulnerable to attack than expected, confronted by looters and armed bandits in Mogadishu and elsewhere. From the start of the operation, journalists raised concerns that the U.S. and UN forces would be inadequate to the task. Still, airdrops of food quickly made the difference between life and death for the starving Somalis, and these immediate results made U.S. officials eager to deliver the food and get out.

As the clock was running out on Bush's term as president, the Somalia initiative he had begun still had a clear, narrow,[57] and concrete goal—to clear and protect the channels for relief supplies—and a timeline: the U.S. forces deployed to Somalia were due to be under American command and control until they were replaced by a UN force early in 1993.[58]

Yet the UN replacements were slow to arrive, and the bandits of Mogadishu had more weapons and ammunition than anyone had realized, putting the troops in the path of guerrilla warfare. Nonetheless, the Bush administration continued to insist that its role was limited to providing relief, not disarming warring factions. Early in January, Democrat John Murtha led a congressional delegation to visit U.S. troops in Somalia and sharply criticized the UN. "The UN is doing nothing but dragging their feet," he observed. "I'd like to see the Americans out of here as soon as possible, because the longer we're here, the more involved we get."[59]

Repeated efforts were made to establish a cease-fire. On January 4, Boutros-Ghali gathered leaders of the fourteen factions together in Addis Ababa, Ethiopia, and called on them to put aside their rivalries. Some criticized the inclusion of all fourteen leaders (especially Aideed, who was blamed for much of the violence), but a top UN official commented that there was no alternative. "These are the power brokers. It is like dealing with [Serbian president Slobodan] Milošević in Yugoslavia." After many negotiations, they agreed on a truce—which did not hold.[60]

The U.S. special envoy to the meeting, Robert Oakley, was the most effective mediator and facilitator on the scene, in part because he could back his commitments with American resources. "The United States is in here big time with a lot of resources, whereas the United Nations is not," Oakley told *Washington Post* reporter Keith Richburg.[61] Another, perhaps more important, reason for Oakley's effectiveness was that he worked hard to promote agreements among local leaders.

A NEW ADMINISTRATION

As Bill Clinton prepared to take office, the situation in Somalia was relatively stable. Humanitarian relief was being delivered, and leaders of the warring clans had agreed to attend a peace conference. U.S. forces were scheduled to withdraw within three months, troops from other nations would arrive, and the operation would be transferred to UN supervision. On the day of Clinton's inauguration, U.S. Marines began leaving Somalia.

At the Bush administration's recommendation, Boutros Boutros-Ghali had appointed retired U.S. admiral Jonathan T. Howe as his special representative to oversee the military operations in Somalia. The authorized force was increased from 4,219 in UNOSOM I (the original UN force) to approximately 28,000 military personnel in UNOSOM II and a civilian staff of 2,800 in UNITAF (the U.S. force). Like everyone else, Howe faced unexpected problems when he arrived. "[I]t took a concerted guerilla campaign to demonstrate unmistakably the significant limitations of the type of UN multinational force deployed in Somalia," he later said.[62] Very quickly, some American officials realized that restoring a stable government to Somalia might be an impossible task for an international operation.

On March 1, 1993, Oakley urged the UN to assume responsibility for the Somali effort promptly, as it had promised. He told the *Washington Post* that the operation was costing the U.S. military $30 million to $40 million a day. It was planned that most of the seventeen thousand U.S. troops still in Somalia would withdraw as soon as possible, leaving about five thousand U.S. troops and troops from more than twenty-five other countries. Oakley himself was scheduled to depart on March 3.[63] On that day, he said that the U.S. mission to "stop the killing from war, famine, and disease . . . has largely been accomplished."[64] Later, Oakley would be criticized by the UN for "attempting to give a somewhat false impression of security" to help hasten U.S. departure from the area.[65] It was true that he pushed hard for quick action, stating flatly that American support for other UN peacekeeping missions was being jeopardized by the perception that the UN was stalling in this one.[66] But it was also true that the UN forces were slow in arriving.

As the date neared for the withdrawal of U.S. troops, relations between the secretary-general and the United States grew strained. Boutros-Ghali acknowledged that U.S. forces had helped restored a degree of peace in Somalia, but said that "a secure environment has not been established."[67] In his report to the Security Council, he said:

> [T]he effort undertaken by UNITAF to establish a secure environment in Somalia is far from complete. . . . [T]he threat to international peace and security . . . is still in existence. . . . UNOSOM II will not be able to implement the . . . mandate unless it is endowed with enforcement powers under Chapter VII of the Charter.[68]

Member states accepted this extraordinary claim, and Boutros-Ghali acquired command and control authority by claiming it—first in Somalia, then in Bosnia, then elsewhere. He encountered no effective opposition in the Security Council, some of whose members objected to U.S. command and saw the United Nations as the only alternative.

The secretary-general would later claim that the United States had made commitments that Bush said he had never made. Still later, President Bill Clinton would complain that the United States had signed on to an expanded mission without having had the opportunity to understand what was involved.[69] In Congress, Republicans and Democrats alike would complain that they had been given no chance to judge the commitment of U.S. forces to a mission that had all the characteristics of war.

But Bush *had* understood the critical difference between delivering food and establishing a government, and he had signed on only to deliver food. The United States and its allies had successfully fostered democracy in Germany and Japan after World War II, but those countries had been defeated in war and were occupied by conquering armies for years. Moreover, both were already modern states with modern political cultures. Somalia had none of these attributes.

In his December 7, 1992, address to the nation, Bush had said: "This operation is not open-ended. We will not stay one day longer [in Somalia] than is absolutely necessary. Let me be very clear: Our mission is humanitarian, but we will not tolerate armed gangs ripping off their own people, condemning them to death by starvation." Bush understood that creating a modern state in Somalia would take decades, if it were possible at all. While he supported large-scale, short-term humanitarian aid, he had never approved occupying this remote society for the purpose of reconstructing it.

A New Team in Washington

When Clinton became president, Boutros Boutros-Ghali sought a mandate for the larger U.S. mission that Bush had rejected. This time he got it. In his March 3, 1993, report to the Security Council, he argued that the central goal of the Somalia operation was "to assist the people of Somalia to create and maintain order and new institutions for their own governance." He made the case for the development of a national police force capable of maintaining order:

> There is still no effective functioning government in the country. There is still no organized civilian police force . . . no disciplined national armed force . . . [and] the atmosphere of lawlessness and tension is far from being eliminated.[70]

The Somali people needed assistance, he said, in "rebuilding their shattered economy and social and political life, reestablishing the country's institutional structure, achieving national political reconciliation, recreating a Somali state based on democratic governance and rehabilitating the country's economy and infrastructure."[71]

Boutros-Ghali, who had arrived in office with a sweeping vision of what needed to be done in Africa and what the UN could do, offered a plan for Somalia. The Security Council (including the United States) commended him on his plan for the establishment of a Somali government, including police, penal, and court systems, and formally recognized national reconciliation as the top priority for UNOSOM II in Security Council Resolution 814.[72] In later accounts, Boutros-Ghali and Bill Clinton would be blamed for the expansion of the mission, but some of that responsibility should be shared by the Bush team—as emphasized by reporter Michael Maren in *The Road to Hell*:

[T]he policy path Bill Clinton wandered down during his first year in office was the logical extension of the direction in which the Bush initiative was heading.

Somalia is not a story of how a humanitarian mission became a military adventure. It's about how the people running a humanitarian mission became so dedicated to their cause that they started to see strafing, bombing and killing as humanitarian acts.[73]

Yet Maren understates the clarity with which Bush himself saw the mission and the firmness with which he held to his vision. Maren also fails to see the conceptual break between the Bush administration's idea of a large-scale but limited humanitarian operation and the Clinton administration's drift into a full-scale international adventure in military operations and nation building.

At first, the Clinton administration's Somalia policy continued on the path laid out by Bush. After several days of rioting and fighting in Mogadishu in late February, Clinton announced that the United States would continue with its scheduled withdrawal. Yet instead, in the months that followed, the Clinton team gradually gave U.S. forces a new mission, sent new troops, and placed them under UN command—all of which largely escaped the notice of the American people.[74] At the same time, the ambiguity and unprecedented nature of the situation obscured the implications of Boutros-Ghali's requests for expanded powers, including command of forces. Although these new powers were a marked departure from previous practices, the Security Council, including the United States, acquiesced.

By late February 1993, danger signals were already multiplying on the ground. At first Aideed had welcomed the arrival of American forces, but on February 24 he accused the United States and the UN of allowing a rival to grab territory after forces loyal to Aideed's archenemy, Mohamad Said Hersi Morgan, slipped past U.S. and Belgian forces and seized several blocks of territory in Kismaayo from Aideed's ally Omar Jess.[75] A mob attacked the Egyptian embassy, the French embassy was fired on, a hotel was attacked, and three U.S. Marines and two Nigerians were wounded. A U.S. military official noted that Somalia was too unstable for UN peacekeepers, who have a reputation for passivity.[76] It was clear that the conflict among the factions had continued to grow despite the presence of the peacekeepers. The response in New York and Washington, however, was to enlarge the UN mandate—while reducing the forces on hand to carry it out.

New Administration, New Mandate

The new administration in Washington endorsed Security Council Resolution 814, which gave Boutros-Ghali most of what he wanted regarding the mission in Somalia. The resolution, passed on March 26, 1993, substantially altered the UN mandate in Somalia, now called UNOSOM II. It gave UN forces expanded military goals, including disarmament of the country. It vested command and control in the UN, and called for the "Secretary-General, through his Special Representative, to direct the Force Commander of UNOSOM II to assume responsibility for the consolidation, expansion, and maintenance of a secure environment throughout Somalia."[77] Nation building became a primary goal, including the reestablishment of national and regional institutions and civil administration; the reestablishment of a Somali police force; and the removal of land mines.[78]

Boutros-Ghali cautioned, as if as an afterthought, that UNOSOM II should not be expected to substitute itself for the Somali people. "Nor can or should it use its authority to impose one or another system of governmental organization," he added. "It may and should, however, be in a position to press for the observance of United Nations standards of human rights and justice."[79]

The Security Council called on the secretary-general, through his special military representative, Admiral Howe, to help with the repatriation of refugees, promote "political settlement and national reconciliation," restore law and order, re-create civil society, and undertake the establishment of the new operation and the transfer of security responsibility from the U.S.-led UNITAF to the UN-led UNOSOM II.[80] At the time, Kofi Annan of Ghana held the top peacekeeping job at the Secretariat. Annan said, "This will be the first time the United Nations has had command and control of an enforcement action under Chapter VII."[81] It was also the UN's first experience in nation building.

The Clinton team was euphoric about the operation in Somalia. For the principals of the administration, peacekeeping provided a new solution to an old problem—and a fig leaf for a president whose campaign had included charges of draft dodging.[82] Like the president himself, many in the administration came from the Vietnam generation and viewed the U.S. armed forces with attitudes ranging from mild distrust to outright contempt. As Colonel Kenneth Allard, author of *Somalia Operations: Lessons Learned*, later commented, "I think much of the atmosphere was poisoned by the . . . deep-seated distrust, often expressed quite publicly, between the people [who] constituted the administration at the second and third tiers and the professional military."[83]

Ambassador Robert Oakley also noted this underlying hostility and its effect on the Somalia operation: "I remember one . . . newspaper article that reported a female employee of the Clinton administration . . . said to one of our top generals, 'We don't appreciate people in uniform in the White House. Please get out if you're wearing your uniform.' " In Oakley's view, the U.S. military in Somalia "felt very keenly that they were . . . not appreciated by the Clinton administration."[84]

For some leaders of the new administration, the "new peacekeeping" provided a new theory of national security under which the military could be used to further a liberal agenda, namely, the rehabilitation of failed third world states. Somalia was an exciting first venture in nation building. Les Aspin, Clinton's secretary of defense, said, "We went there to save a people, and we succeeded. We are staying there now to help those people rebuild their nation."[85] Aspin's sentiments were seconded by Madeleine Albright, the U.S. permanent representative to the United Nations: "[W]ith this resolution, we will embark on an unprecedented enterprise aimed at nothing less than the restoration of an entire country as a proud, functioning and viable member of the community of nations."[86] Aspin and Albright seemed to believe that the UN force would be welcomed as surely as U.S. troops had been—before the peacekeeping troops became involved in the conflict among the clans.

On March 27, 1993, the day after the Security Council passed Resolution 814, the warring Somali clan leaders at the Addis Ababa Conference on National Reconciliation agreed to reestablish some semblance of a government.[87] Despite this agreement, however, there would be little progress toward a rebuilt Somalia through the next year.

Clinton administration spokespeople later claimed that it never considered the transformation of the Somalia mission at the highest level. In her book *On the Edge*, Elizabeth Drew noted that the first principals meeting on Somalia came after the October 1993 ambush in Mogadishu. "The Clinton Administration had inherited an un-thought-through mission to Somalia," she wrote. "Among the un-thought-through questions were the varying responsibilities of the United States and the UN, which was to take over from the U.S. in May 1993."[88]

In fact, the Clinton team had arrived in office with a clear commitment to support the rapid expansion of multinational UN-sponsored peacekeeping operations around the world, to upgrade the size and professionalism of the UN headquarters staff, and to provide troops for operations carried out under UN command and UN rules of engagement. The assumption was that the end of the cold war had freed the United States to deal with the world's problems, and that this could best be done through collective action. Albright called the new approach "assertive multilateralism." At least some of its architects understood that the kinds of involvement and the uses of force they advocated constituted a revolutionary policy that would change the way the Department of Defense—and Americans in general—thought about security. The policy included new conceptions of national interest, national security, and national defense, defined now to include strategic humanitarian emergencies, democracy building, nation building, regional threats, and a new attitude toward the use of force.

The Clinton administration would pursue a policy of global engagement and would not shrink from the use of force. "There is a growing realization," Morton Halperin and David Scheffer wrote in their influential 1992 book *Self-Determination in the New World Order*, "that collective use of military force can be a legitimate means to achieve legitimate ends."[89] For these authors, and for national security advisor Anthony Lake, a UN-based multilateral use of force was more legitimate than the unilateral use of force. Multilateral force—under the guise of "peacekeeping"—could be used to deal with a wide array of problems around the world.

Halperin—who became a Pentagon adviser himself—foresaw that this revolutionary shift in policy would find an intellectual home in the Department of Defense, where civilian appointees would reorganize U.S. force structure and prepare American forces for their new global roles. The civilians sought at once to downsize and transform the military, and to develop UN military, intelligence, and command and control capacities. The convergence of the end of the cold war, Boutros-Ghali's selection as UN secretary-general, and Clinton's election as president resulted in an explosion of Security Council resolutions, plans, and authorizations for peacekeeping and led directly to what happened in Somalia and Haiti.

UN peacekeeping was seen as the centerpiece of a sweeping change in the theory and practice of national security. The concept of "peacekeeping" was stretched to include the management of all phases of conflict—from diplomacy to war—carried out by multinational forces under UN command. Halperin described the concept in a memo to Secretary of Defense Les Aspin:

> Peacekeeping may become the key to preventing virulent conflicts, from causing regional explosions, destroying hopes for democracy, and creating grave humanitarian crises. . . . The president is serious about exploiting new opportunities to bolster international peacekeeping efforts and organizations . . . You should help Americans understand what is, in fact, a revolutionary policy, and what you are doing at DOD to make it happen.[90]

U.S. support would be crucial, Halperin explained. It would require "preparing our forces and our thinking to engage in peacekeeping; paying for peacekeeping so that we have forces ready to fight; and strengthening international peacekeeping organizations and practices so that peacekeeping will be truly international."[91] U.S. forces would be trained to participate in international peacekeeping without earmarking any particular units for this function. The U.S. military would be introduced to the new thinking and new practices, including the idea of serving as requested by the Security Council and under non-American commanders. The new program would include joint training and peacekeeping exercises with other NATO members and with the Russians. The UN's military capabilities would be reinforced by the United States, which would contribute intelligence, planning, communications, and other kinds of help.

William Perry, then deputy secretary of defense, made similar proposals in an essay that appeared after his elevation to secretary. In the essay, Perry anticipated that core activities of international security would be centralized in the UN, and that "preventive diplomacy, conflict resolution, war, postwar management would be carried out as multinational operations under UN command and control." He said that a chief function of DOD would be to prepare Americans for such service and provide equipment, technology, and advice to the UN.[92]

Perry's plan assumed a level of harmony and cooperation among nations that has never existed in human history. It was based on the premise that the United States had an interest in peace, preventive action, and conflict resolution everywhere, and that we should be ready to risk American lives to achieve ambiguous goals in remote countries with which we have no significant ties. Perry's plans further assumed that a large number of other countries would be willing to join the United States in the disinterested use of force; that the putative beneficiaries would desire this international "help"; and that the UN would be competent to exercise command and control, and willing and able to use the intelligence and technology he planned to transfer to them. As Senator Richard Lugar (R-IA) later commented, "The Clinton administration was not comfortable with the use of military power and simply hoped it wouldn't have to be used. Or, if it did, it would be shared responsibility with others, and that there might be some overall legitimacy through the United Nations or some international command."[93]

Underlying the approach was a conception of force that differed sharply from the one that had prevailed up to and throughout the cold war. Gone, too, were the Weinberger-Powell principles: Perry, Lake, Halperin, and others associated with the Clinton administration were less reluctant to use force than the Reagan administration, perhaps because there was less danger of provoking a major war. They were also less concerned about having a preponderance of force when they did use it, instead advocating limited force for limited purposes.[94] The Clinton team and the UN secretary-general seem to have underestimated the dangers of peacekeeping, assuming that peacekeepers would seldom face armed conflict or confront serious adversaries. This benign conception of the new peacekeeping explains why Clinton's Department of Defense leadership was casual about denying requests for additional armor in Somalia, and about intelligence and reinforcements. The president and his top advisers had not seriously considered the dangers involved in committing men to conflict under Chapter VII of the UN Charter, under UN command, or under Boutros-Ghali's rules of engagement.

FROM UNITAF TO UNOSOM II: U.S. FORCES UNDER UN COMMAND

Phase I of the Somalia intervention, the Bush phase, ended on May 4, 1993, when U.S. Marine Lieutenant General Robert B. Johnston handed command of the force to the UN force commander for UNOSOM, the widely respected Turkish lieutenant general Cevik Bir. U.S. Major General Thomas Montgomery served as deputy force commander. Forty-seven hundred U.S. troops remained to ease the transition. The goals of the mission changed again. The Security Council requested the secretary-general—through Admiral Howe—to assist in political reconciliation and rehabilitation. When the four thousand remaining U.S. troops were transferred to UN command, American participation in UNOSOM II got under way.

On July 1, 1993, President Clinton wrote to congressional leaders: "At the height of the U.S.-led Unified Task Force (UNITAF) operations, just over twenty-five thousand U.S. Armed Forces personnel were deployed to Somalia. Consistent with U.S. policy objectives, the current smaller U.S. contribution of approximately forty-four hundred personnel reflects the increased participation by other UN Member States." [95]

In a letter to Senator John Warner (R-VA), Walter Slocombe, deputy undersecretary of defense for policy, explained the planned command and control structure:

The United States is participating in the UN Operation in Somalia (UNOSOM II) with two basic types of forces: support forces for logistics purposes and a quick reaction force (QRF) of combat troops. All U.S. forces . . . remain under the command authority of General Hoar, the Commander of USCENTCOM. He exercises his authority over those American troops through his representative, MG Thomas Montgomery, in his role as the commander of U.S. Forces in Somalia (USFORSOM). MG Montgomery is also dual-hatted as the Deputy Commander of the UNOSOM II Force Command.

The U.S. support forces that provide logistic services to the UNOSOM II Commander, LTG Bir, are under the UN's operational control. General Hoar retains operational control over the QRF but has delegated tactical control to MG Montgomery in specific instances. . . . The UN directs the military actions of the UNOSOM II Force Commander but, although Howe is in the UN's military chain of command, he cannot order the QRF into action. That authority remains within U.S. channels.[96]

At its peak, UNITAF (the original force commanded by the United States, consisting mainly of Americans) had just over 25,000 U.S. troops distributed over 40 percent of Somalia.[97] By August 17, 1993, the secretary-general reported that the UNOSOM II force included 20,707 troops from twenty-seven countries. The primary contributors were Pakistan (4,973), the United States (2,703), Italy (2,538), Morocco (1,341), and France (1,130). The United States also contributed 1,167 troops for the Quick Reaction Force (QRF). Both Force Headquarters Command and total strength had been progressively built up, and about 8,000 more troops were expected, including 5,000 from India, bringing the force strength to the authorized level of 28,000—though these forces would prove neither well armed nor integrated.[98]

Meanwhile, the situation in Somalia had become more dangerous. Warfare among the clans had intensified, and attacks on UN troops had multiplied. The changes in the composition of UN forces and in the command structure affected the relationships among Somali combatants, the UN, and various national groups. As Robert Oakley noted, the relationship between Aideed and the United Nations was deteriorating. "Aideed made it clear he didn't trust Boutros-Ghali in particular or the United Nations in general," he observed. "We kept saying, 'You have to work with him' . . . I think the administration at the top level did not fully appreciate the significance of what was going on." [99]

The Massacre of Pakistanis

Aideed's forces possessed heavier weapons than anticipated and were more aggressive and effective than expected, especially after they had become convinced that the UN favored their rivals. On June 5, 1993, Aideed's men ambushed and killed at least twenty-three Pakistani soldiers. Ten were reported missing and fifty-four wounded [100] in two attacks—one as Pakistani forces left a radio station they had been sent to investigate, the other on lightly armed Pakistani soldiers who were attempting to deliver food. On June 6, the Security Council passed a U.S.-drafted Resolution 837, which condemned the murder of the Pakistani soldiers and announced that UN forces would identify, arrest, and punish the responsible parties. Oakley again noted the lack of communication between the White House and the Pentagon:

> Mrs. Albright felt very, very strongly about [Resolution 837]. As best I can figure out, the resolution was drafted over a weekend between New York and the White House, I'd say, and done in a hurry. General Powell tells people that the first thing he saw about the resolution was when he read about it in the newspapers Monday morning. Therefore, the Pentagon was not properly brought into the loop, in terms of measuring the consequences of this resolution, which declared those responsible to be the enemy in Somalia. [101]

Boutros-Ghali appointed Tom Farer, a professor of international law at American University, to investigate the massacre of the Pakistanis. Farer's report concluded that only Aideed had the "requisite means, motive, and opportunity" to carry out these attacks. Farer was impressed by the military sophistication, planning, firepower, discipline, and first-class weapons employed by Aideed's forces. [102]

Mohammed Farah Aideed saw himself as a principal leader in Somalia—a future ruler and a legitimate competitor for primacy—and he came to see the UN as an obstacle to his status and power. Western opinion about Aideed varied widely. Oakley believed it was important to "play on his pride rather than to try to humiliate him." [103] He observed later that the environment was tense because "the Somalis are very xenophobic, aggressive people." He saw Aideed as "a cunning man of violence," especially volatile when offended. "I would treat him as if he were a vial of nitroglycerine that could go off in my hands." General Anthony Zinni, UNITAF's director of operations, had a more positive view of Aideed's status and role. He saw Aideed as "probably the most successful general in the Somali army, the only one who had tactical successes in the war with Ethiopia. . . . He was very bright; he was an ambassador from Somalia to India. He served in the cabinet there; he was respected by his clan. There was a very strong case to be made on his behalf by his followers." [104]

On the basis of the Farer report and Resolution 837, the immediate mission in Somalia changed again. The Clinton administration and Boutros-Ghali decided that UN forces must capture Aideed, whom they blamed for the escalating violence.[105] But after the attacks on the Pakistani troops, other forces became progressively more unwilling to leave their safe havens and risk danger.

The new mission to capture Aideed, and the flawed logic of the Clinton administration's assertive multilateralism and nation-building theories, put U.S. forces increasingly at risk. Some of the other national units in UNOSOM II defied the U.S.-led mission to capture Aideed and made their own separate peace with him. One of Aideed's officers, Captain Haad, later said, "The Italians were not happy about the war the Americans were fighting against us. We knew that the UN forces were too many for us to confront. So what we did was to concentrate our attacks on the Americans and the forces who were taking their orders from the Americans, such as the Pakistanis. We had an understanding with some UN contingents that we would not attack them, and they would not attack us." [106]

One U.S. Ranger thought the Aideed-Italian "arrangement" far more sinister: "The Italian compound was on the far end of the airfield from us. Every time the profile flights would take off, you would see lights flash. And what we perceived to be going on was that they were signaling the Somalis that we were coming. It definitely left you wondering which side they were actually on." [107]

Targeting Aideed

On June 17, 1993, Admiral T. Howe called for Aideed's arrest, offering a reward of $25,000. U.S. commander Thomas Montgomery later said on *Frontline*, "I didn't have a problem with putting a price on his head . . . because Aideed was a real tyrant, a very, very dangerous man."[108] General Zinni, who had no role in the decision but knew the situation well, disagreed: "I think that the resolution to declare Aideed a criminal and put a price on his head was . . . ridiculous," he said. Such a move meant that the UN forces were "no longer in a peace enforcement or peacekeeping [role]," Zinni felt, but in "a counterinsurgency operation or in some form of war."[109]

Determined to make Mogadishu more secure, UN forces began a systematic drive to collect weapons. Factional retaliation led to the deaths of more UN peacekeeping troops—five Moroccan soldiers on June 17, two Pakistanis on June 28, three Italians on July 2; and then four journalists were murdered. Most of these casualties occurred in confrontations with Somali militias that shielded themselves behind large, disorderly crowds of women and children bearing guns and grenades. Although UN forces retaliated with guns, ships, and planes, the armed Somalis grew increasingly aggressive and the casualties mounted.

Howe urged Washington to send the Delta Force to hunt for Aideed. The request initially met with resistance from Secretary of Defense Aspin and Joint Chiefs chairman Colin Powell, both of whom were reluctant to send more troops and doubted that this assignment was right for the elite force. But Powell changed his mind and convinced Aspin, and U.S. Rangers were dispatched with orders to capture or kill Aideed.[110]

UN intelligence was clearly inadequate, with crucial delays in getting information from collectors to consumers. According to Gene Cullen, the CIA officer in charge of intelligence coordination in Somalia, information had to be transmitted back to headquarters, who would "determine what would be disseminated and what would not"—a process that could take from twelve to seventy-two hours.[111] There were also questions about the reliability of Somali intelligence sources. According to Abdi Hassan Awaleh, Aideed's defense minister, the Rangers "did not have good intelligence for locating General Aideed. They never came close to him. . . . We used the same Somali informers that they used. We knew who they were."[112]

From the time they arrived, U.S. commanders in Somalia had requested heavier armor, tanks, and other vehicles—requests that were turned down by Aspin himself on the grounds that it made no sense to build up and build down simultaneously. But the violence continued to escalate, and events quickly demonstrated why more armor was needed.

On August 8, four American soldiers were killed when their jeep was blown up by a sophisticated remote-controlled mine. Then seven Nigerian soldiers were killed. On August 22, President Clinton himself issued an order to capture and try Aideed.[113] In August and September, the Rangers staged several raids and captured a number of Aideed's aides. (During one raid, they "captured" the astonished staff of a UN development program, who were released with apologies as soon as they were identified.) Curiously, at the same time that Clinton declared Aideed a criminal and ordered his capture, he secretly opened an initiative to negotiate with Aideed, with former president Jimmy Carter as an intermediary.[114] General Montgomery, the deputy UN commander at the time, said that his command "had no idea" about the back-channel effort. "I wish . . . that somebody . . . had told the military chain of command to cease and desist this effort to bring Aideed to justice," Montgomery complained.[115]

Congress sensed the growing danger in the situation. On September 9, the Senate voted 90 to 7 in favor of an amendment to the Defense Authorization Bill that would require the administration to report to Congress on the Somalia operation no later than October 15 and seek congressional authorization no later than November 15. However, before that deadline arrived, there was more action on the ground.

By the beginning of October, the Rangers were closing in on Aideed and his staff. They had identified their meeting place and arrested twenty-four United Somali Congress/Somali National Alliance (USC/SNA) leaders. Yet Aideed had determined that helicopters were the Rangers' chief vulnerability, and on October 3, his forces targeted two helicopters in Mogadishu. The first helicopter was hit by a rocket-propelled grenade (RPG) and crashed; the second, a Blackhawk Super 6-1, which had provided cover for the evacuation of some of Aideed's officers, was also hit. The two pilots were killed, and their bodies and others were pinned in the helicopters. The Rangers, committed to never leaving a fallen comrade, took heavy casualties as they fought their way to the helicopters to extract the bodies.[116] No contingency plans had been made for reinforcements.

In describing the events of that day, General Montgomery described a dangerously cumbersome command structure: "I had a base officer from General Garrison at my side; I had my hand on communications to talk to the QRF and to the Rangers; I had the ability to talk to UN forces." He needed men and tanks. First he sent in the QRF, the Rangers, and the Delta Force. There were only four operational tanks in Mogadishu, and they were Pakistani tanks without night vision. The Malaysians had old but functional armored personnel carriers (APCs) and provided drivers. For eleven hours, the Rangers, the QRF, and the 10th Mountain Division fought their way into and out of the crash scene.

Ultimately, two U.S. helicopters were downed, eighteen Rangers were killed, seventy-five wounded, and one went missing. Cheering Somalis dragged a dead American soldier through the streets of Mogadishu, and a videotape was later distributed showing Somalis capturing and mauling another wounded soldier.

In Washington and Somalia, spokesmen for the Rangers sought to make clear that they had not been overwhelmed. Major General William Garrison, commander of the Joint Special Operations Command, explained later to the Senate Armed Forces Committee that, despite heavy losses, "what we had were helicopter pilots that were trapped inside of the helicopter, and we were going to stay with them until such time as we could extract them. We were not pinned down; we could have fought our way out anytime we chose to do so."[117]

But Congressional impatience with the ambiguity of the Somali mission and its remote relationship to U.S. interests had been growing even before this incident, and impatience turned to outrage when word arrived in Washington that eighteen U.S. servicemen had been killed and more than seventy wounded in an ambush in the Somali capital. Outrage turned to fury when it was learned that the ninety Rangers and others were pinned down for nine hours, during which at least one Ranger bled to death before they were finally rescued by an unprepared, uncoordinated force.

In Washington, information about the circumstances surrounding the Mogadishu debacle seeped out slowly. Mounting evidence suggested that the Rangers had been inadequately equipped and assigned to an inappropriate mission, and that U.S. vehicles lacked adequate armor—in part because the secretary of defense himself had refused a request for less vulnerable tanks and fighting vehicles.

The Rangers took heavy casualties because they would not abandon their trapped comrades. It had not been easy to find troops to help rescue the Rangers, because there were no contingency plans, no backup troops, and no heavy armor with which to barrel through. Language problems complicated planning and execution. After the ambush, General Montgomery's request for more and heavier armored vehicles (which had been turned down at the Pentagon's highest level only a month earlier) was immediately granted. The Pentagon quickly announced that heavier arms and armor would be sent to protect and reinforce peacekeepers in Somalia.[118]

"It is very unusual for the United States to be in a position where we cannot really rescue our own forces in a situation like this," observed Senate Armed Services Committee chairman Sam Nunn (D-GA). Another Senate Democrat, Robert Byrd of West Virginia, said, "Americans are paying with their lives and limbs for a misplaced policy on the altar of some fuzzy multilateralism."[119]

In the PBS documentary *Ambush in Mogadishu*, some of the Rangers who had been trapped in the firefight described having to walk out:

PFC DAVID FLOYD: The thirty or so of us, I guess, that weren't wounded, there was no room for us, so now we were going to—we're going to leave out on foot.

SPEC. MIKE KURTH: I got a sinking feeling there. I was just like, "This is going to be worse than yesterday, because they know exactly where we're at. They know exactly where we want to go."

SGT. JOHN BELMAN: We all got lined up inside our little courtyard there and just went out into the street. And essentially, there's this long column of people running out on either side of the road alongside these Malaysian armored cars.

SGT. KENI THOMAS: It's like, "How in the hell has it come to the point where we have got to run out of this city on our own?" [120]

In testimony before the Armed Services Committee in May 1994, it became clear that more armor would have made an important difference. Senator Nunn pressed the Rangers' commander: "It was my understanding . . . if you had had armor, you would have been able to get to the forces in the original mission that were pinned down sooner. Is that right?" General Garrison replied, "That's correct." [121] General Montgomery also took heat: "As commander," Nunn asked, "did you ever say to any of your superiors . . . 'Look . . . the UN is asking us to have a much broader mission of disarmament, to some extent nation building.' . . . At the same time, the major power—the U.S. power—was being reduced. Did you ever say, 'This makes me uncomfortable'?" [122]

Larry Royce, the father of one of the fallen Rangers and himself a retired army officer, wrote that the deaths in Somalia were "brought about by weak and indecisive amateurs in the Clinton administration . . . To put [the Rangers] into combat with no way to reinforce them is criminal." [123]

Mohamed Sahnoun, UN special representative for Somalia, commented:

It is unfortunate that members of the Security Council tend to rely solely on reports submitted by the secretary-general. Except for inputs and instructions they receive from their own countries—none of which has an embassy in Mogadishu—they look at no other sources of information. Why does the Security Council not hold hearings where Ambassador Robert Oakley, the U.S. representative, and other distinguished diplomats and scholars could provide useful evidence to complement and check what it is being fed to them by the secretary-general? [124]

The effort to apprehend General Aideed was the most dangerous task U.S. forces had attempted in Somalia, and it was clear in retrospect that they lacked the forces, weapons, intelligence, and rationale to carry it out. Senator Strom Thurmond (R-SC) asked the U.S. commanders how they came to attempt such a dangerous task with inadequate forces. "When we had twenty-five thousand troops in Somalia," he pointed out, "our operations were mostly limited to facilitating humanitarian activities. Then, after the transition to UNOSOM II, when we had withdrawn most of our forces and had only about four thousand troops in the country, with only about two thousand of those being combat troops, we became more heavily involved in combat operations—force protection, disarming the Somalis, and trying to capture Aideed." [125]

When Thurmond pressed Montgomery to describe how American forces had become engaged in unanticipated combat operations, Montgomery pointed out the distinction between the original UNITAF mission and the UNOSOM II stage. Resolution 837, which targeted Aideed, "changed the nature of the Somalia mission," he said. "We were under attack from June the fifth, increasingly after that, by a hostile militia force that engaged in, essentially, guerilla warfare." [126]

The testimony of administration officials made it clear that the difficulty and danger of the new mission had been underestimated from the start. On October 20, 1993, Madeleine Albright told the Senate Foreign Relations Committee, "Clearly, the difficulty of apprehending those thought responsible for killing the Pakistanis was underestimated. . . . While the UN was increasing military pressure, the targets of that pressure were gaining strength." [127] She added, "If UNOSOM had had more robust military capabilities last summer, better military results might have been achieved." [128]

Peter Tarnoff, undersecretary of state for political affairs, offered a somewhat different explanation of what the United States was doing in Somalia, how the debacle of Mogadishu happened, and why it would not be repeated. Tarnoff was more candid than most about the sweeping U.S. goals in Somalia:

Our goals are humanitarian. We seek to support UNOSOM in its efforts to help the Somali people help themselves in fashioning a lasting political solution to their civil conflict, and to produce a secure environment to enable the free flow of humanitarian aid. The United Nations ... has taken on a broad mission in UNOSOM II—to help Somalia develop basic political institutions and to assist in establishing a judiciary and police force so Somalis can keep order in their country.[129]

Tarnoff assured the committee that nearly all U.S. military forces in Somalia would leave within five months, by March 31, 1994, though, it was clear that the nation-building goals would not have been achieved by then.

In a 1995 review of the Somalia debacle, former UN special representative Howe wrote that the American people and Congress had never accepted a U.S. role in Somalia beyond delivering humanitarian assistance. He observed that deterring nuclear war and responding to regional aggression (as in the response to the invasion of Kuwait) were widely accepted by Americans as national responsibilities, but the kind of intervention we had ventured in Somalia (and, later, Haiti) was not.[130]

In Mogadishu, American servicemen had risked their lives for an ambiguous cause in a remote place under unreasonable rules of engagement. It was a mission that had never been approved by Congress, a mission in which U.S. forces were expected to coordinate operations among ad hoc multinational units, with troops from other countries who spoke other languages, using incompatible equipment, and without adequate supplies and support. These conditions had set the scene for disaster.

In the wake of Mogadishu, President Clinton abandoned the effort to hunt down Aideed as a mistaken policy. Having come to power enthusiastic about the potential to replace war with peacekeeping, they had discovered how easily peacekeeping could slide into war. Americans would never again see peacekeeping as social work.[131][142] (And yet, on the same day that dead Americans were being counted in Mogadishu and a battered American captive was being displayed on Somali television, Madeleine Albright, the U.S. representative to the UN, voted in the Security Council to send a peacekeeping force to Rwanda.)

The Clinton administration and the UN officials had operated under the assumption that unified forces and a unified command actually existed. In a June 1993 CNN interview with Charles Bierbauer, Albright had described the Somali mission this way:

> This is one of the most interesting and complicated of the United Nations peacemaking operations. And it is what the United Nations is very much into these days—working to keep the peace and to rebuild societies. . . . There are now twenty nations that are contributing forces to UNISOM and more than that are participating in the civilian aspect of rebuilding the Somalian society. . . . [I]t will require . . . sustained multilateral action to try to bring some peace and security to the international community.[132]

But the violence and casualties at Mogadishu were not at all what the proponents of collective multilateral action had in mind. Now the question was how to get U.S. troops out. On October 14, 1993, Clinton wrote to Senator Byrd that he would pull U.S. forces out of Somalia before the end of March 1994, "if at all feasible." The same day, he told a press conference that the casualties in Somalia "would make me more cautious about having any Americans in a peacekeeping role where there was any ambiguity at all about what the range of decisions was which could be made by a commander other than an American commander."[133] (And yet, twelve hours later, he deployed U.S. warships to enforce sanctions against Haiti.) Clinton selected a new commander, Brigadier General Carl Ernst, for the American task force in Somalia and appointed Ambassador Robert Oakley as a special representative who would report directly to the U.S. government rather than to the UN. Planning for evacuation of U.S. troops was put into high gear. On October 15, the Senate voted 76 to 23 for a compromise that did not call for immediate withdrawal of U.S. forces from Somalia, but required U.S. forces to be under command of U.S. commanders.

A MULTILATERAL ENCOUNTER RECAPITULATED

Nothing had worked as intended in Somalia, where Boutros Boutros-Ghali and the UN Secretariat had hastily cobbled together forces from more than two dozen countries with diverse traditions, languages, and levels of development. These forces were never adequately coordinated and equipped. It proved impossible to overcome different priorities, training, weapons, values, goals, languages, habits, and military traditions. Incompatible cultures complicated every action. Disorganization, disagreements, lack of political will, uneven command competence, and wasteful and inefficient administration combined with language difficulties and a general lack of discipline in a number of national contingents.

Everyone associated with the mission had ideas about what went wrong.

First, the problems were very difficult. It was a violent internal conflict in a society without a government, in which several contenders for power made war on one another with murderous weapons. Factional leaders had negotiated but had not implemented a cease-fire and had hindered the delivery of humanitarian assistance.

Second, the operation suffered from progressive "mission creep." What began as an effort to deliver emergency humanitarian assistance was redefined until it became a far-reaching mission to establish a secure environment for the delivery of humanitarian aid; curb lawless clan leaders and disarm warring factions; achieve political reconciliation, stability, and law and order; rebuild the Somali economy and institutional infrastructure; arrest Aideed; repatriate refugees; enforce an arms embargo; and create a new state. Force levels and equipment were not equal to the changed mission.

In May 1994, General Montgomery testified on the purposes of the operation. "The mission of UNITAF was limited. The objective was very clear. Disarmament was undertaken for the purpose of ensuring that relief would flow. . . . [W]ith the clarity of retrospective [view] . . . [i]t was not possible to disarm Somalia totally." [134]

Not until the Mogadishu ambush did the public realize that the mission had slipped from being a humanitarian effort into nation building and war. [135]

The third major problem was the complicated chain of command. During the first phase of UNITAF, U.S. forces operated under American command; under UNOSOM II, almost all U.S. and other forces were placed under the command of the UN, with the secretary-general as commander in chief. The respected (Turkish) Lieutenant-General Cevik Bir had operational command.

General Montgomery, commander of U.S. forces and deputy UN force commander in Somalia from March 1993 to March 1994, described the UN chain of command in a *Frontline* interview: "As deputy UN commander, I was General Bir's assistant and we worked for Admiral Howe, who was the special representative of the UN secretary-general." [136] Montgomery was also commander of the U.S. forces in Somalia. In this position, he reported directly to the commander in chief of the U.S. Central Command, General Joseph P. Hoar, who was in Tampa, Florida. Montgomery explains:

> The . . . Quick Reaction Force was under my tactical control [but] not my operational control. There was a memorandum of agreement on how that force could be employed—basically, only for emergencies or beyond the capabilities of the UN forces. . . . As the U.S. commander, I had control of that. Anything else that force did . . . required the [prior] approval of USCINCENT.[137]

As retired colonel Kenneth Allard noted, "If it takes longer than ten seconds to explain the command arrangements, they probably won't work." In UNOSOM II, Allard said, "You had essentially three chains of command running. You had one that was going back to New York to the United Nations; you had one that was very clearly going back to Washington, DC; and you had another one that was being exercised by the unified command itself, the United States Central Command. . . . [T]hat is precisely the wrong way to do a command control." The right way, he said, is to "have one person in charge." He added, "I think General Schwarzkopf said it very well during Desert Storm: 'When you get off the plane, you work for me.' . . . The command arrangements should be the thing that enables effective command control, not an obstacle to it." [138]

Some Pentagon officials, such as Assistant Secretary of Defense Walter Slocombe, acknowledged the problems with the command structure. "In my opinion," Slocombe later told a Senate hearing, "the UN as such does not currently have the capability to conduct Chapter VII peace enforcement operations which entail serious combat, or the potential for

it. . . . It can do these things only with the leadership of a strong nation . . . or an effective international military organization [such as]
NATO. The complexity of these operations currently exceeds the UN's
capabilities."[139]

The lack of a central authority over national contingents was yet another source of difficulty—a fact that became especially clear as governments began to make unilateral decisions about withdrawing their
forces. In September 1993, Italian forces declined to follow orders regarding the capture of Aideed and threatened to leave because of disagreements with UN methods and goals. The UN command lacked the
authority to fire the Italian commanders. Eventually, after the massacre at
Mogadishu and the return of Robert Oakley as President Clinton's special adviser, the United States withdrew its support for the capture. When
Clinton announced the planned U.S. withdrawal, shortly thereafter, the
governments of Belgium, France, and Sweden followed suit; in the following months, Germany, Greece, Italy, Norway, Turkey, Korea, Kuwait,
Morocco, Saudi Arabia, Tunisia, and the United Arab Emirates all withdrew their forces.[140]

A formerly confidential investigation conducted by the Zambian
chief justice, a retired UN peacekeeping commander from Ghana, and by
Finland's chief of staff into the heavy casualties suffered by Pakistani,
American, and other forces in Somalia reached very different conclusions than the Farer report about the causes of the deaths of twenty-five
Pakistanis and eighteen Americans. While the Farer report had blamed
Aideed, this investigation divided the "blame" among Somali factions,
UN commanders, troop-contributing countries, and the Security Council. This investigation concluded that UN forces had overstepped their
mandate, interfered in internal affairs, and taken sides in the internal
conflict, creating "virtual war situations."[141]

Most observers agreed that the secretary-general bore some responsibility for the failures and casualties, because he had assumed unprecedented powers that he was unable to exercise in a professional manner.
The "traditional United Nations peacekeeping culture that often dis-
dains military solutions or even military expertise" created additional
complications.[142] Boutros-Ghali's embrace of that culture was apparent
in his appointments and decisions. Chapter VII operations were consid-

ered peacekeeping engagements, to be carried out under the peacekeeping rules of engagement. Those rules were inappropriate for the situation in Somalia, as they called for geographic balance in forces, minimum armaments, and minimum use of force for passive self-defense only.

Inadequate Force: How Much Is Enough?

The Clinton administration shared the UN attitude of disdain for military solutions, including a reluctance to supply the necessary means. This attitude became an increasing problem as the UN turned more frequently to military solutions. Richard Haass, George Bush's representative to the Security Council, noted that "the greatest failures come from approaching a mission as one of peacekeeping when it in fact is much more." [143]

General Montgomery said that if the UN's Pakistani troops had had Bradley tanks, Bradley APCs, or M1-A1 Abrams tanks, they could have made a speedier rescue in Mogadishu. The general had repeatedly requested such reinforcements, of both personnel and weapons. In August 1993 he had requested naval battle tanks, a mechanic task force, a cavalry troop, and more intelligence capability, but these requests were rejected in Washington, where the priority was to downsize forces, not strengthen them. The responsibility for this decision lay with the secretary of defense and the upper levels of the Clinton national security team.

General Garrison, commander of U.S. forces in the Mogadishu battle, wrote to President Clinton that "the authority, responsibility, and accountability for the Op[eration] rests here in MOG with the TF Ranger Commander, not in Washington. . . . A reaction force would have helped, but casualty figures may or may not have been different." [144] But others were less ready than Garrison to absolve the administration of responsibility. In *Black Hawk Down*, Mark Bowden wrote the following:

> It seems fairly obvious that a light infantry force trapped in a hostile city would be better off with armored vehicles to pull them out, and few aerial firing platforms are as deadly effective as the AC-130 Spectre. Many of the men who fought in Mogadishu believe that at least some, if not all, of their friends would have survived the mission if the Clinton administration had been more concerned about force protection than

maintaining the correct political posture. Aspin himself, before he stepped down, acknowledged that his decision on the force request had been an error. The 1994 Senate Armed Services Committee investigation of the battle reached the same conclusions. The initial post-mortem on the battle was summed up in a powerful statement to the committee by Lieutenant Colonel Larry Joyce, U.S. Army retired, the father of Sergeant Casey Joyce, one of the Rangers killed. "Why were they denied armor, these forces? Had there been armor, had there been Bradleys there, I contend that my son would probably be alive today, because he, like the other casualties that were sustained in the early phases of the battle, were killed en route from the target to the downed helicopter site, the first crash site. I believe there was an inadequate force structure from the very beginning." [145]

Other deficiencies contributed to the catastrophe in Mogadishu. As Garrison observed, no plans had been made for reinforcing the Rangers in Mogadishu; there was no advance coordination with Pakistani and Malaysian forces, who were not on standby alert and did not understand that they were to be available as reinforcements in an emergency. The Rangers even lacked proper night-vision equipment. [146]

Some UN commanders in Somalia thought the number of troops was inadequate to the task. The secretary-general supposed that more involvement of regional organizations and more troops had been needed. The U.S. military commanders in the Pentagon thought the operation lacked adequate firepower, adequate clarity about its mission, and a clear chain of command. After the massacre in Mogadishu, Secretary of Defense Les Aspin announced that the Pentagon was sending four M1-A1 tanks, twelve APCs, and about two hundred more troops. He used a familiar Vietnam-era excuse that Americans had "unfortunately" become "involved in day-to-day operation in Mogadishu" in the effort to find a military solution to a political problem. [147]

Interoperability

Several participants, including Admiral Howe, emphasized that the approximately twenty-nine-nation UN force had serious problems of coordination and lack of interoperability from the start. Troops from different nations often could not understand one another's languages or methods of operating, and had different ideas about their mission.[148] Sometimes they literally did not understand what they were doing.

Howe observed weaknesses of multinational UN operations that were especially obvious in Somalia: "When troops must work closely in mutual defense to combat organized opposition, practical problems of interoperability, insufficient armament, and widely varying states of training come into sharp focus."[149] He cited inadequate force cohesion; lack of unity and clear command because of donor countries' tendency to micromanage their troops; and differences in the quality of weapons, transport, and training. He emphasized the importance of unity of command to provide forces with the resolve to tackle difficult tasks. He said, "Ducking inevitable conflicts associated with carrying out a UN mandate will quickly make a Chapter VII force ineffectual."[150]

Howe was more realistic than some about the problems. Oakley, who played a critical role, has emphasized that U.S. forces were under U.S. command at all times, and Bowden, in *Black Hawk Down*, reiterated that Task Force Ranger, which was tasked to capture Aideed, was wholly an American production.[151] But while the forces and the commanders were American, the context in which they operated included all the problems inherent in a multinational UN operation.

The forces of some countries (for example, Italy) refused to accept orders from UN commanders to attack Aideed's forces. The Italian commanders believed (with reason, given Italy's colonial experience in Somalia) that they knew the country better than the UN commanders did. Even when the individual national forces did follow UN orders, their coordination with others was poor. And the intelligence on which the forces acted was often inadequate.

CRITICAL DIFFERENCES BETWEEN THE BUSH AND CLINTON EXPERIENCES

There were crucial differences in how the Bush and Clinton operations in Somalia were conceived and commanded. President Bush limited the mission to emergency humanitarian relief and steadfastly refused a larger military role in internal matters. He did not put U.S. forces into a conflict inadequately armed, commanded, and reinforced; in all the operations carried out under his presidency, U.S. forces were adequately trained and armed, and their numbers were sufficient.

Because the Bush administration limited its mission in Somalia—and because Bush's team had a better understanding of the political culture and terrain—the United States avoided head-on conflicts with any Somali clan or leadership group during his term in office. Bush refused to put U.S. forces under UN command and control, and he did not endorse the use of American forces for nation building. These were the differences between success in a limited mission and failure in a broad one.

George Bush, Colin Powell, and the other major players in the Bush administration recognized that modern armies are complex, sophisticated organizations whose management cannot be cobbled together in ad hoc arrangements; that military commanders are not interchangeable parts; that tactics must be adapted to forces, weapons, and terrain; and that armies are not skilled in overhauling societies. The Bush team always provided U.S. forces with the strength they needed to carry out their missions.

The Clinton administration failed to see the dangers of sending U.S. forces into a war zone under an ambiguous command, without reinforcements, without adequate intelligence or weapons, and under peacekeeping rules of engagement (which are far more constraining than is generally understood). Calling a mission operating in a war zone a "peace operation" does not make it peaceful—or safe. For President Clinton, Somalia would prove a difficult first lesson on the use of force.

GETTING OUT

With Rangers dead and wounded, Americans demanded to know how U.S. forces had been committed to a mission that was not authorized by Congress and that did not seem to relate to our national interest. After the ambush in Mogadishu, Clinton was eager to withdraw U.S. forces, but he wanted to make it clear that we would not be scared out. On October 5, he announced that he would temporarily reinforce UNOSOM II with seventeen hundred troops, thirty-six hundred marines, and a ten-thousand-man carrier force.

The next day, Aideed (with whom Oakley had conducted successful negotiations) called for a cease-fire with American forces, and the Security Council lifted the order for Aideed's arrest. As the fighting slowed, hopes rose for an end to civil strife. On December 15, France pulled one thousand troops out of Somalia. A few days later, Germany and Italy announced that their troops would withdraw when the U.S. forces did. The Security Council announced that the UN force would revert to its original mission of protecting humanitarian aid and expediting its delivery. There was no longer any discussion of forcibly disarming Somalis or overhauling Somali institutions.

On March 4, 1994, the last major combat unit of U.S. troops shipped out, and on March 25 the last U.S. Marines departed. The United States warned the still-feuding Somali clans that all remaining U.S. personnel, including a U.S. liaison mission of military advisers and civilians, would withdraw by mid-July unless the Somali leaders reached a settlement. The warning was answered by a Somali attack in which five Nepalese peacekeepers were killed. Peace talks among the Somali factions broke down.

On June 9, the Security Council passed a resolution limiting the UN mission to four more months—less if no progress was made. Two weeks later, the United States sent an amphibious force of two thousand marines to help close the U.S. embassy and evacuate the diplomatic contingent. In the next two weeks, seven Indian peacekeepers were killed and nine ambushed, and three Indian military physicians died in an attack on a field hospital. As the Somalia operation unraveled, the United States agreed to assist in the evacuation of 18,900 UN peacekeepers, who the Security Council had decided should be withdrawn by March 31, 1995. In January 1995, four U.S. warships carrying twenty-six hundred marines departed to help in the final evacuation of UN peacekeepers.

Fighting among Somali clans escalated as the last U.S. Marines departed from Mogadishu in March 1995, two years after Secretary-General Boutros-Ghali had proposed that the mandate of UNOSOM II cover the whole country and include Chapter VII enforcement powers for rebuilding Somali society. Phase 2 of this ill-conceived and poorly executed international adventure in Somalia had ended. General Hoar wrote of the UNOSOM II experience, "The application of decisive rather than sufficient force can minimize resistance, saving casualties on all sides." [152] In Somalia, reinforcements were too little, too late. Achieving the UN goals would have required peace, and establishing peace was impossible. It would have required the concurrence and effort of Somalis, which was not available, and it would have required a century.

Somalia quickly became a nonevent at the White House and the Pentagon, where officials were busy writing rules of engagement for a different venture in a different country. The word *Somalia* did not cross Bill Clinton's lips when he spoke to the UN General Assembly in September 1994; rather, he spoke of challenges in Haiti, Rwanda, and other troubled societies. But the ghosts of Mogadishu haunted the streets of Port-au-Prince and Cap-Haïtien. [153]

Even as U.S. forces moved into Haiti, the UN shut down the billion-dollar-a-year "peace operation" that was to have brought national reconciliation to Somalia's warring clans and restored the rudiments of government. The departing UN peacekeepers needed U.S. protection from the marauding gangs that had again begun to roam the streets of Mogadishu—looting, threatening, attacking, and killing.

Somalia was neither a futile mission nor a failed one. The emergency food and medicine that had been provided before George Bush left office saved hundreds of thousands of Somalis from starvation. But the second mandate—to bring peace and restore the nation's infrastructure—was quietly abandoned. There was no mention anywhere of the UN Secretariat's ambitious plans to train national police and create a court system. As U.S. ambassador to Somalia Daniel Simpson made the final preparations to shut down the American embassy, he said, "There's no more Somalia. Somalia's gone. You can call the place where the Somali people live 'Somalia,' but Somalia as a state disappeared in 1991." [154]

There is still no national government and no order in the interior or in Mogadishu, where gangs prey on one another, on UN forces, and on UN-protected targets, stealing what can be stolen. No one expects that the United States will recover the huge quantities of heavy equipment leased to the UN forces. General Aideed, after successfully eluding U.S.

forces, succumbed to a wound inflicted by a rival. His son, Hussein Farah Aideed, who had served as a U.S. Marine, returned to Somalia to assume his father's leadership role.

Most postmortems on Somalia have emphasized the incompatibility of the country's indigenous social structure with the requirements for building a modern nation-state. Most observers agree that it is impossible to keep the peace without the cooperation of warring parties. The attention of the "international community" moved on—to Haiti, Bosnia, Kosovo. There was no indication that Clinton or anyone in his administration thought much about the failure of the team's first venture into peacekeeping and nation building, although the problems encountered in Somalia foreshadowed many that would be encountered in subsequent peacekeeping ventures.

Most of these troubled societies have much in common with each other and with other African societies—Angola, Mozambique, Burundi, Sudan, Liberia. All are failed states—that is, states in which conflict and civil war are endemic, and postcolonial state structures have succumbed to the pressures of indigenous social organization. UN Security Council authorizations to use "necessary force" to solve the problems of these failed states are already on the books, but no one seems to know how to graft peace, order, and modern government onto these fractured societies.

Sobered by the casualties in Mogadishu, President Clinton was ever after cautious and reluctant about committing U.S. troops to UN command and into harm's way. Nonetheless, he recommitted the United States to the "sacred mission" of building a new world, promising to encourage democratic governments to "help civil societies emerge from the ashes of repression." [155] The main problem with this scenario is that no one knows much about how foreign forces can help civil societies or modern states emerge in very different cultures. No one knows how to harmonize hostile elites, end violent behavior, or induce respect for law and restraint in the use of power in another culture without a larger commitment of personnel, money, and time than any president or any administration is prepared to make.

In the spring of 1995, while Congress debated whether it would be possible to balance the budget without raiding the Social Security fund or denying Medicare to aged dependents, U.S. Marines were sent back to Somalia as part of a fourteen-thousand-man "extraction mission" to cover the retreat of twenty-five hundred UN troops. The last phase of the failed UNOSOM II mission got under way in March 1995. The mission had cost 30 Americans killed in action, 175 wounded, and about $1 billion a year. The evacuation alone was expected to cost about $50 million. As with most U.S. contributions above our assessed 31.7 percent share of UN peacekeeping costs, most of this sum was simply diverted from funds authorized and appropriated by Congress for conventional defense activities, such as training and providing spare parts.

This mission was to be carried out with nonlethal weapons—barbed wire, rubber bullets, pepper grenades, wooden pellets, and sticky foam—and under UN rules of engagement. Secretary of Defense William Perry explained that he wanted to make certain that no one on either side was hurt in the operation. The spirit of this policy was captured in Perry's comment that the marines would enter Somalia with their guns pointed backward. However, he did not want to leave the impression that the marines would be defenseless. A DOD briefing explained that "authorization to use lethal force for self-defense against deadly threats would be unaffected by the use of nonlethal weapons for achieving mission objectives."

The very idea that marines would need specific authorization to use regular weapons in self-defense in a war zone told us that they were operating under UN rules of engagement, which permit the use of force only if the peacekeeper's life is directly and immediately threatened. These are the rules UN peacekeepers operated under when, at the airport in Sarajevo, Bosnia, on January 9, 1993, they allowed Hakija Turajlic, the deputy prime minister of Bosnia-Herzegovina, to be murdered in cold blood after having promised to keep him safe. French peacekeepers explained to a reporter that they had not drawn their guns because their own lives were not directly threatened. Such rules of engagement are incompatible with unit cohesion, force security, and morale.

American military personnel are both fierce and restrained in the use of force, because they are disciplined professionals. Trusting the discipline of well-armed forces preserves their credibility and their confidence. When the Clinton administration sent marines into danger armed with silly putty and hot pepper, it sent the message that it was as concerned about the safety of the adversary as it was for the safety of U.S. forces. But American officials have a primary and overriding responsibility to the forces they command and the taxpayers who support them—a responsibility not to unnecessarily endanger lives or waste money, and not to take on missions that are not likely to succeed.

UNITAF, George Bush's mission to deliver food and medicine to a starving Somalia, was successful. UNOSOM II, undertaken by the Clinton administration and the UN, and aimed at nation building, was a predictable failure that was abandoned after unexpected casualties. Like the phrase "Vietnam syndrome," "Mogadishu massacre" had different meanings for different commanders. To some allies, it meant that Americans could not take casualties. To some, it referred to the need to avoid becoming involved in internal conflicts among factions.

All aspects of the Somalia experience had violated both the traditional Hammarskjöld rules of peacekeeping and the lessons of Vietnam. Soon American troops would depart for Bosnia, where UN forces were applying some—but not all—of the lessons learned in Somalia.

3.

HAITIANS' RIGHT
TO DEMOCRACY?

The U.S. military intervention in Haiti resembled that in Somalia. Both were multilateral military interventions undertaken for humanitarian purposes in failed states. In both countries, the Clinton administration started with the George H. W. Bush administration's narrow objective of providing food to alleviate starvation, then expanded it to the broader goals of fostering peace and democracy.

The concept of the failed state came into prominence at about the same time as the crises in Somalia and Haiti and the arrival of the Clinton administration. In an influential article in *Foreign Policy*, Gerald B. Helman and Steven R. Ratner described the failed nation-state as a disturbing new phenomenon: an underdeveloped state characterized by "civil strife, government breakdown, and economic privation" and "utterly incapable of sustaining itself as a member of the international community."[1] Refugee flows, political instability, and random warfare spread within these countries and across borders. The Clinton administration agreed that something should be done to help states that had fallen into violence and anarchy.

In the case of Haiti, the U.S. government adopted the notion that democracy is a human right, and that the United States is responsible for protecting or restoring it around the world, regardless of the costs or whether American interests are at stake. This experience showed the danger of assuming—naively, with insufficient planning and resources—that democracy can be imposed on a historically lawless and chaotic nation.

In a later *Foreign Affairs* article, Secretary of State Madeleine Albright identified four categories of countries: those that were "full members of the international system; those in transition, seeking to participate more fully; those too weak, poor, or mired in conflict to participate in a meaningful way; and those that rejected the very rules and precepts upon which the system is based."[2] Haiti fell into the third category. Its poverty, internal divisions, and chronic violence disrupted civil order so severely that it could barely survive, much less play a role in the international system. "We are trying to help Haiti overcome divisions and build its young democracy," Albright wrote.[3] Whether Haiti actually was, or could even become, a "young democracy," was by no means clear.

A series of developments at the end of the cold war had encouraged many to expect the rapid spread of democracy. In 1991, the Organization of American States (OAS) approved the Santiago Commitment to Democracy, which called for an automatic meeting of the OAS Permanent Council "in the event of . . . the sudden or irregular interruption of the . . . legitimate exercise of power by the democratically elected government in any of the Organization's member states."[4] This was the latest in a series of declarations by the OAS stating that representative government was the only legitimate government in the Americas. But neither Cuba nor Haiti was a democracy. Cuba was a one-party dictatorship, and Haiti had virtually no experience with democracy, or even with the rule of law. Severely underdeveloped, it more closely resembled a West African francophone country than its Latin and British Caribbean neighbors. It had won independence in 1804 through a revolt of its original inhabitants, most of whom were slaves.[5] In the decades since, it had suffered harsh dictators and chronic disorder, instability, and violence. In the words of a 1998 World Bank report, "Historically, Haiti's state has been essentially a personalized system of authoritarian rule based on vertical power relations centered on the chief executive. There has been only limited institutional development and few functions beyond the maintenance of power, patronage, and the extraction of wealth."[6]

From 1957 through 1971, the country was ruled by the infamous

François "Papa Doc" Duvalier. He was succeeded by his son, Jean-Claude "Baby Doc" Duvalier, who was president until he was ousted in 1986 after an uprising against his dictatorship. A series of strongmen served as president for brief periods until 1990, when Jean-Bertrand Aristide, a radical Roman Catholic priest, was elected in a landslide.[7]

With Aristide's election, many observers concluded that the global trend to democracy sweeping Eastern Europe and South America had reached the Caribbean.[8] But events quickly demonstrated that Haiti's transition to democracy would not be so smooth. Democracy requires free speech and freedom of assembly. It requires periodic elections in a context of personal security, competition, toleration of opposition, and orderly, transparent procedures. Democracy requires the rule of law, including politically neutral police and honest courts. Haiti did not have the political culture, traditions, or institutions associated with democracy. Papa Doc and Baby Doc had thoroughly politicized the judicial system and the police; they did not respect citizens or protect personal security.

In addition, democracy thrives where there is a reasonably good living standard and a substantial middle class, and Haiti had neither. It was the poorest nation in the Western Hemisphere. The same World Bank report noted that "the overwhelming majority of the Haitian population are living in deplorable conditions of extreme poverty."[9] The country's per capita income of $250 was less than one-tenth the Latin American average.[10] Haitian unemployment was as high as 80 percent; in rural areas, where two-thirds of the population lived, more than 80 percent of Haitians lived below the poverty line.[11] An International Monetary Fund (IMF) report pointed out that Haiti's "social indicators are . . . comparable to those of sub-Saharan Africa."[12] Life expectancy was only fifty-four years, and adult literacy just 43 percent.[13]

Haiti's extremely weak economy, widespread poverty, extreme disparities in wealth, political polarization, and habits of violence bred chronic disorder, the effects of which could be observed in the governments of the Duvaliers and then Aristide. U.S. trade accounted for 61 percent of Haitian imports and 87 percent of its exports;[14] by 1994, these figures had dropped by nearly half.[15] Between 1990 and 1995, the country's GDP contracted by about 6.5 percent annually.[16] The GDP of

Haiti's neighbor, the Dominican Republic, grew by approximately 3.9 percent annually during the same period.[17]

To this day, Haiti possesses few of the political requisites of democratic government. Mobs have threatened and attacked members of the opposition. Political murder remains shockingly common and is rarely prosecuted. Corruption is widespread.[18] Turnout for elections is usually very low. In the elections of November 2000, boycotted by the opposition, it was less than 10 percent in many areas.[19] Between 1987 and 2002, Haiti had thirteen governments—nearly one each year.[20] This chronic instability has persisted since the fall of the Duvalier dictatorship in 1986, and it was rampant in the early 1990s, when an aborted election brought it to worldwide attention.

A FALSE START FOR DEMOCRACY

In December 1990, Jean-Bertrand Aristide was elected president of Haiti; he was inaugurated on February 7, 1991. He promptly named René Préval as prime minister and General Raoul Cédras as commander in chief of Haiti's armed forces. Both men would play critical roles in the evolution of Haiti's new government, although not in the roles to which they were appointed. Less than eight months after his inauguration, Aristide was deposed by a military coup that left General Cédras in charge and sent Aristide into exile in Washington. The country was once again ruled as a military dictatorship, as it had always been except when it was governed by civilian dictators, notably the Duvaliers.

It was clear from the beginning that "Father Aristide" lacked the experience and temperament to be a constitutional ruler. Among those who observed him firsthand, many judged him to be seriously unstable, and there was speculation that he suffered from bipolar disorder.[21] He displayed a marked attraction to violence, and his behavior suggested that he brought to the presidency a fanatical disposition, intolerance for opposition, and a habitual disregard for law. His political style was not conducive to debate, compromise, or peaceful settlement of disputes. As journalist Mark Danner wrote, "even in Haiti's long and colorful history of delirious emperors, mad kings, and paranoid dictators, [Aristide] stands out as an extraordinary political phenomenon."[22]

Before his inauguration, Aristide had been expelled from the Roman Catholic Salesian Order for preaching liberation theology and advocating violent class war. During his first seven months as president, he repeatedly incited crowds to violence and threatened his opponents. He showed no interest in discouraging violence when his supporters attacked his opponents.

On September 27, 1991, Aristide delivered his infamous "Père Lebrun" speech from the steps of the presidential palace, inciting a crowd to burn his opponents alive. Père Lebrun is the Haitian term for "necklacing," a brutal form of murder in which the arms of a victim are hacked off and a gasoline-filled burning tire is put over his head. (The name comes from a well-known importer of tires in Haiti.) Aristide told his supporters, many of whom were carrying tires, that if they saw "a faker who pretends to be one of our supporters . . . just grab him. Make sure he gets what he deserves . . . with the tool you have now in your hands. . . . You have the right tool . . . the right instrument. . . . It smells good and wherever you go, you want to smell it." [23] [Translation from Creole; Aristide's words were recorded on videotape.] Two days later, Silvio Claude, a leading democrat, human rights activist, and political opponent of Aristide, was burned to death by a mob in front of the parliament building. Several other Aristide opponents had been killed by necklacing during his first months in power, and he had incited action against the papal nuncio, who was forced to flee for his life. The nuncio's assistant barely escaped; both of his legs and his jaw were broken by a mob of Aristide's followers. [24]

Aristide's increasingly strident speeches and threats contributed to the fear and hostility that led to the coup. Charles Lane observed in *Newsweek* that "Aristide's pro–Père Lebrun speech was the last straw for a military already furious over the president's creation of a French-trained, fifty-man presidential security detail answerable only to him." [25] Aristide's efforts to intimidate political opponents and encourage vigilante tactics were chronicled regularly by correspondents in the country.

Aristide also violated Haiti's laws and constitution. In the months before the coup, he appointed judges without parliamentary consent, failed to provide public accounting for funds, signed blank warrants for arrests, replaced newly elected mayors with committees of Lavalas supporters, and sought to establish one-party rule. [26] (Aristide had built the loosely organized Lavalas—"the flood"—movement during his campaign; he was nominated for the presidency by the National Front for Change and Democracy but separated from the party before taking office.[27])

Over two days, September 29–30, General Cédras led a coup that forced Aristide and his government into exile in the United States. Georges Fauriol, one of the best-informed observers of Haitian affairs in Washington, wrote the following:

> The army led the effort, with sympathy and initially open support from many in the nation's small economic elite and right-wing factions. . . . During the three-year political impasse [that followed], Cédras not only became the key figure but was assigned the central responsibility for the 1991 coup. U.S. diplomatic language encompassed this view as it argued its case for intervention in 1994.[28]

Fauriol is more skeptical that Cédras directed the coup. He cites evidence suggesting that NCOs and mid-level officers (majors and some lieutenant colonels) were involved in the initial phases of the coup at Aristide's personal residence and later at army headquarters, where he was almost killed. He was saved by the intervention of the French and U.S. ambassadors (by some accounts, also by Cédras) and ultimately escorted out of the country on a Venezuelan air force plane dispatched by President Carlos Andrés Pérez.[29] Aristide was accompanied by an entourage of associates, who undertook an intensive lobbying effort to support his return to the presidency.

Outrage in America and the International Community

The Bush administration and the international community were outraged that an elected government had been deposed and replaced by a military dictatorship. On September 30, 1991, the OAS Permanent Council met in emergency session and condemned the coup. The same day, the UN Security Council assembled informally at the request of Haiti's ambassador, who was representing the overthrown government, but took no action; most delegations viewed the coup as an internal matter that did not constitute a threat to international security.[30] The UN secretary-general, Pérez de Cuéllar, stated that he was "disturbed at the grave threats posed to democracy" and hoped that "the democratic process will resume in accordance with the constitution."[31] David Malone, in *Decision-Making in the UN Security Council: The Case of Haiti, 1990–1997*, notes that the Security Council president, French ambassador Jean-Bernard Mérimée, took the "unusual move" of expressing his personal support for the secretary-general's statement. This "was seen by some at the UN as a precedent in placing on the record Security Council concern over the preservation of democracy."[32]

President Bush declared himself in favor of collective action in response to the Haitian coup. "[T]his hemisphere is united to defend democracy," Secretary of State James Baker told the OAS ministers on October 2. Refusing to recognize the "outlaw regime" that had seized power, Baker said it was "imperative" that the OAS "act collectively to defend the legitimate government of President Aristide. . . . Until President Aristide's government is restored, this junta will be treated as a pariah throughout this hemisphere, without assistance, without friends, and without any future. . . . [T]his coup must not and will not stand."[33]

On October 3, the OAS passed a resolution calling for the diplomatic isolation of those who had seized power in Haiti.[34] The UN Security Council agreed to receive Aristide in a formal meeting and, although it did not adopt a resolution, Mérimée condemned the coup and voiced support for the OAS resolution. On October 4, President Bush stated in an executive order that the Cédras coup constituted "an unusual and extraordinary threat to the national security, foreign policy, and economy of the United States."[35] He suspended $90 million in nonhumanitarian aid to Haiti. At a news conference that day, he said he hoped that democracy could be restored without using force; however, he did not rule out U.S. participation in a multinational force:

> The United States has been . . . wary of using U.S. forces in this hemisphere. . . . I would like to see [this OAS mission] succeed without having to . . . put together such a force, to say nothing of using it.[36]

According to James Baker,

> [N]o serious consideration was given to the use of such force to restore Aristide to power in Haiti. In our view, the national interests of the United States clearly did not require risking American lives and expending billions of dollars in a full-scale military invasion and occupation. And history had taught us that it could not have been done *without* an extended occupation—something our successors now know.[37]

The OAS delegation met with General Cédras and others on October 6 and 7. By then Haiti was on the edge of chaos, and the ministers determined that outside assistance was necessary to restore constitutional order in the country.[38] Some members of the delegation were roughed up by Haitian soldiers.[39] On October 8, the OAS adopted a resolution recommending that member states impose a trade embargo (exempting humanitarian aid) against what was already being called the "Cédras regime."[40]

On October 11, the UN General Assembly passed a resolution condemning "the attempted illegal replacement of the constitutional President of Haiti"; demanding the "immediate restoration of the legitimate Government of President Jean-Bertrand Aristide, together with the full application of the National Constitution and hence the full observance of human rights in Haiti"; and appealing to member states to "take measures in support of the resolutions of the Organization of American States" concerning the condemnation and isolation of the regime in Haiti and the trade embargo.[41]

Embargo Against Haiti

Experts believed that the Haitian economy, already the poorest in the Western Hemisphere, could be paralyzed in a month by comprehensive international trade sanctions. The country's acting prime minister, Jean-Jacques Honorat, warned that "Haiti cannot withstand an embargo for more than three days. It is a country that produces nothing."[42] In late October 1991, the U.S. government announced a strict ban on "all commercial trade with Haiti, both exports and imports of goods and services."[43]

Haiti quickly ran out of fuel—it soon lacked gasoline for buses, kerosene for lamps, and electricity for factories and hospitals. It also ran out of money. At least sixty-five thousand Haitians lost their jobs in the first month of the embargo.[44] Farmers could not get their bananas, coffee, and cacao to market. Food became scarce, and medical supplies were nearly exhausted within a few months. "We don't even have fuel for our generators to operate our blood banks," said Haitian Red Cross director Dr. William Fougere in late November.[45]

People began taking to the sea in small boats; about half were picked up and returned, and the other half drowned. The local director of Project Care told the *New York Times*, "People . . . are starting to eat roots and things that animals eat." An international relief worker said, "The international embargo . . . has increased the pain of poverty in ways that are likely to lead thousands more people to set out for the U.S., regardless of the perils of the sea journey."[46]

The embargo was designed to force the restoration to power of Haiti's elected president. But the effects were economic, not political, and they fell on the poorest people and on the country's small middle-class commercial and business sectors. The embargo made life even more difficult for the many Haitians who normally lived on the razor's edge of subsistence, pushing the country toward a tragedy on a massive scale. Medicines, fuel for basic services, and foodstuffs such as wheat, sugar, rice, flour, and cooking oil were permitted, but these commodities were not provided. Haiti's economy was extremely feeble and heavily dependent on America for basic necessities. It could be argued that, in imposing the embargo, Washington was responsible for creating famine in the poorest nation in the hemisphere.

The OAS and the Bush administration used the embargo to demonstrate their collective commitment to democracy in the hemisphere, but it was a strange decision. You do not promote democracy by denying food to desperate people, destroying a rickety economy, or pushing a society over the edge to anarchy and chaos. The U.S. government could not solve all of Haiti's terrible problems, but, as I argued at the time, we could stop pretending that we did not bear the lion's share of responsibility for the stream of refugees making their way toward Miami and that Haiti's problems could be resolved simply by returning Aristide to power.[47]

Non-OAS states continued to send trade shipments to Haitian ports, and within a few months many OAS members, including the United States, began to carve out substantial loopholes in the embargo. Tankers delivered fuel during the winter and spring of 1992. The U.S. Government Accounting Office (GAO) reported that "at least a dozen countries routinely ignored the embargo."[48] In February, the United States announced unilateral exemptions to the embargo on a case-by-case basis for American-owned companies in the assembly industry in Haiti. The administration explained that the shift in policy was intended to put as many as forty thousand Haitians back to work and keep U.S. businesses from leaving the country.

Still, the embargo continued to take a serious toll, and sanctions would be tightened repeatedly over the next two years, often at the urging of Aristide. In May 1992, the OAS reiterated its call for member states and all other countries to respect the embargo.[49] President Bush complied, banning vessels that traded with Haiti from U.S. ports. In a May 28 statement, he explained:

> Our actions are directed at those in Haiti who are opposing a return to democracy, not at the Haitian poor. We are continuing to provide substantial, direct humanitarian assistance to the people of Haiti and are working to intensify those efforts. Our current programs total 47 million dollars and provide food for over six hundred thousand Haitians and health care services that reach nearly two million. While tightening the embargo, we will continue to encourage others to ship food staples and other humanitarian items to those in need. The action that I have directed will not affect vessels carrying permitted items.[50]

Meanwhile, the boat people continued to take to the sea. Contemplating the waves of Haitian refugees heading for American shores, international lawyer Lori Fisler Damrosch noted: "Ironically, the economic sanctions that became the chosen instrument of international involvement turned out only to exacerbate that problem for the United States, while bringing the crisis no closer to resolution."[51]

If Bush and Baker did not see a vital U.S. interest (or, at least, one sufficient to merit military action) in Haiti, they couldn't mistake the U.S.

interest in stanching the flow of Haitian refugees pouring into Florida. After the military coup, that flow had increased from two to three thousand per week. Under a 1981 agreement with the Duvalier regime, U.S. policy was to intercept refugees and return them to Port-au-Prince. The Bush administration sought a regional solution to the new refugee problem, but only four countries expressed interest. On November 20, Bush announced that the United States would grant asylum only when petitioners could substantiate claims of political oppression. By this time, so many people were leaving Haiti that the U.S. Coast Guard began taking refugees to a camp at the American naval base at Guantanamo, Cuba, for processing. When Guantanamo reached its maximum capacity of 12,500 refugees in May 1992, Bush authorized the Coast Guard to repatriate the refugees without processing them. But 1992 was a presidential election year, and Bush's decision would have political costs: Bill Clinton made Bush's policy an issue, criticizing the president's decision and calling for immediate and unconditional political asylum for the refugees—a policy he reversed soon after his inauguration.

THE CLINTON ADMINISTRATION ADOPTS A NEW POLICY

Clinton had strong political ties to the Congressional Black Caucus (CBC), whose members strongly advocated action on Haiti, and he arrived in the White House ready to implement a more activist policy. In February 1993, less than a month after his inauguration, a joint UN/OAS international civilian mission (MICIVIH) to Haiti was established to monitor human rights practices and investigate violations.

In June 1993, the UN Security Council unanimously adopted Resolution 841, imposing a mandatory oil and arms embargo against Haiti. Lori Damrosch, an expert in international law, observed at the time that "the Haitian sanctions resolution goes farther than any other to date in applying universal, mandatory, and severe economic sanctions to influence a domestic political crisis over democratic governance. Its cautious wording (stressing more than once the 'unique and exceptional' circumstances) cannot hide its precedential significance."[52]

It was widely assumed—on the basis of virtually no evidence—that imposing economic sanctions on Haiti would hasten the return to democracy by putting pressure on the ruling junta. But it did not. It only made the Haitian people poorer. In a careful study of the effects of the sanctions, Elizabeth Gibbons, a former UN representative in Haiti, described the harsh impact of the sanctions; while the efforts of the international agencies and the strength of the Haitian people prevented outright famine, she concluded, they could not protect Haitians from the ravages of the embargo.[53] The embargo had devastating effects on the poorest Haitians, especially children, and on public services. A measles epidemic killed or debilitated thousands of children. Conflicting goals created cross-cutting pressures that blocked effective humanitarian work by UN agencies.

International pressure to restore Aristide to the presidency led to the Governors Island Agreement of July 1993. Drafted by Dante Caputo, Special Envoy for Haiti of the Secretaries-General of the UN and the Organization of American States (OAS), and signed by Aristide and Cédras in July 1993, the agreement called for the retirement of Cédras and the restoration of Aristide by October 30, 1993; dialogue among Haitian political parties to allow parliament to resume normal functions; and international technical and financial assistance to build a new police force, modernize the army, and implement administrative and judicial reform. It also called for the appointment and confirmation of a new prime minister, and for amnesty for the coup leaders. Aristide selected Robert Malval as prime minister, and the selection was ratified by the Haitian parliament. In late August, the UN passed Resolution 861, suspending the embargo, and Resolution 862, deploying a UN mission in Haiti.

The Governors Island Agreement was Clinton's preferred instrument for dealing with Haiti. As late as July 23, 1993, he said, "I'm excited about this process. There's a major potential for a victory for democracy here."[54] Clinton pledged economic aid to Haiti, and he offered to send U.S. military personnel to retrain the nation's army and police force. Ultimately, though, neither Cédras nor Aristide honored the agreement, and it was never implemented. Nor was the Malval plan to restore democracy, which was put forth by the State Department after the Governors Island plan failed. Generous funding eliminated an important incentive for Aristide to cooperate: the U.S. government provided him with

$1 million a month, and the State Department authorized payments of $5 million a quarter.

While the Clinton administration waited to see whether Aristide and Cédras would honor the Governors Island Agreement, the U.S.S. *Harlan County* made its way to Haiti. The ship, carrying about two hundred American and twenty-five Canadian military and police trainers armed with light weapons, was slated for a six-month stay in connection with the UN mission. Although the Governors Island commitments were not being kept, National Security Advisor Anthony Lake, his deputy, Sandy Berger, and Secretary of State Warren Christopher prevailed over Secretary of Defense Les Aspin in allowing the *Harlan County* to proceed. When the ship arrived in Port-au-Prince on October 11, it was met by a flotilla of small boats filled with several hundred armed attachés, who sought to prevent it from docking.[55] The ship retreated. (This incident occurred just days after the battle in Somalia in which eighteen U.S. servicemen were killed.)

The ensuing flood of criticism got the attention of the top leaders of the Clinton administration, who did not like appearing incompetent. President Clinton immediately called for the reimposition of sanctions; the Security Council agreed, passing Resolution 873 on October 13. Three days later, after Aristide pushed to strengthen the sanctions, the Security Council voted to impose a naval blockade on Haiti. Aristide continued to press for tougher sanctions against the Cédras regime through the spring and summer of 1994, although the Haitian parliamentarians and the State Department did not agree. That spring, the Congressional Black Caucus—which supported Aristide—proposed the Haitian Restoration of Democracy Act, which would tighten the economic embargo, block financial assets held in the United States by Haitian nationals, sever commercial air links, and reverse the summary repatriation of refugees. The act was never brought to a vote, but most of its provisions were eventually adopted as administration policy.

In April 1994, Aristide terminated the 1981 treaty allowing repatriation of Haitian boat people, and Randall Robinson, chairman of TransAfrica (a leftist organization that focused on foreign policy issues of concern to African Americans), went on a hunger strike to protest the U.S. repatriation policy. The administration reaffirmed its support for

Aristide and agreed to look into strengthening the Dominican Republic's border with Haiti against sanctions violations. Clinton caused a stir in Washington when he attempted to distance himself from his own administration's policy by voicing support for Robinson's protest. His comments triggered a new wave of refugees. On May 8, the administration agreed that Haitians would be interviewed to determine whether they were political refugees and would not be summarily sent back to their country.

The Clinton administration's actions were not without consequence. A *Washington Post*-ABC News poll released in mid-May 1994 confirmed that public confidence in President Clinton's handling of foreign affairs had declined sharply, with 40 percent approving, 53 percent disapproving, and only 13 percent saying they thought the president had a clear foreign policy.[56] The clamor for stronger action against Haiti was rising on the left of the Democratic Party, a faction to which the president was especially sensitive. Clinton's former envoy to Haiti, Lawrence Pezzullo, described the administration's policies as irrevocably headed toward military intervention to restore the ousted Haitian president.[57]

On May 6, 1994, the administration joined in UN Security Council Resolution 917, which imposed a comprehensive mandatory commercial embargo and an international naval blockade unless the junta departed Haiti by May 21. The blockade included seven U.S. naval warships, frigates from Argentina and Canada, and 650 Marines aboard the U.S.S. *Wasp*. Two members of the Friends of the secretary-general for Haiti,[58] France and Canada, announced that they would not participate in or support a U.S.-led invasion.

The sanctions' cumulative impact on Haiti's economy was overwhelming and devastating. Before October 1991, for instance, there were 145 garment factories; by January 1994, only 44 remained. Before the coup and the sanctions, the assembly industry employed forty-four thousand workers; by May 1994, that number had dwindled to eight thousand.[59] But still the Clinton administration moved to tighten the screws. Former president Jimmy Carter later revealed his negative opinion of the sanctions: "I told [General Cédras] that I was ashamed of my country's [embargo] policy."[60]

During the summer of 1994, Clinton and his special adviser on Haiti, William Gray, tried with little success to find other countries that would accept Haitian refugees. The effort dramatized Haiti's isolation in the region. Panama backed out of an earlier agreement to accept ten thousand Haitians. Country after country found reasons not to accept the displaced people; they feared the refugees' competition for scarce resources, jobs, and public assistance, and they feared AIDS, which was believed to be more widespread in Haiti than in most other Caribbean and Latin American countries.

Aristide asked the UN for action to restore democracy in Haiti. On July 31, 1994, by a vote of 12 to 0 (Brazil and China abstained, and Rwanda was absent), the Security Council passed Resolution 940, authorizing a multinational force and clearing the way for U.S.-led military action. On August 5, members of the U.S. Senate engaged in a heated debate on Haiti policy. Majorities in both parties opposed a U.S. invasion, and Congress was sharply divided on whether the president had the authority to undertake military action without congressional authorization. A U.S. force had never been used to "restore democracy" or been sent anywhere without congressional authorization.

A clear majority of the American public was opposed to such action, and criticism of Clinton's handling of foreign affairs grew. The deepening U.S. involvement in Somalia made the proposed military commitment in Haiti even less attractive. Clinton spelled out for Congress and the American people several reasons that restoring Aristide to power was in the U.S. national interest: to stop the brutal atrocities that threatened the lives of tens of thousands of Haitians; to secure our borders and prevent a mass exodus of Haitian refugees; to preserve stability and promote democracy in the Americas; and to emphasize the reliability of U.S. commitments.[61]

Nonetheless, a September 17 ABC News poll showed that 60 percent of Americans opposed U.S. military action in Haiti, and only 31 percent agreed that the situation was a vital American interest.[62] The *New York Times* also reported that 60 percent of Americans opposed sending or keeping American troops in Haiti, and 56 percent said that the United States had no responsibility to restore democracy in Haiti.[63]

AN ENTITLEMENT TO DEMOCRACY?

For several months, the Clinton administration seemed to be preparing for war (although they called it a "peace operation"). By whatever name, what they were planning was an invasion, intended to depose Haiti's unconstitutional government and restore democracy in the form of Aristide. Already the administration had tightened the embargo in an effort to remove Haiti's military government from power. Already it had reinforced Haiti's long, rugged border with the Dominican Republic. Obviously, U.S. military forces would have no problem subduing Haiti's small army, establishing control of ports and government buildings, and installing Aristide in power. But what then?

Would restoring democracy to Haiti mean trying to ensure democratic government, with the rule of law and respect for rights of citizens? Would it mean accepting responsibility for Aristide's use of power? Did the Clinton administration plan to become directly involved in governing Haiti—as UN secretary-general Boutros Boutros-Ghali had proposed that the United States do in Somalia?

The operation proposed for Haiti would not just replace one ruler with another; it would also aim to build a modern democratic state. But nation building requires a long-term commitment, intimate familiarity with the country, and deep cultural affinities. The United States and the Clinton administration clearly had few of the fundamental requisites for successful nation building in Haiti.

I believed, further, that an invasion of Haiti would be incompatible with American interests. Haiti was not a menace to the United States or the hemisphere. It was not then a center of Caribbean drug trafficking (though that would later change). It did not provide a base for a hostile power. It did not export subversion and revolution. It had not declared open season on Americans, as Manuel Noriega did earlier, nor held Americans hostage, as had Grenada's revolutionary Committee of Safety. It had not engaged in terrorist plots against Americans, as Libya did.

There was only one conceivable ground for invading Haiti, and that was to implement some sort of "Brezhnev doctrine" for democracies, invoking the principle that no democracy could be overthrown. But Haiti had only the weakest claim to ever having had a democratic government in the first place, and the prospects were not bright that Aristide would govern by constitutional means if he returned to power. During his first term, he consistently undermined the rule of law, violated constitutional practices, ignored established institutions, shut down parliament (where he did not have a majority), and relied on mobs and a private gang of enforcers—on the model of Papa Doc's Tonton Macoutes—to attack his opponents. By violating democratic norms and human rights standards, Aristide had forfeited his claim to constitutional rule. As George H. W. Bush noted at the time, restoring Aristide might not be the same as restoring democracy—he could prove as difficult as Haiti itself to deal with. On the other hand, the military rulers had violated Haiti's constitution and had no claim to legitimacy.

One of the fundamental questions at issue was the existence of a right to democracy, or a right to be governed democratically, by rulers chosen in free elections. Did Haiti have such a right? The Clinton administration thought so, and tried for many months to rouse support in the international community for action to depose the military government and restore Aristide. These efforts, and the political skill of chief U.S. delegate Madeleine Albright, produced Security Council Resolution 940, authorizing "the use of all necessary means" (that is, force) to achieve this end. But was this force to be provided and paid for exclusively by the U.S. government?

Weeks of effort to persuade other governments to contribute netted little. Canada turned down an appeal to join the expeditionary force, but offered to send peacekeepers. France and Venezuela, both in the Group of Friends, declined to participate in the multinational force (MNF). Among European allies, only the United Kingdom, Belgium, and the Netherlands agreed to participate in the military phase. Other countries agreed to contribute troops, including Argentina, Australia, Bolivia, Israel, Jordan, and a number of Caribbean and Central American nations. Eventually, twenty-eight nations would contribute about two thousand peacekeeping troops, police monitors, and translators to the U.S.-led MNF.[64] The largest contribution, of five hundred troops and one hundred police monitors, came from Bangladesh.[65]

The Clinton team justified its plan to invade Haiti on grounds that force was required to restore democracy. It offered other supporting arguments, mainly that the Cédras government had violated the civil rights of Haitians and had failed to carry out the decisions of the UN Security Council and the provisions of the Governors Island Agreement. But the fundamental justification for the use of force was a postulated "right to democratic government" of which Haitians had been deprived.

Resolution 940, the first action of its kind, constituted a significant expansion of the Security Council's jurisdiction over the internal affairs of member states. The idea of a right to democracy, which could be imposed by force, was a dramatic departure from previous theory and practice. International lawyers, notably Thomas Franck, whose work was an important source for the ideas and arguments of Morton Halperin and other Clinton administration officials, had written of an emerging democratic entitlement and right to democratic governance.[66] Franck argued that the end of the cold war, the collapse of the Soviet Union, and the disappearance of Marxism-Leninism as a competing paradigm for understanding the world and legitimizing political action, had resulted in a global move toward democracy and a new global ethos in which all persons enjoy all democratic rights and under which only democratic governments are legitimate.

If democracy is viewed as a human right shared by all persons, and the world community has an obligation to use force when necessary to protect this right, then it is also appropriate to use force to depose any government that achieves power by force and violates its citizens' rights. By acting against the government of Haiti on these grounds, we would logically be committed to act again if the Haitian government did not respect the rights of Haitians. As it turned out, Aristide did not respect the rights of Haitian citizens, but too little attention was paid to that threat at the time.

Resolution 940 built on the precedent established in the mandatory economic sanctions imposed on Haiti in June 1993. It justified ousting Haiti's military rulers, basing its argument on domestic conditions. (The nation's humanitarian situation had deteriorated; violations of civil liberties had increased; the condition of refugees had deteriorated; a UN team monitoring human rights had been expelled.) The resolution implicitly endorsed a right to democracy, and made clear that this right overrode the prohibitions in the UN Charter against the use of force (except for self-defense and collective self-defense), and against intervention in the internal affairs of states.

This idea of a right to democracy was appealing to Americans, many of whom believe that all people should govern themselves democratically. But should we overthrow governments because we disapprove of them? Should we risk American lives? And why Haiti? Why not Cuba? Presidents justify their policies, especially decisions to use force, in terms of the nation's basic values and established practices. The Clinton administration prepared Americans for military action in Haiti by emphasizing that an illegitimate government must be forced out of power and democracy restored. But the case Clinton made for U.S. intervention would seem to have applied far more clearly to Cuba than to Haiti.

Thomas Franck foresaw a day when the global community would guarantee democracy as a legal entitlement. But he also believed that "the collective use of military force to protect the people's right to democracy is an extremely remote bridge which need not be crossed at present."[67] This was precisely the bridge Clinton and Christopher were ready to cross; the mystery was what they intended to do when they reached the other side.[68]

Justifying Intervention

It was disturbing, in that summer and fall of 1994, to hear Clinton administration spokespersons call the planned military operation a "police action" rather than a war; they seemed to be evading the constitutional requirement for congressional consent and treating the Security Council resolution as sufficient grounds to spend half a billion dollars and risk American lives. It also seemed cynical for a government that endlessly sought a negotiated peace in Bosnia-Herzegovina to speak of having exhausted all alternatives to the use of force in Haiti. In fact, the U.S. government had discouraged the efforts of Venezuela and other Latin American governments to resume talks, even though the Cédras government repeatedly indicated its willingness to negotiate. Administration officials even cited the U.S. military action in Grenada as a precedent, an especially inapt and objectionable analogy. The Reagan administration's 1983 military action in Grenada was conducted under the treaty of alliance with the Organization of Eastern Caribbean States (OEC), and in the face of a clear and present danger to more than six hundred American students held prisoner by a band that had already murdered Grenada's prime minister and cabinet. In fact, a comparison of the situations—Grenada in 1983 and Haiti in 1994—is a textbook illustration of the difference between problems that do justify the use of force and those that do not.

The situation in Haiti posed no urgent threat to the lives of Americans. In Grenada, the American students were trapped in an atmosphere of extreme violence. On October 19, 1983, Grenada's Marxist prime minister, Maurice Bishop, and five members of his cabinet were shot in cold blood by Bishop's Cuban-trained deputy, Bernard Coard. Coard imposed a round-the-clock shoot-on-sight curfew and closed the airport, trapping some one thousand American citizens, including several hundred American medical students who were held incommunicado under guard. The U.S. government obviously had an urgent interest in bringing this crisis to a swift and safe conclusion.

Haiti had no strategic importance to the United States. The situation in Grenada was quite different: the country had been transformed into a base for the projection of Soviet and Cuban military power in the Caribbean. The largest airstrip in the Western Hemisphere was nearing completion under Cuban auspices, and the flow of military traffic to and from Grenada was causing widespread anxiety among Grenada's island neighbors. Their fears were confirmed by the discovery after the invasion of eight hundred armed Cuban troops and six warehouses filled with advanced Russian weapons. In addition, secret treaties were discovered between Grenada and the Soviet Union, Cuba, and North Korea that included plans to make Grenada a major base for guerilla operations in the Caribbean basin.

Haiti posed no threat to the peace and security of the region, but the United States and most Caribbean nations perceived Grenada as a threat. The prime ministers of Jamaica, Barbados, and Dominica had appealed to the United States and the United Nations for help. With eloquent, data-rich presentations, they described the destabilizing effects on the region from the extension of Soviet-Cuban power to Grenada. The weapons caches and secret documents discovered there offered ample justification for these concerns.

Moreover, surprise was not a necessary element of the Clinton plan for deposing the government of Haiti, but secrecy and dispatch were needed to save the American students in Grenada. The Reagan administration did not consult with Congress or seek UN authorization to invade Grenada. (There was never any question of going to the UN. It was the height of the cold war, and the Soviet Union would certainly have vetoed any action against a communist state.)

The Grenada invasion was applauded by a large majority of the American public but denounced in the UN by the Soviet and nonaligned blocs, and in Congress by liberal Democrats. The CBC, which later urged action in Haiti, strenuously denounced the liberation of Grenada—even after the American students shared their terror on national television. Top officials of the Clinton administration expected that once the military landed in Haiti, the American public would rally around, and the action would prove to be a source of new political support for President Clinton.

But the American people have an acute sense of what is and is not in the national interest. A majority opposed risking American lives to restore Aristide to power, because they knew Haiti was no threat to the United States. While few Americans approved of the military government in Haiti, most knew that Haiti had had a series of bad leaders, and they did not want the United States to assume responsibility for the quality of government in Port-au-Prince. Most Americans oppose the use of military force except when a vital U.S. interest is at stake, and no one had made a persuasive case that Aristide's return to the presidency was such an interest. When the administration decided to use military force in Haiti, it was only the latest example of a growing habit of discounting public views and values in making foreign commitments. The public had shown little support for committing American forces and resources to the various peacekeeping activities, yet the Clinton team remained enthusiastic about pursuing them.

The administration's plans to invade and occupy Haiti also failed to meet traditional standards of international relations and law. The UN

Charter—which is to international law what the U.S. Constitution is to American law—explicitly prohibits "the use and threat of force against the territorial integrity or political independence of any state" (Article 51) and forbids intervention in the internal affairs of states except in self-defense, collective self-defense, or where there is a serious threat to international peace and security. In keeping with the traditional view of many liberals on the unacceptability of the use of force, the prohibition against using force in international relations was the heart of the UN Charter, the crucial norm, the very basis of a civilized world. International lawyer Louis Henkin spoke for many when he wrote, "The UN Charter—an epitaph to Hitler—is not neutral between democracy and totalitarianism, between justice and injustice, or between respect for human rights and their violation. . . . [But] those fundamental goals are not to be pursued by force." No one had ever suggested that violation of the Universal Declaration of Human Rights or the International Covenant on Civil and Political Rights created grounds for using force.

In Iraq, intervention to provide humanitarian assistance to the Kurds and Shiites threatened by Saddam Hussein's forces was justified because these massive violations of human rights—on the borders of Iran and Turkey—constituted a serious threat to international peace and security. The government of Haiti constituted no serious threat to Haiti's only close neighbor, the Dominican Republic, and the human rights violations in Haiti did not threaten international peace and security.

Military intervention also violated the charter of the Organization of American States. The Santiago Declaration, adopted by the OAS on June 5, 1991, was sometimes said to justify intervention to preserve democracy because it defined representative democracy as "an indispensable condition for the stability, peace and development of the region." But the Santiago Declaration did not authorize force to restore democracy, in Haiti or anywhere else. The states of the Western Hemisphere have always been extremely sensitive about their sovereignty, and they have a special aversion to U.S. intervention, with which they have had extensive experience. In fact, Haiti endured U.S. military occupation from 1915 to 1934. Major Latin American states had already made it clear in the UN and the OAS that they opposed U.S. military intervention to return Aristide to power. France, a permanent member of the Security Council, also opposed military intervention.

The Clinton administration's Haiti strategy had its origin in views Clinton's appointees brought with them to the White House, the State Department, and the Department of Defense. These views concerned broad issues of sovereignty and its limits, peacekeeping and war making, cooperative security and collective restraint, as well as democracy and how to expand it. These ideas constituted a new theory of international relations and global engineering, but because the concepts were discussed in specialized journals and introduced one or two at a time rather than as a doctrine—and because they did not fit familiar Right-Left, liberal-conservative categories—they drew little attention or scrutiny, even as they exerted a major influence on U.S. foreign policy.

Those who subscribed to this new doctrine put their faith in a set of interlinked ideas: the decline of sovereignty, the doctrine of failed states, an interventionist imperative, and the legitimacy of multilateral force. The core principle was the conviction that sovereignty was an outdated concept, no longer appropriate to an interdependent world in which many important problems transcend state boundaries and states were "collapsing" at an unprecedented rate. In 1992, Secretary-General Boutros-Ghali wrote, "[T]he time of absolute and exclusive sovereignty . . . has passed; its theory was never matched by reality."[69] Global realities, it was argued, required more activist international approaches. If the notion of sovereignty was outdated, for instance, it was not necessary to respect Article 2.7 of the UN Charter, which asserts, "Nothing contained in the present Charter shall authorize the United Nations to intervene in matters which are essentially within the domestic jurisdiction of any state." (A 2002 *Foreign Affairs* article by Gareth Evans and Mohamed Sahnoun illustrates the further development of thinking on the limits of sovereignty.[70])

Hand in hand with the decline of sovereignty came the rise of the doctrine of failed states. The failed state was seen as a threat to its neighbors and to international peace and security.[71] Failed states "imperil[ed] their own citizens and threaten[ed] their neighbors with refugee flows, political instability and random warfare" and "their problems tend[ed] to spread," thus making them appropriate objects of UN military action.[72] This postulated link between failed states and international peace was used to justify the Security Council's claim of jurisdiction over the internal problems of states and its authorizations of international force to deal with them. Because democratic governments do not engage in aggressive wars and so must be preserved, this new doctrine held, the UN (or the United States) should be allowed to force as a necessary strategy to protect international peace and security.

Haiti was a perfect match for this new approach to third world policy: a failed state to which the international community could restore democracy. The proposed mission, which won multilateral authorization, was an occasion for the altruistic use of American power in a location close to home—an operation that would not be too dangerous or too expensive. The convergence of the Clinton team's doctrine and the problems in Haiti—along with the enthusiasm of key Democrats— paved the way for intervention and suggested how easily this new ideology of international altruism could be used to justify a military action in virtually any less-developed country.[73]

THE CARTER-JONASSAINT DEAL AND OPERATION UPHOLD DEMOCRACY

Clinton's top priority was to remove General Cédras from the presidency and restore Aristide. During the summer of 1994, pressure for U.S. military action grew on the left and in the CBC. Even as he was preparing for war, however, Clinton rather reluctantly asked former president Jimmy Carter to head a mission that included Senator Sam Nunn (D-GA) and General Colin Powell to try to resolve the problem and avoid a military confrontation.

On September 17, Carter, Powell, and Nunn arrived in Port-au-Prince to meet with Haiti's leaders. They persuaded Cédras to agree to the Carter-Jonassaint Accord (signed by Emile Jonassaint, Haiti's military-appointed civilian president). The accord provided that the military troika (Cédras, army chief of staff General Philippe Biamby, and Lieutenant Colonel Michel François) would take "honorable retirement" by October 15; that the government of Haiti would cooperate with U.S.

armed forces and hold legislative elections promptly; and that UN sanctions would be suspended without delay.[74] The dictators told Carter that they would leave and permit Aristide to return to power. Clinton's last-minute diplomatic effort had narrowly averted a military conflict.[75]

By the time Clinton called off the planned invasion, paratroopers of the 82nd Airborne were already on their way to Haiti. The *New York Times* described the reaction aboard the U.S.S. *Eisenhower*:

> [S]enior commanders said Sunday night and throughout the day that the plans for what the troops would do were extremely fluid, indeed uncertain. The situation, in fact, was changing so swiftly that the [ship]'s skipper, Capt. Mark A Gemmill, ordered his helmsmen to make continual adjustments so the ship's antenna could pick up the sometimes flickering CNN broadcasts. Many young soldiers and junior officers gathered in the First Brigade's headquarters shook their heads as they watched President Clinton's speech announcing that an agreement had been reached for the Haitian military rulers to leave, and that American troops would have a security role rather than go in as an invasion force.[76]

On September 19, U.S. troops began to land in Haiti as an occupation force rather than an invasion force. They encountered no resistance. By sundown, almost three thousand troops from the U.S. Army's 10th Mountain Division were on the ground, the lead elements of the twenty-eight-nation multinational force in Phase 1 of what was dubbed Operation Uphold Democracy. On September 21, Clinton dispatched one thousand military police to supervise the Haitian police. Before the end of the week, more than twenty thousand U.S. troops, including eighteen hundred marines, were in Haiti.

As in Somalia, the Haiti mission consisted of two distinct phases. Phase 1, carried out by U.S. forces and a few troops (266) from nearby Caribbean islands, removed General Cédras and his administration and replaced them with President Aristide. That mission was carried out with virtually no bloodshed, thanks in significant measure to the negotiations by Carter, Nunn, and Powell.

Phase 2—restoring democracy—was undertaken by a multinational force assembled by the secretary-general.[77] Participants in the MNF in-cluded approximately two thousand peacekeeping troops, police moni-tors, and translators from Antigua, the Bahamas, Barbados, Dominica, Guyana, Jamaica, Panama, St. Kitts-Nevis, St. Vincent, Trinidad, Ar-gentina, Bolivia, Belgium, Britain, the Netherlands, Bangladesh, Israel, Ghana, and Belize.[78]

Congress reacted badly to Clinton's decision to send troops to Haiti, whatever their assignment. In a joint resolution, the House and Senate declared that the president should have sought congressional approval for military operations in Haiti and urged the "prompt and orderly with-drawal" of U.S. troops.[79] But the troops would not withdraw for some months. The tasks assigned to the American forces—maintain order and provide security for Aristide—were problematic. It was hard to keep supporters of the military and supporters of Aristide from engaging in acts of violence against one another, and difficult to provide stability and order in a country where political violence was entrenched.[80]

NATION BUILDING IN HAITI

Establishing a new government is always difficult, especially after a vio-lent rupture with a previous regime. Should those responsible for the vi-olence be punished for mass murder, rape, or torture? At what point should the slate be wiped clean? These questions were being asked in Haiti, as in Rwanda, Bosnia, Somalia, and Russia. French scholar Ray-mond Aron says that it is always prudent to "prefer the limitation of vio-lence to the punishment of the presumably guilty party." Clinton took this approach when he accepted the Carter-Jonassaint Accord. Carter, Nunn, and Powell recommended trading the pleasures of revenge for a chance at national reconciliation, understanding that even against an ad-versary as small and weak as Haiti, war was never to be undertaken lightly.

Moving the country toward democracy was the main challenge. Haiti had virtually no experience with democracy; its traditionally per-sonalist political system has made it difficult for Haitians to lay down arms against one another. Haiti was not a Latin regime with a strong army and a tradition of military rule, like the Dominican Republic. It had no inherited democratic tradition, like the British Caribbean states. Haiti's politics had been dominated by a series of strongmen who ruled as long-term dictators. Some were generals; others, like Papa Doc, were chosen in elections that lacked most of the characteristics of democratic elections. Papa Doc's was no military government. As soon as he was elected, he disarmed the army and kept its weapons locked in an armory to which he alone had the key. He relied on the violence of the Tonton Macoutes, which ran the country like a protection racket, bullying, beating, and collecting "taxes" at will. And Aristide continued the tradition of corrupt rule. Though he was originally elected in a reasonably fair and free election, no sooner did he take office than he began to bypass parliament and the army and start using his network of private gangs to terrorize citizens.

Haiti's political culture emphasized violence, voodoo, and vengeance, and many Haitians saw politics as a winner-take-all zero-sum game. This did not breed trust, confidence, or consensus. The only hope for a viable government was to break the cycle of vengeance and rule by force. The Carter-Jonassaint Accord sought to end the cycle by permitting Cédras and his associates to step down with dignity, letting Aristide assume the presidency peaceably, and protecting the population against uncontrolled violence. It was a prudent course that increased the chance that democracy might take root, but the prospects were still not good. Democracy makes complex demands on people. It requires both participation and restraint, active citizenship, inclusive policies, and a modicum of good faith. These complex demands have proved hard for Haiti to meet, and the country has failed to make much progress toward a functioning democracy in the years since 1994.

The immediate outcomes of the Carter-Jonassaint Accord and Operation Uphold Democracy were not very impressive. The world's only superpower had managed to land forces in one of the world's smallest, poorest, and least developed countries. Those forces succeeded in destroying some arms caches and arresting some attachés who had formerly terrorized the population. The military leaders left Haiti, with Cédras going into exile in Panama. The U.S. government arranged for Aristide's return and the repeal of the economic embargo, and made commitments of economic aid.

It was announced that all judges would be fired and that most police and a large part of the armed forces were unfit to serve in the new regime. The situation was less dangerous than that in Somalia—Haiti was so small and unarmed that hostile persons were less able to cause problems for U.S. and UN troops. But the country lacked virtually all the requirements for a democratic government: rule of law, an elite with a shared commitment to democratic procedures, an educated populace, a sense of citizenship, a decent standard of living, and habits of trust and cooperation. Although no one could explain how this venture contributed to the U.S. national interest, the Clinton administration and various journalists termed it a success.[81]

On March 31, 1995, the multinational force was officially disbanded, and control passed to the UN Mission in Haiti (UNMIH), which was about six thousand strong and included more than twenty-three hundred American troops.[82] UNMIH undertook the training of the Haitian police. Training the new recruits to respect citizens and protect personal security proved to be exceedingly difficult, as old patterns of corruption and violence remained.

The Clinton administration had much riding on the success and integrity of Operation Uphold Democracy. The plan had faced opposition in the Security Council and the OAS, but the administration had pushed forward. To avoid a failure of this operation comparable to the failure in Somalia, Clinton urgently pursued the campaign for democracy in Haiti.

THE RESTORATION OF ARISTIDE AND THE 1995 ELECTIONS

Parliamentary, municipal, and local elections were to be held within a few months of Aristide's return to power, but once again he flouted the law. Instead of holding elections by December 1994, he postponed them until February 4, 1995, when the terms of incumbent parliamentarians and local officials expired; this included the entire 83-seat Chamber of Deputies, two thirds of the 27-member Senate, 137 mayors, and 565 town council members. Aristide then ruled by decree.

As for the judicial branch, Aristide attempted to pack the Haitian high court with members of his Lavalas movement, without Senate approval. He forced the minister of justice, Ernst Malebranche, to resign after he criticized Aristide for replacing appointed-for-life judges with political allies. As a Clinton aide wrote in March 1995, "The judicial system [in Haiti] is not only unjust; it barely exists."[83] And soon after returning to office, Aristide circulated to supporters a secret "watch list" of thirty persons charged with unspecified crimes against humanity. Some observers were concerned that the list heralded a campaign of repression against opposition candidates in the upcoming parliamentary elections.

Instead of strengthening the rule of law, Aristide's policies under the American occupation only further destabilized the country. He eliminated the existing structure of police and military forces in Haiti, and reduced the army to one-quarter of its previous size. In February 1995, he dismissed all officers above the rank of major.[84] He also tried to put in place a civil security force (consisting of followers who had committed human rights abuses in his earlier tenure) but was pressed by the United States to revise this politicized recruitment plan. This move was part of Aristide's general reliance on private militia—a practice with a long tradition in Haiti.

The U.S. government was relieved when the UNMIH took over at the end of March 1995. After several postponements, elections were finally held on June 25 under the protection of the UNMIH, which had approximately 6,000 members from twenty-one nations as well as 850 civilian police to reinforce the national police (roughly 3,500 in number).[85] These forces should have been able to provide security for candidates and voters, but they were outmatched, and assassinations multiplied in advance of the elections.

Once again, Aristide's best-known opponents were among the targets. Mireille Durocher-Bertin, an articulate woman and determined opponent of Aristide, was shot dead on March 28.[86] Her killer was never identified. Matsen Cadet, a candidate for mayor and an Aristide critic, was shot twice at a campaign rally. A Senate candidate and critic of Aristide was wounded and his driver killed. On June 27, Jean-Charles Henoc, an opposition candidate for a deputy seat, was assassinated. Another opposition candidate for the legislature, Duly Brutus, was arrested, jailed, and severely beaten in Port-au-Prince; he was released after he managed to pass written appeals to UN special envoy Lakhdor Brahimi and others.

In October, after Brutus testified in Washington, DC, his house was attacked by a mob some three hundred strong. Violence continued to spread, often carried by mobs shouting "Aristide or death." On September 12, Ambassador Colin Granderson, chief of the UN/OAS Civilian Mission in Haiti, said there had been at least twenty commando-style killings of political figures in Haiti in 1995.[87]

Shortly before the June elections, the Haitian Provisional Electoral Council (CEP), which Aristide had staffed with loyalists, announced that a million voter registration cards had been stolen. As the Carter Center reported: "[I]nstead of building confidence in an electoral process, the government and the CEP—by their words, actions and inaction—eroded confidence. Instead of building bonds with the political parties by listening to their concerns and complaints and responding in an expeditious and helpful manner, the CEP stiff-armed the parties and never responded to their complaints."[88] Repeated delays in publishing voter and candidate lists meant that few voters knew much about the elections. All suggestions for reforming the CEP were stonewalled. The Carter Center's Robert Pastor said that "Aristide rejected any change" in the system, and instead he "appointed figures that were viewed as even more biased in favor of Lavalas," effectively making the CEP an arm of the party.[89] On June 24, the day before the elections, the International Republican Institute issued a report concluding that "the pre-electoral process and environment in Haiti has seriously challenged the most minimally accepted standards for the holding of a credible election."[90]

About 30 percent of registered voters participated in the polling on June 25.[91] As Pastor reported:

> [T]he level of irregularities was so high, and the vote count so insecure that virtually all of the parties except Lavalas condemned the legislative elections, called for annulment, and threatened boycotts if their concerns were not addressed. President Aristide met with opposition leaders, but they could not agree on which parts of the election should be accepted and which parts should be rerun.[92]

Reruns were finally held on August 13, and runoffs followed on September 17. Lavalas won all the Senate races, giving the party seventeen of twenty-seven seats in the Senate, and sixty-six of eighty-three seats in the Chamber of Deputies.[93] Voter turnout for the runoffs was 14 percent—indicating that electoral legitimacy had suffered.[94]

Nonetheless, the Clinton administration initially sought to cast the elections in a positive light. National Security Advisor Anthony Lake commented, "What an extraordinary act it is to have conducted elections in Haiti on time when a year ago if a Haitian expressed freely a political view, he or she risked having his or her face cut . . . And today, millions of Haitians expressed those political views in safety."[95] U.S. Agency for International Development (AID) director Brian Atwood, who headed the administration delegation observing the elections, called them "a major step forward for democracy in Haiti." He added, "There are problems, but we are confident that they will be resolved."[96] In a State Department briefing on June 27, Atwood elaborated:

> We believe that there was a very significant breakthrough for democracy in this election, that people were voting for the very first time without fear of intimidation by the military. . . . For the most part, there was very little violence. . . . the security situation was excellent. We didn't see any systematic effort to commit fraud in this election. . . . [W]e do not believe that the electoral council tried . . . to influence the results of this election in a way that would favor a single party. . . . I am confident that this election is going to bring political stability to Haiti that has not existed heretofore.[97]

U.S. embassy spokesman Stan Schrager described the elections as "the most free, most complex, and least violent" in Haiti's history.[98] The secretary-general of the OAS issued a statement declaring that "from all indications, electors were able to exercise their franchise freely."[99] Officials from the OAS, the UN, the Clinton administration, and labor groups were unanimous in pronouncing the elections free and fair.[100]

Yet in the weeks that followed the OAS, the UN, and the Carter Center all issued reports detailing extensive election irregularities—from the CEP's disqualification of candidates without explanation, to mistakes on the ballots (including candidates' names or symbols being left off the ballot), fraud in counting the votes, and the burning of a polling office.[101]

In December 1995, a presidential election was held. The constitution of Haiti says that the president may not succeed himself, and this provision applied to Aristide, even though the coup had deprived him of three years of his term.[102] His handpicked successor, René Préval, won the December presidential election. Aristide, out of office, continued to wield considerable power over Haitian government.

In his January 1996 State of the Union address, President Clinton declared that the dictators had fled and "democracy has a new day in Haiti." But the administration's effort to restore democracy was running into new trouble. Clinton invoked executive privilege to deny congressional access to forty-seven documents said to concern political assassins connected with the personal security force of the new Haitian president, René Préval. The chairman of the House International Relations Committee, Benjamin Gilman (R-NY), called Clinton's refusal to cooperate in an investigation of the Haitian government's connection to the murder of political opponents "a blatant abuse of power to cover up a massive foreign policy failure in Haiti."[103]

Administration officials and the U.S. ambassador in Port-au-Prince knew that violent agents of the government had prevented the participation of opposition leaders in the 1995 elections, and that President Préval's U.S.-trained security guards took part in political violence during 1996, but the desire to claim success in Haiti was apparently stronger than a commitment to open discussion of the problems.

During Préval's 1996 term, the Haitian Senate passed privatization and administrative reforms. One result of these reforms was that $226 million was released through the IMF. On the other hand, Haiti's failure to form a new government cost the country $162 million from the Inter-American Development Bank (IADB). The reforms displeased Aristide, who reacted by forming a new party, the Lavalas Family (Fanmi Lavalas).

THE ELECTIONS OF 2000

International observers agreed on the importance of the elections of May 21, 2000, to Haiti's future. Préval had dismissed parliament in January 1999 and governed by decree—that is, unconstitutionally. A dispute concerning the composition of parliament dragged on, intensifying after

Préval signed a law annulling the 1997 elections. Hundreds of millions of dollars in economic assistance from international agencies and other governments were held up pending the legislative, municipal, and local elections of 2000. Aristide had given Clinton personal assurances that the elections would be free and fair and would meet international standards.

But they did not. Irregularities and danger signals accumulated in the weeks before the elections, which were postponed three times, confusing parties, candidates, and voters and almost surely lowering the turnout. The most serious "irregularities" were violence and murders. Haitians and international observers alike were shocked by the April 3 assassination in broad daylight of Jean Dominique, journalist and director of Radio Haiti Inter.[104] That no progress was made in finding his killer dramatized the lack of personal security that existed. The OAS recorded seventy acts of violence, leading to the deaths of seven political party candidates and activists.[105]

As the OAS Electoral Observation Mission reported, procedures broke down shortly after the vote:

> Armed groups of men broke into election offices . . . and burned ballot boxes. The receipt of the tally sheets and other electoral materials was extremely disorganized . . . Exhausted polling officials arrived in overcrowded electoral offices and threw their materials on the floor. The following day's newspapers showed [pictures of] ballots and official tally sheets strewn on the street.[106]

The most serious problems involved illegal counting for the senatorial races. Haiti's constitution and electoral law explicitly state that a senatorial candidate must receive an absolute majority to be elected on the first ballot. Otherwise, a second-round election must be held. In late May, the CEP issued preliminary results: Of the seventeen winners declared in the first round, sixteen were Lavalas candidates. Election observers testified that the vote-counting stacked the outcome in favor of Lavalas candidates. The OAS Mission calculated that if the votes had been counted accurately, ten of the sixteen races would have required a second round.

The situation had serious consequences. Most of the opposition members of the CEP resigned at the request of their parties, and the president of the council fled the country rather than certify the false calculations. The CEP refused to correct the counts. By July, the OAS Mission had determined that "the results are biased and had a major impact on the number of senatorial candidates elected in the first round, and thus cannot be the basis for a credible and fair electoral process."[107] In a July 13 report, the mission concluded that "the highest electoral authority of the country violated its own Constitution and electoral law."[108]

The United States, Canada, France, the UN Security Council, and the OAS appealed to the government of Haiti to correct the fraudulent results of the May 21 elections, but to no avail. Haitian authorities would not budge. As a result, the runoff elections on July 9 were not observed by international monitors and were considered to be as flawed as the originals. The opposition then boycotted the presidential election in November 2000, with the result that voter turnout was under 10 percent and Aristide won with 92 percent of the votes.

In some ways, Aristide's second run for president was different from his first. No longer "Father" Aristide, he was married and had two daughters. No longer poor, he lived with his family in a large house in an expensive neighborhood. To be sure, he won the election, but it is difficult to interpret the results. Aristide had no real opposition—most other candidates had withdrawn—and there were very few foreign observers. The OAS and the United Nations declined to send observers, as did France, the European Union, Canada, and the United States—all "special friends" of Haiti. Anticipating violence, American Airlines and Air France canceled all flights to Haiti on the day of the election.

By now, preelection violence was a familiar phenomenon in Haiti. Murders and drive-by shootings multiplied. The chief of Haiti's national police, Pierre Denize, seemed to concede that law enforcement was beyond his control: "The last elections were the same thing. I don't think there is too much we can do about this except go through the elections and get it over with."[109] On November 17, the State Department issued an advisory to Americans that their safety could not be guaranteed in the week leading up to November 26, and noted that the tone of the dialogue among candidates and government officials had become "distinctly anti-American." Members of the so-called popular organizations supporting Aristide were responsible for sporadic violence, threats, and fires.

Democratic Convergence, the fifteen-party opposition alliance, boy-cotted the election and refused to recognize Aristide as the legitimate president-elect; they vowed to create a shadow or alternative government unless an agreement could be reached with him to rectify the problems of the 2000 elections. Threats against the opposition intensified; its leaders were told to drop their plans to form an alternative government or suffer extreme consequences. On January 9, 2001, groups claiming to support Aristide and the Fanmi Lavalas summoned Haitian reporters to St. Jean Bosco (the church where Aristide had preached as a liberation theologian) and read the names of opposition figures whose "blood will serve as the ink and their skulls the inkwells for writing Haiti's second declaration of independence."[110] The list included most of the leading opposition figures.

Representatives Porter Goss (R-FL) and Benjamin Gilman (R-NY) denounced the threats. "The long list of political assassinations in Haiti is proof enough to believe these are not idle threats," they said in respective statements. "Instead of keeping his promises to President Clinton, Mr. Aristide is condoning by his silence thuggish acts of violence in his name."[111] Most dramatic were the efforts of purported Lavalas followers to kill opposition leader Evans Paul, the leader of the Espace de Concertation (Space for Dialogue) party. Haiti's police remained passive during these attacks, but arrested, beat, and jailed Aristide's opponents after the elections. These patterns of violence continued despite repeated appeals by France, Canada, the United States, the OAS, and the UN Secretariat for Aristide to denounce the violence.

The political opposition continued to protest the outcome of the November election, and Aristide's followers protested the protests. Nonetheless, on the basis of the May and November elections, Aristide and a new parliament were inaugurated on February 7, 2001, with few international observers.[112] Aristide and his followers claimed that the elections were free and fair, but virtually all opponents and observers declared them neither free nor fair. While Aristide was being sworn in as president, the Democratic Convergence Party denounced Aristide's election and named its own alternative president, Gérard Gourgue. This government became the principal target of pro-Aristide factions, which have kept up violence in the streets ever since.

It is not easy to describe the process by which the "new" Aristide government came into being. Aristide was declared president after a so-called election, and a new parliament (consisting entirely of his followers) was "elected." He and his followers ran virtually without opposition after violence and threats of violence had driven out or silenced potential opponents. Haiti's elections were mired in force and fraud. Violence was endemic; corruption universal, and turnout extremely low.

Aristide and his Lavalas Family constructed a one-movement state that ruled on the basis of intimidation, fraud, and fear. Nonetheless, they sought to reap international support for having been "chosen" in democratic elections and tried to collect the $600 million from international agencies and other governments that would have been available to a fairly elected government.

The many violations of normal democratic rules discredited the new government before the governments of the world and the leading international organizations, including the UN and the OAS. In a report to the General Assembly in November 2000, Secretary-General Kofi Annan charged that Haitian authorities had disregarded all calls for rectification of the May 2000 elections, and he recommended that the UN close its mission (MICAH—the International Civilian Mission in Haiti) because it could not function in a "climate of political turmoil." [113] (MICAH closed on February 6, 2001, the day Gérard Gourgue was named provisional president by the opposition and the day before Aristide's inauguration. Since the May 2000 elections, Annan noted, "Haiti's political and electoral crisis has deepened, polarizing its political class and civil society." [114] The first report of the OAS mission on the elections reported: "[T]he aftermath of the May 21 elections . . . exacerbate[d] an existing political and democratic-institutional crisis in the country rather than beginning to resolve it, as it had been hoped. The sense of the urgent need for political dialogue now coexists with extremely serious doubts about whether such a dialogue is possible." [115]

OAS assistant secretary-general Luigi Einaudi traveled to Haiti twenty times over the next two years to mediate a dialogue between the government and the opposition. A number of proposals were advanced, but the OAS generally found a mutual lack of trust and an atmosphere that was not conducive to negotiations. Meanwhile, the political polarization and security climate worsened. In March 2001, the State Department issued a statement that said, in part, "The United States is deeply concerned by escalating political violence in Haiti. Opposition demonstrators began peacefully on March 14. However, anti-opposition protests by 'popular organizations' have turned increasingly violent in recent days, with incidents of tire burning, rock throwing, roadblocks, and shootings that have resulted in several reported casualties."[116] Gourgue went into hiding after the Senate passed a unanimous resolution calling for his arrest.

In May 2001, Aristide advanced a proposal that included the resignation of the seven senators whose seats had been contested and a commitment to appoint a new CEP, which would set dates for elections for those seven seats and organize early elections for the rest of the parliament. The Democratic Convergence argued that this proposal would allow the Aristide government to secure the release of the blocked foreign aid and still avoid holding immediate elections for all offices. The OAS worked to build on the initiative and continue dialogue.[117] On July 18, the two parties agreed to hold new legislative and local elections, although they could not agree on election dates.[118]

But violence flared once again. On July 28, armed men in military uniforms attacked two police stations, killing five police officers and injuring fourteen. A *Los Angeles Times* article reported that the attacks had lasted sixteen hours before police commando units were dispatched to restore order. The Aristide administration denounced the attacks as an attempted coup and claimed that the armed men were former members of the Haitian armed forces. Government officials took to the airwaves and urged Haitians to mobilize against any plots. The opposition denied involvement and suggested that the administration had arranged the attacks to derail progress on resolving the political impasse. OAS secretary-general César Gaviria urged the parties to continue negotiations,[119] and the U.S. embassy in Port-au-Prince called on the Haitian government to put a stop to arbitrary arrests and killings in the wake of the attacks. (By August 8, more than forty people, most of them members of the opposition, had been arrested in connection with the attacks. All were released by mid-September.) Not surprisingly, these events largely undid the progress toward a political compromise that had been made in July.

This became a familiar pattern. Under OAS auspices, talks between the government and the opposition coalition were resumed in October, only to collapse a day after they started. The opposition said, "We cannot accept the unacceptable. . . . Last year's so-called elections were an electoral coup d'état." [120] Meanwhile, in a letter to President George W. Bush, the CBC pleaded for a change in U.S. policy, in particular its refusal to allow the IADB to release loans to Haiti.

Einaudi returned to Haiti in December to restart the dialogue, but was defeated in the attempt by another resurgence of violence. Two days before his scheduled arrival, radio journalist Brignol Lindor was hacked to death with machetes—he had received death threats the week before for inviting opposition supporters onto his talk show. Violence flared at his funeral. On December 17, armed commandos stormed the National Palace. Seven were killed before the attack was thwarted. Once again government officials condemned the attack as an attempted coup, and their supporters took to the streets and burned down opposition headquarters and the homes of several opposition leaders. Gérard Gourgue again went into hiding, saying "I don't know what happened at the National Palace, but it has become a pretext to massacre the opposition." [121]

In the spring of 2002, the OAS appointed a Commission of Inquiry to examine the acts of violence that occurred on December 17 and created a Special Mission for Strengthening Democracy in Haiti, which would work in four core areas: security, justice, human rights, and governance. In July, the OAS issued a report stating that the December 17 attack was not a coup attempt "but rather an outbreak of violence connected to the general breakdown of law and order." [122] Einaudi was sent to Haiti to recommence negotiations, but again little progress was made. The OAS negotiator commented that the "government was *not* assuming its responsibilities" vis-à-vis the negotiations. [123] An Aristide spokesman claimed that the opposition would not budge until Aristide resigned and general elections were held. [124] In August, anti-Aristide protests broke out in Gonaïves and surrounding areas.

In September 2002, after more than two years of failed attempts to broker an agreement between the government and the opposition to resolve the political crisis stemming from the May 2000 elections, and citing "the continuing deterioration of the socioeconomic situation in

Haiti, the ongoing suffering of the people, and its potential for humanitarian disaster," the OAS passed a resolution calling for the "normalization of economic cooperation between the Government of Haiti and the international financial institutions," which would mean allowing the release of the $500 million in blocked international aid and loans.[125] The Bush administration signed on to the resolution, which laid out goals for the Haitian government that would allow the aid to be unblocked. However, the government failed to meet a November 4 deadline to appoint a new CEP, improve security, punish gangs, and disarm the population.[126] Einaudi commented, "The long and short of it is that the key actors have been unwilling to rise above entrenched personal positions in terms of allowing for an end to the fragmentation and paralysis that are leading the country as a whole toward disaster."[127]

Meanwhile, Haitian boat people continued to head for Miami, causing disputes in the United States over what to do about the refugees.[128] In late October 2002, more than two hundred Haitians arrived in a rickety freighter, many jumping overboard to reach land. In Haiti, protests against Aristide broke out in mid-November. The more than $500 million in aid money that Haiti sorely needs to get its economy moving is still frozen.

New Democracies, New Problems

Military intervention in another country to promote democracy is a relatively new idea; in fact, many nations still consider it an illegitimate interference in the internal affairs of another state.

The world has learned a great deal about how democracies fail. In his book *The Breakdown of Democratic Regimes*, political scientist Juan Linz provides a brilliant analysis of how new democracies establish the legitimacy, authority, and effectiveness necessary to survive, and how they can lose it. Much of what Linz says is directly relevant to problems that exist today in countries in Eastern Europe, Central America, the Philippines, and elsewhere. So far, the United States is the only country that has attempted to systematically design policies that promote democratic practices, and its efforts have been extremely controversial.

The United States has embraced the idea that human rights and democracy should be important concerns of American foreign policy, but we have been inconsistent in the pursuit of these goals. Concern about human rights and democracy have been factors in U.S. policy toward the former Soviet Union and in Latin America, where Congress has demanded democratic and human rights reforms in return for U.S. aid. There has been broad public approval of making aid to the Philippines, Nicaragua, El Salvador, Guatemala, and other countries conditional on the development of democratic governments.[129]

The occupation of Haiti, in contrast showed that our government had become too casual about using military force, deploying U.S. troops, and assuming open-ended obligations. Large majorities of Americans feel very cautious about direct military interventions abroad, and a majority opposed sending U.S. forces to Haiti. The primary concern of the largest number of Americans was discouraging immigration from Haiti.

The Clinton administration was committed to success in Operation Restore Democracy, although opposition existed in both the UN Security Council and the Organization of American States. In his effort to "restore democracy" in Haiti by restoring Aristide to power, President Clinton exercised his new approach to foreign policy,[130] involving the United States deeply in Haiti's political development—at great expense. The American presence in Haiti to put Aristide back in power and subsequent involvement with and assistance to Haiti cost U.S. taxpayers about $3 billion.

Once the intervention had occurred, I believed we should follow through on our investment for three reasons: (1) in the interest of human solidarity; (2) because imposing the embargo, which further weakened Haiti's fragile economy, gave us some responsibility and created an increased flow of refugees; and (3) because we had imposed a military occupation on the island. Having intervened so far into the internal affairs of Haiti, the United States had a responsibility to foster the implementation of democratic practices, and the provision of nonpolitical police and judiciary systems.

But the U.S. government could not solve Haiti's terrible problems. No one knows exactly how foreign forces can help civil societies or mod-

ern states emerge in very different cultures, and the influence of an external power on the process of democratization is limited. I suggested that
the United States concentrate on what we could do well—improving literacy rates and providing police and job training—rather than aiming
for the total renovation of Haiti, which would be an inappropriate lapse
into neocolonialism. As we now know, even the assistance for judicial reform and police training we provided in the mid- to late 1990s produced
scant improvement, as corruption and government indifference created
obstacles to reform.

Literacy can be taught from outside a society. Vocational training
can also be effectively taught by foreigners. Obviously, efforts to help
with literacy and vocational training should respect the traditions of
Haiti—its language, culture, and educational system. By playing an active role in programs to promote literacy and technical education, the
United States could guard against efforts to politicize these programs
and ensure their effectiveness.

There were a good many highly educated Haitians, although many
fled the country in these years. Haiti's small commercial and business
sectors, along with tourism, constituted the primary sources of foreign
exchange, which is why the embargo was so devastating. The development of larger, more skilled working and middle classes could help create
the foundation for economic and political development.

But the key to assisting Haiti on the road to democracy is to establish
reliable law and order—and since the United States restored Aristide to
the presidency in 1994, neither he nor René Préval, who ruled between
Aristide's first and second terms, brought about the rule of law that has
been missing throughout Haiti's history. Aristide holds a dubious distinction. He was ousted from power twice, again in February 2004 and
forced back into exile. A rule of law might have prevented Aristide's second coup and would encourage Haitian exiles to return from the United
States, France, and South America, and would contribute greatly to political and economic development.

I was pessimistic in the 1990s about the prospects for democracy in
Haiti. Today, I am not much more sanguine—not when the cultural, social, and economic foundations of genuine democracy are still missing

there. As sociologist Anthony Maingot concluded a grim article on poverty and corruption in Haiti by saying: "All that has changed are some of the actors. The play is a tragedy and in Haiti, as in theater, the outcome of a tragedy is predictable; it invariably ends without solutions and with many deaths." [131]

4.

THE BALKAN WARS: MAKING WAR TO KEEP THE PEACE

[Bosnia was] the historic boundary between East and West, Islam and Christianity. Most of all, it represented the limits of European integration, of humanitarian concern, and of political interests.[1]

—JAMES GOW, *Triumph of the Lack of Will:*
International Diplomacy and the Yugoslav War

Many Western observers were impressed with the skill of Josip Broz Tito in preserving Yugoslavia's independence from Joseph Stalin's boundless appetites, from the end of World War II to his death in 1980. Only after Tito died and wars of secession broke out did it occur to Americans that his greatest achievement had been holding together the diverse peoples of Yugoslavia. It was ironic that the Communist autocrat who ruled this one-party state understood the importance of ethnic identity and decentralization to the preservation of Yugoslavia. Tito (who was half Croat, half Slovene) respected ethnic diversity but rigorously suppressed separatism; he was especially sensitive to manifestations of Serb nationalism.

After his death, Serb nationalism asserted itself in the person of Slobodan Milošević.

The story of U.S. involvement in Bosnia is an interesting counterpoint to the experience in Iraq/Kuwait, Somalia, and Haiti. Here the problem was not overreaching but hesitation in the face of tyranny and mass murder on European turf. It shows the danger of allowing political sensitivities, both domestic and international, to stand in the way of strong action. The first President Bush was reluctant to act because he was in an election. Clinton wanted to act, but the Europeans would not agree, the UN peacekeepers were subject to too many limitations, and the UN secretary-general insisted on having the final say. The failure to organize a timely and effective response to Slobodan Milošević demonstrated a global lack of seriousness about a new world order, the Clinton administration was forced to retreat in the face of the lack of support, and NATO stepped in to make the difference where the UN had been merely obstructive.

SETTING THE SCENE

Under Tito's governance, Yugoslavia adopted new constitutions in 1946, 1963, 1968, and 1974. The 1974 constitution provided for an eight-member collective presidency, with one representative from each of the six republics (Slovenia, Croatia, Serbia, Bosnia, Montenegro, and Macedonia) and the two autonomous provinces (Kosovo and Vojvodina). Each region had its own parliament, as well as representatives in the federal parliament. The presidency of the collective executive rotated annually among the eight states, with the commander in chief of the armed forces presiding and the Communist Party reinforcing unity.

In 1989, as Communist governments throughout Central and Eastern Europe began to break apart, opportunities for free expression and self-determination encouraged the development of separatist movements. Democratic reforms in the neighboring countries of Eastern Europe encouraged reformers in all the Yugoslav states; they also hardened the determination of nationalists everywhere, including Milošević, who encouraged the Serbian Communist government to resist reform and preserve its power.

Ambition, political skill, and control of the Communist Party organization quickly made Milošević a powerful source of trouble in Serbia and in the Yugoslav federal government. He purged the Communist Party of Kosovo in 1988 and promised to "make Serbia whole again" by depriving Kosovo and Vojvodina of autonomy and votes, increasing Serb (and his own) power in the collective bodies. With Serbia effectively exercising the votes of Vojvodina and Kosovo, it would control three of the eight votes in the federal presidency. It could generally count on Montenegro as well, giving it control of four of the eight votes.

In January 1990, the Yugoslav Communist Party voted to give up the monopoly on power it had held for forty-five years and permit the formation of other parties. In the fall of 1990, the first free elections since 1945 were held in Macedonia, Bosnia-Herzegovina, and Serbia. Nationalist political parties triumphed in the first two states; in Serbia, Milošević won 60 percent of the votes. In December, he won the presidency again (after winning in 1989), and Serbia's Communists won 194 of the 250 seats in parliament.

Milošević's reelection had been credited, in part, to his successful rallying of Serbian nationalists, who rewarded Milošević for unilaterally abolishing the provincial legislature in Kosovo and extending Serbia's control—first limiting, then revoking the 1974 statutes that had made Kosovo and Vojvodina autonomous provinces with powers nearly equal to those of the six Yugoslav republics. He confiscated the weapons of Kosovo's territorial defense forces and moved to disarm the defense forces of Yugoslavia's other states.

Then, in short order, Milošević began his policy of "ethnic cleansing" in Kosovo, expelling ethnic Albanians (that is, Muslims) from their jobs, homes, and farms and forcing their relocation. The Muslims' jobs and property were quickly taken over by Serbs from other provinces. From the start, teachers were special targets of the policy, as were trade union leaders, who were also targeted for expulsion, beatings, arbitrary arrests, imprisonment, and shooting. Virtually all institutions in Kosovo were sucked into the purges, which became progressively more brutal.[2]

Slovenian, Croatian, and Bosnian officials watched the evolution of events in Kosovo and in the federal presidency. Milošević's determined effort to gather all power into his own hands increased the interest of these republics in secession from the federal state of Yugoslavia, an interest that was heightened when Milošević blocked the scheduled accession of the Croatian representative Stipe Mesic to the rotating presidency.

Janez Drnovsek, a member of the collective presidency who later became prime minister of Slovenia, provided a day-by-day account of the deepening political split between those who desired economic and political reforms with multiparty elections and free markets and those who saw reform as a threat to the Communist system.[3] In his memoir, Drnovsek asks whether it might have been possible to prevent the breakup of the Yugoslav nation. His answer: "Yes, the tragic, violent development could have been prevented—by prompt and helpful action from outside Yugoslavia."[4] But the needed "political and economic aid, swift incorporation into the processes of European integration, and . . . rapid results"[5] to overtake the mounting hostility and disorder were not forthcoming. Western Europe was busy with its own integration.

As Slovenia and Croatia leaned toward independence, Milošević and other Serb leaders insisted that independence was not an option because Serbs were spread throughout Yugoslavia. Members of the collective presidency tried to draw up a balance sheet of each state's debts and credits with the national government. Again and again, Milošević drove differences to a climax and refused to recognize the authority of the federal executive body. Citizens rallied in the street, calling for his resignation.

Beginning in May 1990, Milošević used the Yugoslav National Army (JNA), which he controlled, to try to disarm the militias of each of the republics, demanding that all arms be handed over to the JNA, which was largely controlled by Serbia. Slovenia's militia refused to hand over its weapons, which enabled the republic to declare independence in 1991 and face the national army.

Together with Croatia, Slovenia proposed that Yugoslavia become a loose confederation with a democratic multiparty system. In December 1990, 88 percent of Slovenes voted to separate from Yugoslavia. The Croatian referendum on independence passed with 94 percent of the vote. Milošević announced again that if component states of Yugoslavia became independent, Serbia would insist on bringing the 8.5 million Serbs, scattered in several areas, into a single state and would demand territory commensurate with their numbers.

In June 1991, Croatia and Slovenia declared their independence. Immediately, fighting broke out between Croatia and the largely Serb JNA. Brutal from the start, the war included all the elements of ethnic cleansing: siege, murder, confiscation and destruction of property, rape, beatings, and forced relocation of whole populations.

The European Community (EC) called repeatedly for an end to the fighting and announced that it would recognize no unilateral declaration of independence. The Bush administration was willing to defer to the EC on political matters in Central and Eastern Europe, and the EC and its Luxembourgois president, Jacques Poos, were ready to take charge. The EC appointed a troika to deal with the problem: Poos and the foreign ministers of the Netherlands and Italy, Hans van den Brock and Gianni le Michelis. They thought they had achieved their goal of preserving Yugoslavia with the Brioni Agreement, a cease-fire pact signed on July 7, 1991. The agreement stipulated that the Slovenes and Croatians would take no action toward independence for ninety days, the JNA would return to its barracks, and Stipe Mesic would assume the presidency of the federal executive. The agreement came unstuck as soon as it was adopted, but Poos insisted, "If anyone can do anything here, it is the EC. It is not the U.S. or the USSR or anyone else."[6]

The Americans had no desire to assume responsibility. James Baker, then secretary of state, commented in his memoir, "The Bush administration felt comfortable with the E.C.'s taking responsibility for handling the crisis in the Balkans. The conflict seemed to be one the E.C. could manage. Yugoslavia was in the heart of Europe, and European interests were directly threatened. . . . our vital national interests were not at stake."[7]

The EC asked the British foreign secretary, Lord Peter Carrington, to undertake the task of making peace. Carrington convened the first meeting of The Hague Peace Conference on September 7, 1991. He had concluded that Milošević and Franjo Tudjman, Croatia's prime minister, were prepared to split Bosnia between them, which would surely lead to war. The only way to avoid war, Carrington believed, was to devise a settlement that was acceptable to all. For a brief period, it looked as if he had secured an agreement on a loose association of independent republics; arrangements for protecting minority communities in all republics, including human rights guarantees; and no unilateral changes in borders. The agreement was quickly approved on September 25, 1991, by the UN Security Council.

Lord Carrington wrote of this plan, "It seemed to me the right way to do it was to allow those who wanted to be independent to be independent, to associate themselves with a central organization as far as they wanted to. Those who didn't want to be independent, well, they could stay within what had been Yugoslavia. In other words, you could do it . . . à la carte."[8] Croatia, Slovenia, and Bosnia quickly agreed, but Serbia refused. Milošević insisted that it was essential for all Serbs to live in one state, not in a number of independent republics. The Carrington plan called for equal rights for all minorities in all states, but this provision was vetoed by Milošević, who was interested not in unity or fairness but in a greater Serbia. To that end, he pressed Serbia's claim to sovereignty over Kosovo, whose population was 90 percent ethnic Albanian Muslim.

The violence worsened, even as Carrington continued to seek an agreement in which all minorities had equal rights in all states. Once again, all states except Serbia agreed. Milošević was unyielding and this newest attempt also fell apart. Meanwhile, the Bush administration quietly suspended all economic assistance to Yugoslavia, including assistance from international financial institutions, effective May 6, 1991.[9]

The new UN secretary-general, Boutros Boutros-Ghali, proved unhelpful in the process. Boutros-Ghali complained that Carrington should have sought his approval of the EC plan before it was submitted to the Security Council; he could have told them that the UN lacked the resources to implement the cease-fire agreement. The secretary-general stated bluntly that activities relating to peace and security should be managed by him and his staff. This was by no means the last time Boutros-Ghali would claim powers for himself that the UN Charter clearly vests in the Security Council, not in the secretary-general.

The United States and most EC member states favored preserving Yugoslav national unity over self-determination for the republics. Only

Austria and Germany (which had the closest historical association with the Croats and Slovenes) expressed sympathy for the aspirations of those who wished to secede. In June 1991, as Slovenia and Croatia moved toward independence, James Baker made a hurried visit to Belgrade, where he called for Yugoslav unity, discouraged secession, and rebuked Serb leaders for their repression of Kosovo. He emphasized that diplomatic recognition by the United States would not be forthcoming if any of the republics unilaterally declared independence. At the same time, Germany warned the Belgrade government not to send troops to fight Slovenia and Croatia, and said that Bonn would recognize them as independent states. But the Serb attacks spread. That same year, Bosnia-Herzegovina also declared its independence.

Two top officials of the Bush administration had long personal experience in Yugoslavia. Lawrence Eagleburger, who served as deputy secretary of state and then as secretary of state, had been U.S. ambassador to Belgrade in the late 1970s and knew Milošević well. Brent Scowcroft, Bush's national security advisor, had served as an attaché in Belgrade and had written his doctoral dissertation on Yugoslavia. Both men had headed Kissinger Associates after their assignments in Belgrade and had done business with Yugoslav enterprises and leaders during their tenures there. Both were very concerned about the effects of the collapse of central authority. Before Baker traveled to Belgrade, he was briefed by Eagleburger and Scowcroft. Eagleburger emphasized Yugoslavia's history of violent conflict: "We believe the only solution to these internal differences in Yugoslavia is an open, multiparty democracy throughout the entire country which protects individual rights."[10] The new U.S. ambassador to Yugoslavia, Warren Zimmerman, a talented Foreign Service officer who had already served two tours of duty in Belgrade, also strongly opposed Slovenian independence and the breakup of Yugoslavia.

American and European diplomats sought to prevent the disintegration of Yugoslavia, to prevent Croatia and Slovenia from seceding, to prevent the outbreak of violence and war, and, after war had broken out, to prevent it from spreading and prevent the establishment of independent states. They failed to achieve any of these goals.

Baker and President Bush made trips to the region, where they made strong statements discouraging secession. Baker urged Yugoslav leaders to accept two basic realities: (1) that they needed to negotiate their differences, not act unilaterally, and (2) that under no circumstances would the international community tolerate the use of force.[11] His message to Milošević went further: "[W]e regard your policies as the main cause of Yugoslavia's present crisis. . . . If you persist in promoting the breakup of Yugoslavia, Serbia will stand alone. The United States and the rest of the international community will reject Serbian claims to territory beyond its borders."[12]

Bush's top advisers were united in the belief that the United States should avoid becoming bogged down in a protracted civil war; they feared that a Yugoslav breakup might encourage present and former republics of the Soviet Union to consider secession. This concern inspired Bush's famous "Chicken Kiev speech," in which he declared that "Americans will not support those who seek independence in order to replace a far-off tyranny with a local despotism. They will not aid those who promote a suicidal nationalism based upon ethnic hatred."[13] General Colin Powell, chairman of the Joint Chiefs of Staff, argued that the United States had "no clear military goals in Bosnia" and stressed that "the solution must ultimately be a political one."[14] Bush, Baker, and Powell expressed no sympathy for the aspirations of the Croatians, Slovenes, or Bosnians or their desire for independence. Nor did they speak of self-determination. Scowcroft, Bush's national security advisor and good friend, wrote,

Eagleburger and I were the most concerned about Yugoslavia. . . . We tried very hard to prevent the recognition of Slovenia and Croatia. The British and French agreed, but the Germans for the first time really asserted themselves in the Community. The French were very sympathetic to us, but in the end the cohesiveness of the Community was more important.[15]

Both the American and the EC governments were shocked when the Serb-controlled central government used savage force to crush rebellions against the authority it claimed. The U.S. State Department was only slightly behind the British, the French, and the rest of the EC in realizing that this had become the principal problem.

On June 25, 1991, Slovenia declared independence; Croatia followed suit. The next day, national army tanks rumbled into Slovenian towns, but the Slovenes were ready. Forty-four JNA soldiers were killed and 187 wounded. Slovenia's secession was recognized. Yet the alliance between Slovenia and Croatia soon collapsed, and the JNA began shelling Croatia, including the historic, defenseless cities of Vucovar and Dubrovnik. Three months of heavy shelling and siege reduced Vucovar to shambles, its inhabitants to hunger and to hiding in basements. The Croats surrendered on November 20, 1991. Many were killed; others were stuffed into overcrowded prisons.

Tapes of telephone calls between Milošević and Bosnian Serb leader Radovan Karadzic in July 1991 provide clear evidence that Belgrade was making regular secret deliveries of arms to Bosnian Serbs.[16] The tapes also make it clear that Milošević, who was collecting weapons from the republics of Slovenia, Croatia, and Bosnia-Herzegovina, was simultaneously directing the JNA to deliver weapons to the Bosnian Serbs in an attempt to build a Bosnian Serb military force and link Serbs in Bosnia and Croatia with those in Serbia. Vojislav Seselj, a political ally of Milošević, told *Der Spiegel* that Bosnia, Macedonia, Montenegro, and much of Croatia should be transferred to Serbia.[17]

On September 21, 1991, a number of European and American publications reported the invasion of Croatia by the JNA. Serbs in Croatia and Bosnia had designated four "Serb autonomous regions," which then requested protection from the JNA, by then wholly under Serb control. On September 25, at the request of Yugoslavia, the UN Security Council adopted Resolution 713, which imposed an arms embargo that froze in place Serbia's huge advantage in weapons and left the newly independent states nearly defenseless at a time when massacres of their populations were already under way.[18]

At the Conference on Security and Cooperation in Europe (CSCE), President Bush said that the world must stop Serbian terrorism "no matter what it takes." But he later added that the United States would not act without European engagement. The result was that two U.S. warships waited off the Adriatic coast for a European naval force that was incredibly slow in arriving to begin monitoring the embargo.

The arms embargo, enforced by the United States (among others), remained in effect until 1995, when Resolution 1021 recalled previous resolutions,[19] in spite of the fact that the people of Bosnia, Croatia, and Slovenia had voted in favor of independence, and the countries had been admitted to the UN on May 22, 1992. As UN member states, these nations were entitled to all the rights, privileges, and protections in the UN Charter, including "the inherent right to self-defense." The legal interpretation underlying the decision rested on a recommendation of former president Carter's secretary of state, Cyrus Vance, the UN secretary-general's appointed mediator and personal envoy to Yugoslavia.[20] But Vance, the secretary-general, and the Security Council had not considered the full implications of imposing an arms embargo in the context of Serb aggression against the people of Croatia and Bosnia, who lacked the means of self-defense. And the resolution did not take into account the clear contradiction between denying Bosnia and Croatia arms to defend themselves and various other Security Council resolutions that dealt with violations of Bosnia's territorial integrity.

Because the embargo rendered Croatia and Bosnia virtually defenseless, it became the center of a heated and long-lasting debate in the U.S. Congress between those who wanted to provide arms to the Croatians and Muslims so they could defend themselves, and those who believed that such measures would only further inflame the fighting. An intense debate developed in the UN between the United States (which repeatedly threatened to revoke the embargo unilaterally) and its allies, especially the British and French, who had historic ties with the Serbs and eventually had troops in the UN Protection Force for Croatia (UNPROFOR), who they feared would be endangered if arms were provided to the Bosnians. Denying arms to the Croatians and Bosnians laid the foundation for mass slaughter.

Newsday correspondent Roy Gutman provided the first descriptions in English of Serb concentration camps. Later, in his Pulitzer Prize–winning book *A Witness to Genocide*, he noted that the Bosnian Serb army—the old Yugoslav army—began to oversee the burgeoning horror of ethnic cleansing, which featured "arbitrary executions and wholesale deportations." [21]

In November 1991, Serb troops surrounding Dubrovnik broke a multiple cease-fire, negotiated weeks before, and began shelling the beautiful medieval city. On December 6, Dubrovnik was shelled for ten hours; the attack would leave its inhabitants without electricity or water for a month. All of its citizens (including Serbs) were engaged in the defense of the city. [22]

Accounts of brutality filtered out of Croatia. Gutman told the world what he had learned of the ghastly conditions in the Omarska prison camp. He interviewed survivors, who described the unbelievable tragedy. One man said:

> I will tell you about the conditions in the camps. All the grass has been eaten by the people. Every day in Omarska between 12 and 16 people die. In the first six days, they don't receive any food. There is no possibility of any visit. No possibility of packages. No medical help. Two thirds of them are living under open skies, in an area like an open pit. When it rains, many of them are up to their knees in mud. [23]

Gutman found eyewitnesses and former detainees who described death camps where "emaciated men with their heads shaved in an open field" were routinely slaughtered. [24] The first British television photos of Omarska appeared, deeply shocking Europe.

By this time, reports of savage fighting in the former Yugoslavia were appearing in the Western media. The *Washington Post*, the *New York Times*, and other major newspapers published accounts of the destruction of Vucovar, and of tens of thousands of Croats and Bosnians driven from their homes and stripped of their possessions. In Europe, reports in *Le Monde* and the *Frankfurter Allgemeine Zeitung* estimated that five hundred thousand persons had become refugees in Croatia alone in the weeks after the fighting broke out, and that the Serbs were using rape as a weapon.

As Ambassador Warren Zimmerman later wrote, "The pattern of Serb atrocities that continued throughout the war was set in these first few days. Typically, the Serbian paramilitaries would storm a town, killing civilians in the assault, would expel the Muslim population, and would turn the town over to Serbs, who, protected by the JNA, could destroy mosques and other Muslim symbols at leisure. Military-age Muslims were sent to concentration camps or executed." [25] Beatings, rapes, and murder were ubiquitous. [26]

The pressure to act against these atrocities mounted, and blame began to rub off on the Western powers. Some observers believed that Milošević could not have launched the attacks without the acquiescence of the United States and Western Europe, or at least that their passivity could easily be interpreted as acquiescence.

On December 17, 1991—in its first significant unilateral move in foreign affairs since the end of World War II—Germany announced that it would recognize Slovenia and Croatia on January 1, 1992. It was a bold move that reflected foreign minister Hans-Dietrich Genscher's determination to do something for Croatia and Slovenia. But it was a sharp disappointment to Lord Carrington and the U.S. diplomats, who believed that the chances of containing the violence would be far better if no government granted diplomatic recognition to any republics that unilaterally declared independence. Repelled by the Serb offensive, the EC and the Vatican soon followed Germany in recognizing Croatia and Slovenia.

On February 21, the Security Council passed Resolution 743, establishing a small peacekeeping force for Croatia (UNPROFOR), pending a solution of the Yugoslav crisis. The resolution authorized a force of 45,000 troops, but reached a maximum strength of only 39,922. The force was to serve as a buffer—separating the Krajina Serbs and the Croatians, monitoring cease-fires, and facilitating the return of refugees. In Bosnia, the force was also to escort humanitarian convoys and deter attacks on safe areas. Its cost over four years was approximately $4.6 billion. [27] UNPROFOR's size and mandate were repeatedly enlarged as the situation deteriorated throughout 1992. Although Bosnia-Herzegovina had also become a major target of Serb aggression, Boutros-Ghali rejected the appeals of France, Germany, and Poland to deploy a peacekeeping force there. Violence spread in Bosnia, as regular and irregular Serb forces moved with impunity from village to village, pillaging and driving out the inhabitants.

On the last day of February 1992, a referendum in Bosnia produced a Bosnian Serb boycott and a strong Muslim-Croat vote in favor of independence. On April 21, the shooting started in Sarajevo, Bosnia's multi-ethnic capital city. Heavy shelling continued throughout April; but no action was taken to defend the unarmed Bosnians. In late May, James Baker announced that the United States had joined in discussions on adopting Chapter VII sanctions (authorizing the use of force). Still, the ever-cautious Baker added a caveat: "[B]efore we consider force, we ought to exhaust all of the political, diplomatic and economic remedies that might be at hand."[28]

Almost all the parties to this increasingly violent situation were slow to realize that they were truly at war. Nowhere was this clearer than in the extraordinary events surrounding the kidnapping on May 2 of Bosnian president Alija Izetbegović, who was returning to Sarajevo after a day of fruitless negotiations in Lisbon. No one realized that the JNA had taken over the airport. Minutes after Izetbegović's plane landed, he was a prisoner of the Serbs. In the negotiations that followed, he sought to save himself, his daughter, his party, and his position as president, while the JNA sought to secure its control of the airport, Izetbegović, and the Bosnian government by imposing a coup d'etat. UN commander general Lewis MacKenzie described it as "the worst day of my life."[29] In extremely complex conversations, President Izetbegović managed to secure his freedom and that of his daughter, while Bosnia's military commanders ambushed the JNA forces at the airport. By the end of the day, it was clear that the Serbs and Bosnians were engaged in a fight to the death.

In the weeks that followed, the situation worsened. Cease-fires were negotiated and violated. Pledges were made and broken. The Sarajevo airport remained closed. The "safe routes" created for delivery of humanitarian supplies remained tightly shut. Serb mortars pounded Sarajevo's neighborhoods. On May 27, 1992, as the inhabitants of Sarajevo stood in a long line waiting to buy bread, Serbs shelled the bread line, killing twenty-two people and wounding many others.

In the spring and summer of 1992, the UN Security Council passed a series of resolutions calling for the reopening of the Sarajevo airport, which had been closed since early spring. On May 15, 1992, the Security Council passed Resolution 752, demanding that all parties cease efforts to change the ethnic composition of any part of the former Yugoslavia, that Croatians and Serbs who had not lived in Bosnia withdraw from Bosnia, and that all parties cooperate to ensure access to Bosnia's airports and the delivery of humanitarian assistance. The United States joined in Resolution 757, which passed on May 30, 1992, imposing a trade embargo on Yugoslavia (Serbia and Montenegro) and freezing Yugoslav assets abroad and in the United States. Resolutions 758 and 761 banned all military flights over Bosnia-Herzegovina and demanded that the UN and humanitarian organizations be given open access to camps, prisons, and detention centers. But none of these good things happened. Finally, in June 1992, the Security Council passed Resolution 762, belatedly authorizing the deployment of peacekeeping forces to Sarajevo.

The Western organizations did not deal well with this conflict. Neither the EC nor the UN nor the Conference on Security and Cooperation in Europe (CSCE) roused itself quickly to oppose the violence, which had forced 2.2 million Bosnians to flee their homes by mid-July 1992, and reintroduced into Europe the barbaric notion of "ethnic purification" that was thought to have died with Adolf Hitler. The international failure to counter this Serbian aggression could be traced to a diverse set of motives. Some governments were inhibited by their historic ties to Serbia. Some argued that the conflict was a hopelessly complicated ethnic struggle in which outsiders should not become involved, or that blame was so evenly divided among the parties that there were no meaningful moral issues. Some governments (notably France) opposed NATO's involvement, anxious to avoid setting a precedent for a post–cold war U.S. role in Europe. Everyone equivocated.

Reminding the world that leaders make a difference, recently retired heads of state Margaret Thatcher and Ronald Reagan signed public appeals deploring the Serb violence and demanding that Milošević cease military action against Bosnia and the flow of weapons to Serb forces in Bosnia, turn over heavy weapons to an international body, and permit Bosnian civilians to return to their homes under international protection—or else expect NATO airpower to target and destroy Serb military assets.[30]

Secretariat du Prince Sadruddin Aga Khan
PO Box 6—1211
Geneva 3 Switzerland
Telephone: (022) 346 8866
Telex: 129825 MUROCH
Fax: (022) 347 9159

18 November 1992

WORLD LEADERS URGE BOSNIAN ACTION

We are now witnessing in Bosnia a replay of one of the darkest eras of modern history: the invasion of one sovereign nation by another. It is the attempted genocide of people who have lived in peace and tolerance with their neighbors for centuries. The scale of atrocities and the appalling human suffering tell the story.

The savagery can and must be stopped or the tragedy will spread far beyond Bosnia. People of conscience must speak out now.

Every assistance should be provided to help the Bosnian refugees; otherwise countless more will die.

> *The Hon. Gerald Ford*
> *The Hon. Ronald Reagan*
> *Baroness Margaret Thatcher*
> *Ambassador Jeane Kirkpatrick*
> *General Alexander Haig, Jr.*
> *Prince Sadruddin Aga Khan*

In the summer of 1992, François Mitterrand, the oldest of the Western presidents, flew into Sarajevo in a French helicopter, donned a flak jacket, and spent seven hours under continuous mortar and sniper fire. "The people of Sarajevo are truly prisoners, condemned to murderous blows, and I feel an overpowering sense of solidarity with them," he declared.[31] But his trip had few consequences, perhaps because, in spite of his expression of solidarity, he took no formal action.

The shelling eventually became so heavy that General MacKenzie suspended the ongoing airlift of food and medicine, which had barely sustained the city. By this point Sarajevo had little electricity or water.

Many civilians were killed and wounded by mortars and bombs, and the city's hospitals were severely damaged. On August 13, acting under Chapter VII of the UN Charter, the Security Council passed Resolution 770, authorizing states to use "all measures necessary" to deliver food and medicine and fuel to Bosnians trapped in cities under siege, and imposing a ban on military flights over all of Bosnia-Herzegovina, except for flights in support of UN missions.[32] A month later, a follow-up Resolution (786) was passed providing for monitoring (but not enforcing) the no-fly zone.

In an oblique threat, Baker said that the United States would not accept the blockage of the airport and humanitarian relief. He warned Karadzic that the United States might use air and naval units in Europe to attack Serbian artillery around Sarajevo. Baker held intensive consultations with the other members of the Security Council on how to get food and medicine to Sarajevo, which by now had been surrounded by Serb forces for months. In the United States, impatience grew. Democrats in the Senate—including Claiborne Pell (RI), Paul Simon (IL), and Joe Biden (DE)—demanded U.S. participation in delivering food and medicine.

General MacKenzie was scheduled to leave the Balkans in the first week of July 1992, and the Canadian battalion, 850 strong, would be leaving with him. It was rumored that he was being removed at the request of the UN because of the Bosnian government's strong feeling that he favored the Serbs. (A few years later, the secretary-general would request the recall of French generals Philippe Morillon and Jean Cot because the Serbs felt they favored the Bosnians.)

Throughout MacKenzie's last day, urgent appeals for help flooded in from amateur radio operators in the Muslim town of Gorazde, which had suffered months of siege and repeated ethnic cleansing. Bosnian president Alija Izetbegovíc appealed again to President Bush for American air attacks on Serb artillery positions to break the sieges of Gorazde and Sarajevo and to clear the way for new shipments of food and water. Bosnian foreign minister Haris Silajdzic appealed for more decisive action from the Security Council or the United States. But the world stood by—the NATO forces idle—while heavily armed Serbs slaughtered civilians in the heart of Europe.

THE BALKAN WARS 171

Secretary-General Boutros-Ghali's response was unfathomable. Despite urgent appeals, he refused to provide UN peacekeepers for Bosnia when he deployed UNPROFOR troops in Croatia. This elegant francophone Egyptian Copt, who often called himself an African, was preoccupied with Somalia, not Bosnia, which he described as a "war of the rich." It was his stated view that the Bosnian Serbs and the Yugoslav government should not be blamed for the violence in Bosnia.[33] Again and again, he found excuses for those who attacked Bosnian Muslims. Eventually, the evidence overwhelmed his prejudice—but not until two million Croatians and Bosnians had been "relocated" under conditions that resembled the Nazi relocation of Jews: stuffed for days in stifling boxcars without food or water; crowded into prisons and army barracks, where women were raped, men beaten to death, and people "disappeared."

"We have reached the end," Bosnian president Izetbegovíc told Bernard-Henri Levy early in July 1992. "We have no food, no arms, no hope. We are the Warsaw ghetto. Will the world once more leave the people of the Warsaw ghetto to perish?"[34] On July 14, an amateur radio operator in Sarajevo transmitted a similar message: "We are awaiting death together. Please tell the world that we beg them to do something to stop these attacks as quickly as possible."

"Everywhere," Roy Gutman wrote, "Serb ethnic cleansing, the euphemism for murder, rape, and torture, was continuing against Muslims and Roman Catholic Croats. The Serb onslaught had displaced two million civilians and left ten thousand dead. It was the most vicious conflict seen in Europe or nearly anywhere else since World War II. But the steam had run out of Bush's presidency."[35] According to Gutman, at the G-7 meeting early in July 1992, when violence was already widespread in Yugoslavia, Bush had said, "I don't think anybody suggests that if there is a hiccup here or there or a conflict here or there that the United States is going to send troops."[36]

Bush's reaction was no different from that of other Western leaders, all of whom were essentially passive in the face of near genocide in Bosnia. Despite reports of cruelty worse than anything seen in Europe since World War II, the Western leaders equivocated, procrastinated, and offered only the most measured and detached responses. Though conditions in Croatia and Bosnia-Herzegovina worsened throughout 1991 and 1992, the Western leaders responded to Serbian aggression mainly by assuring one another that there was little they could do.

At the end of 1992, Robert Gates, Bush's CIA director, reviewed the war's human costs: "At least nineteen thousand persons have been killed . . . and perhaps another hundred thousand have died as a result of the hardships created by the war. At least three million Bosnians have lost their homes, and more than five hundred thousand have fled to neighboring nations . . . Most of the victims have been Slavic Muslims, who formed 44 percent of the republic's population before the war, while Serbs made up about 31 percent."[37]

The harsh, dehumanizing policies imposed by Serbs in conquered Bosnian territory bore a stark resemblance to those imposed by Nazi occupiers a half century earlier. In September 1992, the Associated Press published an abbreviated version of the declaration issued in the Bosnian town of Celmac.[38] Among the many restrictions, non-Serb citizens were not allowed to move around the town from 4:00 PM to 6:00 AM; to swim in the rivers, fish, or hunt; drive motor vehicles; to gather in groups of more than three; to use telephones except in post offices; or to wear uniforms of any kind. They were expected to do any tasks and work assigned to them.

The behavior of Bosnia's neighbors and the international organizations in ignoring Serbian brutality demonstrated the inadequacy of the elaborate arrangements that had been constructed to deal with international crises. It also called to mind a July 1938 conference in Evian, France, assembled to consider how to deal with Adolf Hitler. In that infamous meeting, a *New York Times* correspondent wrote, "[L]eaders of the three great democracies—the United States, the United Kingdom and France—engaged in a nontruthful poker game . . . in which each of the players refused even to contemplate raising the stakes."[39]

The UN response to Serbia was distorted by a persistent attitude of neutrality toward victim and victimizer, and by the mistaken expectation that an arms embargo would diminish the violence, when it actually empowered the Serbs while denying Croats and Muslims the weapons they needed to defend themselves. U.S. and NATO officials repeatedly told one another that "no one was innocent" in the Balkans, implying that no one was to blame. But in fact Serbia was clearly the aggressor, and Bosnia and Croatia unmistakably victims. The passivity and impotence of the

UN, the United States, the EC, the Western European Union (WEU), and NATO in facing this horror sapped the political and moral foundations of collective action.

The Bush administration, which had taken the lead in Kuwait, offered several explanations for its failure to act, notably the "unsolvable nature" of the tragedy of interethnic violence and the "European" character of these Yugoslav wars. Assistant Secretary of State Ralph Johnson offered this limp excuse: "The bottom line is that the world community cannot stop Yugoslavs from killing one another so long as they are determined to do so. . . . Only the people of Yugoslavia and their leaders can do that."[40] That excuse was a convenient way of blaming the victim and excusing the observer.

Secretary of State James Baker, having concluded that it was in the interest of the United States to prevent this humanitarian nightmare from continuing, emphasized the importance of multilateral pressure to enforce an ending to the conflict.[41] He took a series of diplomatic measures, withdrawing recognition from Belgrade's ambassador to the United States, closing the Yugoslav consulate in Chicago, stepping up consultations with allies on relief operations in Sarajevo. Baker seemed astonished at the Serbians' behavior: "It's hard to believe," he testified later, "that armed forces will fire artillery and mortars indiscriminately into the heart of a city, flushing defenseless men, women and children out into the streets and then shooting them."[42] But his disbelief never moved him to address the crisis effectively.

The Bush administration was not only unwilling to take unilateral action, it was reluctant to take any action at all. The Senate majority leader, Robert Dole (R-KS), called repeatedly for lifting the arms embargo to permit Bosnia and Croatia to secure the means to defend themselves. Senator Don Nickles (R-OK) observed that fifty countries had recognized Bosnia-Herzegovina while the United States had not, nor had it recognized Croatia and Slovenia. In the late summer of 1992, California Democrat Tom Lantos accused Bush of being too preoccupied with campaigning to make Bosnia a priority:

> The problem is that there is an election in ninety days, and this election paralyzes the Administration. . . . What is called for is for the one

remaining superpower on the face of the planet to take a . . . stance in the face of an outrageous, unacceptable mass extermination of innocent civilians.[43]

Through the summer and fall of 1992, conditions in Bosnia grew steadily worse, demands for action by the United States grew more urgent, and protests against the Bush administration grew stronger. Bosnia became a major issue in the American press and within the State Department, where several young Foreign Service officers resigned to protest U.S. policy.[44] But the policy remained the same, and Bush's time was running out.

One might have thought that the EC, which had eagerly claimed jurisdiction over the Yugoslav conflict at its outset, would have requested emergency meetings of the Security Council, or that the Islamic Conference would have acted in solidarity with Bosnia's Muslim population. But the EC limited its efforts to low-key diplomatic démarches throughout 1991 and 1992, and the Islamic Conference was occupied at the UN with its usual vendettas against Israel.

If the United States, the United Kingdom, Russia, France, China, and four or five of their colleagues in the Security Council had decided to stop the slaughter in Bosnia or Croatia, they could have done so. Had the WEU or the NATO countries decided to act, they too could have done so, even though some of the Bush team recognized that there would likely be long-term consequences to such action. "The entire future of our efforts to build our common security based on common values is at risk in the Balkans," warned Lawrence Eagleburger, who became acting secretary of state on August 23, 1992, when Baker left to run Bush's campaign.

Three days later, on August 26–27, Lord Carrington led a conference in London to stimulate action on Bosnia. The London Conference began with a declaration that participants must "reject as inhuman and illegal the expulsion of civilian communities from their homes in order to alter the ethnic character of any area."[45] In his remarks at the conference, Eagleburger said, "[I]t is Serbs who face a spectacularly bleak future unless they manage to change the reckless course their leaders chose for the new nation. . . . The civilized world cannot afford to allow this cancer in the heart of Europe to flourish, much less spread."[46]

Among the urgent tasks for the conference were collaborating in the delivery of humanitarian relief; lifting the sieges around Sarajevo, Tuzla, Bihac, Mostar, Gorazde, and other Bosnian towns; introducing human rights monitors; bringing all forces, including irregulars, under central control; restoring the civil and constitutional rights of the inhabitants of Kosovo and Vojvodina; banning military flights; helping refugees return to their homes; closing detention camps and dismantling concentration camps; securing the release of all detained civilians; and bringing to account those who had committed grave breaches of the Geneva Conventions. Those were the necessary dimensions of a general peace in the Balkans, but they were not achieved.

As the Bush campaign pushed on toward the fall elections, three factors were pushing the United States toward action: (1) the threat that violence would spread to Kosovo, Macedonia, Albania, and perhaps Bulgaria and Greece; (2) the massive human rights violations; and (3) growing American support for action in response to televised horrors. Bosnian civilians were literally starving and freezing to death in villages that had been under siege for months without food or firewood, and reports of their plight anguished Americans (and others), as did reports of camps in which Bosnian men were starved and beaten to death.

By the fall of 1992, many Americans understood that Bosnians lacked the weapons to defend themselves, because the U.S. government and the UN Security Council had imposed an arms embargo on Bosnia and Croatia. There's no real evidence that the Bush administration's shared responsibility for the Balkan tragedy caused his electoral defeat; most politically alert observers understood that the European Community was at least equally complicit. But Bush's electoral fortunes were not helped by the continued reports of destruction and death in Croatia and Bosnia.

Among the most damning such reports were those from Tadeusz Mazowiecki, the first democratically chosen leader of Poland, who was appointed by the UN Commission on Human Rights (UNCHR) as special rapporteur on Bosnia. Mazowiecki's first report, delivered on August 28, 1992, described the pattern of brutality and conquest in half a dozen towns in Bosnia-Herzegovina, including regions of Bihac and Sarajevo; it covered replacement of elected officials, discrimination, dismissal from work, confiscation of property, expulsions and forced population transfers, destruction of mosques, the use of sieges to cut off food sup-plies, and beatings, torture, and summary executions.[47]

This ethnic cleansing was also the subject of Mazowiecki's second and third reports, which described its deliberate and methodical charac-ter. All non-Serbs were potential targets. Serbs who sought to protect or help their Muslim or Croat neighbors were themselves attacked and harshly punished. Neighborhoods and towns in which Serbs, Muslims, and Croats had lived together peaceably for years were swiftly "cleansed" of Muslims. The reports described people being told to leave or die, or stuffed into airless trucks and cattle cars for unknown destinations. Mosques and Catholic churches were burned and bombed. Hospitals were repeatedly attacked. Humanitarian workers and convoys were ha-rassed, attacked, and blocked. Journalists were attacked.

The year 1992 had seen summary executions become commonplace in areas occupied by Serb forces. In March, all the men in the village of Jelec were rounded up and machine-gunned. In April, the first round of organized killing began in Srebrenica, with young and distinguished males as preferred victims. Elsewhere, women and girls were raped; men and boys were strangled, shot, or drowned. In May, carloads of Serbs drove through Zaklopaca, murdering at least one hundred Muslims. Three-quarters of the forty-five hundred inhabitants of six small moun-tain villages were killed when Serbs captured the villages. In Mostar, there were reports of mass graves, of victims shot at close range with automatic weapons. From village after village came reports of Serbs making house to-house searches in each new town they entered, rounding up men and boys, and frequently shooting them in cold blood.

In Mazowiecki's reports, Serbs were most often the aggressors and Muslims the victims. But Croats and parts of Croatia under Serb control were also major targets, and Muslims and Croats were also occasionally guilty of brutal mistreatment. The UNHCR and the International Com-mittee of the Red Cross (ICRC) sought to determine the number and lo-cation of prisoners. By the end of December, more than 10,800 detainees had been registered in fifty places of detention in Bosnia and Herzego-vina. Many more were undeclared. Most of these people were not prisoners of war but civilians who had been seized and detained for possible exchange.

By the end of 1992, Mazowiecki estimated, 810,000 people had been driven from their homes and become internal refugees; another 700,000 had become refugees in other countries that had been part of Yugoslavia. Conditions in the camps, prisons, and other places of detention were inhuman, the treatment of prisoners often ghastly. Starvation, thirst, beatings, torture, and rape were commonplace in towns and camps alike.

After losing the election, Bush joined British prime minister John Major in calling for enforcement of the ban on military flights over Bosnia and threatening drastic action unless Serbs made a "rapid and radical change of policy." The joint U.S.-U.K. statement was triggered by the deterioration of the humanitarian situation and concern about the spread of fighting to Kosovo and Macedonia, a concern that had already prompted the UN to send seven hundred peacekeeping troops to Macedonia and more observers to Kosovo. Major and the United Kingdom were reluctant to use airpower against Serbs, and the British and the UN qualified their threats.[48] Retaliation would not be automatic; it would not involve extreme measures, such as shooting down Serb aircraft, but rather a warning period of fifteen days.[49]

The American response was not much more helpful to the victims of Serb violence. The State Department announced that, in addition to recalling its ambassador, the United States would join in diplomatic efforts and economic sanctions. The United States, Russia, and three EC members—Britain, France, and Belgium—cosponsored Security Council Resolution 787, which threatened more economic sanctions against Serbia. Instead of following the special rapporteur's recommendation and establishing safe areas for the burgeoning refugee population, however, 787 merely "invited" the secretary-general and the UN High Commissioner on Refugees (UNHCR) to "study the possibilities of and the requirements for the promotion of safe areas for humanitarian purposes."[50] These and other toothless resolutions and measures hardly qualified as serious attempts to deal with the murder and mayhem under way in Yugoslavia.

Bush also began to push for the establishment of UN observer missions to monitor the southern regions of Kosovo and Macedonia, and for more serious commitments to enforce action. In a December 1992 letter, he warned Milošević and JNA chief Zivotr Panic that "In the event of conflict in Kosovo caused by Serbian action, the United States will be prepared to employ military force against the Serbs in Kosovo and Serbia." [51]

George H. W. Bush had provided effective world leadership to cope with Iraq's aggression in Kuwait and with starvation in Somalia. He had announced his intention to help usher in a new world order. Why, then, did he avert his eyes from the destruction of Vucovar and Dubrovnik, from the sieges of Sarajevo, Gorazde, and a dozen other Muslim towns and villages? Perhaps because he was preoccupied with the presidential election. Perhaps also because he was frustrated with the decisions of France and Germany to create a military force outside NATO and take an independent role in policy. Bush preferred that the United States be the leader in international actions, as it was in imposing sanctions.

If Bush took no effective action, neither did John Major, François Mitterrand, or the French foreign minister, Roland Dumas. Mitterrand's Sarajevo visit became the center of a public controversy after a French documentary film reported that President Izetbegović had personally briefed Mitterrand on Serb massacres of civilians and on the ghastly atrocities at the concentration camps, where thousands had been deported. Mitterrand never mentioned any of this until he was publicly criticized for his silence. Major and British foreign secretary Douglas Hurd also steadfastly resisted participating in an international response to Serbian aggression. Britain's reluctance to mount air attacks was based on the fear that "they might bring Serb reprisals against the twenty-seven hundred troops it had on the ground helping to deliver humanitarian supplies." [52] To at least one informed observer—historian Brendan Simms, who has made an exhaustive study of British biases in favor of Serbs and against Bosnian Muslims, against robust resistance to Serb aggression, and against an air war—Hurd and British defense secretary Malcolm Rifkind were "the two men who bear the greatest responsibility for Britain's disastrous Bosnian policy." [53]

On February 8, 1993, after the Clinton administration had assumed control of Yugoslav policy, Bush's last secretary of state, Lawrence Eagleburger, issued his own epitaph for the Bush policy in Bosnia. "[W]e failed, from beginning to end," he said. As for the ongoing violence, he said, "I don't know any way to stop it except with massive use of military force."[54]

THE VANCE-OWEN PLAN

As the guard prepared to change in Washington, there was plenty of blame going around. Even before meeting with Clinton's choice for secretary of state, Warren Christopher, EC negotiator Lord David Owen said it was probably the fault of the United States that war still raged in Bosnia.[55] During the presidential campaign, Clinton had delivered strong activist speeches that Owen believed gave hope to beleaguered Bosnian Muslims, encouraging them to resist any peace plan that gave an advantage to the Serbs (as all of them did). But Bosnia's government needed no encouragement to resist these plans. Foreign minister Haris Silajdzic vowed that the Muslims would not negotiate under "continuing genocide"—that is, until the Serbs stopped shelling Sarajevo, Gorazde, and other villages.

In late 1992, Owen and Cyrus Vance unveiled what became known as the Vance-Owen peace plan,[56] which proposed to divide Bosnia into ten ethnically distinct, autonomous provinces. Three provinces would be dominated by Croats (who made up 17 percent of the population); three by Muslims (who constituted 44 percent of the population before the ethnic cleansing drove them from their homes); and three by Serbs (who made up 31 percent of the population).[57] Sarajevo, the capital, would be jointly administered. Under the Vance-Owen plan, the elected government of Bosnia, then headed by President Alija Izetbegovíc, would resign and be replaced by a nine-person interim commission whose members would be equally drawn from each of the ethnic groups.

The Croats quickly accepted the plan, but they were the only ones. The Bosnian Serbs were equivocal; they wanted a corridor connecting Serbian enclaves to Serbia, but on January 20, 1993, the Bosnian Serb Assembly reluctantly accepted the plan in principle. The Bosnian Muslims—the principal victims of Serbian aggression—would be the big losers in the Vance-Owen settlement. Early in his tenure as EC negotiator, Owen had told the London *Financial Times* that "what has happened in the former Yugoslavia must be reversed . . . we are not going to have the Muslims treated like the Jews once were in Europe."⁵⁸ But to the Bosnians and most of their supporters and sympathizers, it seemed clear that the Vance-Owen plan *did* reward ethnic cleansing, by legitimizing Serb control of land from which Muslims had been driven. It also rewarded Croats, who were already moving to consolidate military control over lands assigned to them under the plan, on the grounds that Muslim "extremists" and "fundamentalists" would threaten the national identity of the Croatian community in Bosnia. At the same time, the Croats, who had managed to procure some weapons and training, were fighting successfully to reclaim Serbian enclaves in Croatia.

None of these events caused Vance or Owen to doubt that their plan could end the fighting if it were imposed by the Security Council and enforced by the major powers. Owen believed that President Clinton "should stop all of this loose talk about using force, make it clear to Izetbegovíc that he's got no real alternative to these negotiations . . . then provide American troops as part of a NATO force."⁵⁹ To his credit, Clinton had no stomach for forcing Bosnia to accept a solution that was fundamentally unacceptable. Madeleine Albright, newly appointed U.S. ambassador to the United Nations, had promised that the United States would "try to make a peace settlement which does not punish the victims and does not reward the aggressors."⁶⁰

The Vance-Owen plan not only awarded the Serbs most of the areas from which Muslims had been driven, it also gave them political representation far beyond their numbers. The Bosnian government was dismantled, Bosnia's constitution scrapped. Silajdzic spoke for the Bosnian government when he said, "We, as a member of the United Nations, will never accept the abolition of our constitution, our legality, which are based on free and democratic elections."

Nevertheless, the British and French governments signed on to the plan. The EC and Russia offered encouragement. Initially, Warren Christopher leaned toward accepting the plan, with some changes: tightening the sanctions against Serbia and Montenegro, creating a war crimes tribunal, and enforcing the ban on Serb flights over Bosnia and Herzegovina. Surprisingly, he did not mention lifting the arms embargo or press for enforcement of existing Security Council resolutions. A *New York Times* headline announced that "Clinton . . . Supports Current Bosnia Peace Plan."[61]

On February 10, 1993, Christopher announced that the United States would become actively involved in the negotiations and would not deploy troops in Bosnia for any purpose except to enforce an agreement that was accepted by all parties. Colin Powell, still serving as chairman of the Joint Chiefs, continued to operate on the basis of the Weinberger-Powell doctrine, which called for using force decisively to attain clear objectives.[62]

After studying the Vance-Owen plan further, however, Christopher changed his position, describing it as "legitimizing the ill-gotten gains from ethnic cleansing." With Senator Bob Dole leading Congressional opposition to the plan, Clinton's secretary of defense, Les Aspin, proposed an alternative approach: lifting the arms embargo to permit Bosnians to defend themselves, while using U.S. airpower to enforce the no-fly zones and knock out Serb artillery and airfields. The UN turned down Aspin's proposal.

Vance and Owen had labored long, but their plan was neither fair nor enforceable. It was finally abandoned not because it was unacceptable to Bosnia, but because the Bosnian Serbs put it to a popular referendum, where it was rejected by more than 90 percent of those who voted. In the wake of its failure, a coalition known as the Contact Group, consisting of France, Germany, Russia, the United Kingdom, and the United States, tried to restart the peace process in the spring of 1994 by attempting to resolve the ongoing territorial issues—but that plan, too, failed after it was rejected by the Bosnian Serbs.

In this period, Serbs and Croats often collaborated in attacking Bosnia's Muslims. Milošević and Croat president Franjo Tudjman had previously discussed dividing Bosnia and Herzegovina between them, but Milošević was resistant to any agreement; he and Karadzic were apparently moved only by the prospect of enhancing Serb power, and did not hesitate to ignore cease-fire agreements they had signed. It was clear to all that Milošević had no interest in a peace that established democratic self-government for Bosnia or Kosovo, or protected the minority rights of Muslims. He had no intention of demilitarizing Kosovo, reducing his forces, or implementing a cease-fire. Instead, he escalated his campaign of violence.

Early in 1993, the cold-blooded murder of Bosnia's deputy prime minister, Hakija Turajlic, seriously called into question UN competence and rules of engagement.[63] As the nation's vice president, Ejup Ganic, said, "The assassination was in a United Nations APC, on a road controlled by the United Nations, and under the protection of UN soldiers. . . . The United Nations is responsible for this tragedy."[64] The true responsibility for the murder, of course, lay first and foremost with the Serb soldier who fired seven shots into the deputy prime minister, but it was shared by the UN officer in charge and by the UN itself, whose rules of engagement made it unnecessarily difficult to protect Turajlic. The armored vehicle carrying Turajlic was halted at gunpoint by a force of forty Serbs. One of the UNPROFOR troops assigned to guard him—probably the French officer in charge, Colonel Patrice Sartre—opened the locked back doors of the vehicle, leaving him exposed to the Serb assailants. A peacekeeper was asked if he had fired at the man who killed Turajlic. "Returning fire," he explained, "is not permitted under UN rules of engagement except to save your own life."[65]

In February 1993, the Security Council passed Resolution 808, establishing the International Criminal Tribunal to investigate and prosecute war crimes committed in the former Yugoslavia. But the siege of Sarajevo continued unabated, with no war criminals arrested or charged. In June, air strikes were again authorized but not conducted, and though six Muslim towns were declared safe areas, none were safe. All remained under attack and without food and medicine.

PEACEKEEPING: THE UNITED STATES MEETS THE UN RULES OF ENGAGEMENT

At the start of 1993, three enclaves in eastern Bosnia remained under Bosnian government control. Each was desperately overcrowded and undersupplied, overflowing with refugees from other towns that had been attacked, and each was surrounded by Serb forces. Sniping and heavy artillery cut them off from food, medicine, and supplies. Despite

UN resolutions and the assurances of Serb officials, UNHCR convoys bearing aid were repeatedly denied passage. Inside the towns, hunger became starvation and people died.

In a report to the U.S. Agency for International Development (USAID), humanitarian relief expert Thomas Brennan charged that the United States and the UN were "clearly failing to prevent genocide in Bosnia-Herzegovina, and may actually be facilitating its implementation." [66] Brennan also blasted the UN High Commissioner for Refugees (UNHCR), who had "generally opted for negotiation and appeasement rather than forceful determination to deliver relief supplies to those most in need," and concluded that "immediate U.S. political leadership and military intervention is essential to halt genocide in Bosnia-Herzegovina." [67]

The UN's typical unwillingness to take forceful action unfortunately characterized its response to ethnic cleansing. It was no surprise that the Serbs opposed relief for Bosnian Muslims, whom they were systematically starving and driving out of the country. But the opposition of UN commanders to the airdrops of food and medicine—as if there were some sort of international duty to stand in a neutral pose while Muslims died—was another matter entirely. In Bosnia, to require neutrality was to suggest moral equivalence between the refugees who were forced to flee their homes and those who forced them to flee.

By early 1993, the community of Cerska in eastern Bosnia had been cut off for seven months. People were reduced to eating leaves, animal fodder, and the bark of trees and many froze to death. They lived under constant shelling, with no medical supplies; those who were wounded or ill usually died. Damaged arms and legs were amputated with hacksaws and without anesthesia. Those too old or ill to flee were beaten unmercifully, and amateur radio operators reported massacres of the people remaining at Cerska.

From Zepa, another "safe" area, a shortwave radio operator broadcast "a cry for help from a frozen hell." The twenty-nine thousand people of Zepa were described as "too exhausted to dig graves in the frozen ground." [68] There was no electricity, no running water, no heat. There were no houses, few clothes, no live animals—all had been killed by Serb shelling. The town had been under siege for months. The broadcast continued, "I beg . . . to all humanitarian organizations to help us. My personal statement and from all the people of Zepa is that we don't trust the United Nations any longer. . . . But we have great confidence in the American people. It's only the Americans whom we can trust now." [69] The UNHCR had agreed to consider making emergency airdrops of food to Zepa and two or three other beleaguered cities. Incredibly, UN officials told the Bosnian government that this could only be done if the Serbs agreed. They did not. [70]

Traditional peacekeeping requires the consent of all parties to a conflict, a will to peace among the parties, and peacekeepers committed to strict neutrality. Yet none of these circumstances obtained in Bosnia. Serbia denied that it was an active participant in a war against the Bosnian Muslims and did not consent to the deployment of peacekeepers on its border with Bosnia. And the UN peacekeepers could hardly be neutral when the UN had called for acts in support of Bosnia. In a series of resolutions, the Security Council had authorized safe havens and necessary measures to deliver humanitarian assistance, declared a no-fly zone, and put monitors on the Serb border to stop the flow of vehicles, weapons, ammunition, oil and gas, and food from Serbia into Bosnia. These resolutions were not consistent with neutrality or equal treatment of the parties, but they were followed regardless.

Boutros-Ghali believed that peacekeepers should be studiously neutral, never affecting the balance of power between the contending forces, and in Bosnia the Secretariat restricted its peacekeeping force to the sacrosanct UN rules of engagement. Yet (like UN peacekeeping itself) those rules had no formal standing; they were simply conventions that had developed over time in small conflicts in places like Cyprus, Kashmir, the Golan Heights, and southern Lebanon. The UN forces following these rules in Bosnia failed to contain violence, prevent the shelling of civilians, deliver food to the hungry, or ensure safety in the safe havens, but no authority challenged the appropriateness of these rules to this specific region or conflict.

In Somalia, the Bush administration had declined from the outset to be bound by such rules. The American forces who entered Somalia in Operation Rescue did not have the consent of General Mohammed Farah Aideed. They were not neutral, and they used force. Though some participants in the Somalia peacekeeping operations complained that the use of force increased the dangers in the region, Boutros-Ghali had dismissed the complaints: "[W]e have to use force to disarm the various forces . . . without that we will not be able to promote national reconciliation and offer humanitarian assistance to the populations." This was equally true in Bosnia. In Somalia, UN policies had been flexible and fairly successful. In Bosnia, they were rigid and timid and achieved little. Certainly, they did not help the Bosnian Muslims.

Each effort by the Security Council to modify the terms under which UN forces operated in Bosnia met with resistance from the Secretariat. And personal heroism was not consistent with UN rules of engagement. In March 1993, when the UN commander for Bosnia, General Philippe Morillon, demonstrated extraordinary personal courage to expedite the delivery of food and medicine to Srebrenica, the secretary-general asked the government of France to recall him.

By January 1993, UN troops in Bosnia were being widely condemned for failing to protect the people they had been sent to protect. The director of UN relief operations for Bosnia, José-Maria Mendiluce, reported that convoys bearing food, fuel, and medicine were being blocked or turned back from the area, exacerbating the humanitarian crisis. Mendiluce asked the UN to ask NATO to airlift supplies to the estimated 150,000 Bosnians in danger of starving and freezing,[71] assuming—like almost everyone else—that planes dropping food to starving people under siege would be protected. Yet the British resisted introducing any further military elements into Bosnia, including protection for humanitarian aid.

During his presidential campaign, Bill Clinton had promised repeatedly that he would airlift food to the Bosnians to save them from starvation. Clinton had barely been inaugurated before the UNHCR made a passionate appeal to him to airdrop food and medicine. José Maria Mediluce, director of UN relief operations for over a million and a half Bosnians in "safe areas" such as Zepa, Cerska, and Srebrenica, publicly stated that civilians in these towns were dying of starvation and incessant shelling.[72] In March 1993, without securing Europe's approval, the United States began dropping food into Bosnia. Boutros-Ghali approved the plan only after Clinton agreed that the U.S. planes would fly without military escorts and would make their drops from about ten thousand feet (which greatly reduced their accuracy). Despite these limits, UN professionals and others later agreed that the drops were critical to the survival of the Bosnians.

But the Serbs' merciless attacks on Bosnia-Herzegovina and Croatia continued, and the presence of the severely restricted UN peacekeepers was doing little to stem its tide. Ethnic cleansing was in full force in Banja Luka, Bosnia's second largest city. The Serbs were attacking private homes with grenades, firing non-Serbs from their jobs, looting and seizing Muslim property, and murdering Muslims. Some twenty thousand Muslims and five thousand Croats fled the area in January 1993.[73] UN commander general Lewis MacKenzie of Canada and his successor, General Michael Rose of Great Britain, were repeatedly charged with partiality to the Serbs. Yet the Americans still sent no men and no commanders to the region.

Early in 1993, the Clinton administration moved tentatively toward a more active role in Bosnia. Soon after he took office, Secretary of State Warren Christopher attempted to persuade the U.S. allies to lift the arms embargo and mount air strikes against the enemy. This policy was widely supported in the U.S. Congress, but was opposed by the Europeans, and Christopher made little progress.

The continued violence presented Clinton with an extremely difficult decision: should he commit U.S. troops to peacekeeping under a UN command that seemed incompetent and ineffective, and that imposed rules of engagement that would render them nearly helpless and unable to achieve most of their goals? The Serb forces, which had driven the Muslims out of one Bosnian town after another, were closing in on Srebrenica. By April 1993, they had begun shelling this last Muslim town in East Bosnia—now jammed with refugees. Once again, unarmed inhabitants of a large Muslim town waited helplessly while Serb forces prepared to drive them from their homes.

Now it was Bill Clinton's turn to explain why he was not doing more. "I've done everything that I know to do consistent with the possibilities we have for further action in the United Nations with our European allies and the members of the Security Council," Clinton said. "It is a very frustrating and difficult situation."[74] In many cases, he asserted, the United

States was willing to do more than its allies, who resisted virtually every proposed move to counter Serbian ethnic cleansing of Bosnia. But except for the food drops—undertaken in the face of European and UN resistance—the United States did not act on its own, and Clinton quickly became the second American president who allowed the Europeans to stifle the impulse to action.

There was some truth in the excuses offered for every capitulation, but no adequate explanation of why President Clinton so quickly accepted the limits on U.S. action in Bosnia that candidate Clinton had harshly criticized. "If you believe that we should engage these problems in a multilateral way," he said, "if you believe in what happened in a good way in Operation Desert Storm, then the reverse has to be true, too. The United States has got to work through the United Nations, and all of our views may not always prevail." [75] Apparently, he had forgotten that Desert Storm was conducted under American command.

Was it the search for consensus among the permanent members of the Security Council that led Clinton to abdicate American leadership in the Balkans? Did he see unilateral action as incompatible with U.S. commitments to NATO?

Christopher had promised that the new administration would bring "the full weight of American diplomacy" to a search for a negotiated settlement, after which U.S. troops would help enforce the peace. But it often seemed that he did not fully understand the many obstacles to peace. He assured all that the new administration would take immediate steps to reduce suffering and bloodshed—including enforcement of the no-fly zone under the UN resolution. He called on all parties to expedite the flow of humanitarian aid to relieve suffering. He suggested that American forces might be used to ensure safe passage of food and medicine to areas under siege, and reiterated that the Clinton administration planned to work with Vance and Owen to make their proposals more acceptable to the Muslims. [76]

In March 1993, special rapporteur Mazowiecki made an urgent appeal to the UN Human Rights Commission, then meeting in Geneva. Serb officials agreed to give UNHCR brief access to prison camps and towns under siege. On March 5, General Morillon (the French UN commander for Bosnia-Herzegovina) and the UNHCR representatives arrived in Srebrenica for a fact-finding mission. The facts were not difficult to ascertain. Serb forces temporarily diminished their shelling, but resumed it as soon as Morillon left. Many people who tried to escape behind his convoy were ambushed and murdered. Among those who stayed, many died under the intensified shelling.

On March 12, General Morillon arrived again in Srebrenica, where he announced that he would remain until Serb officials permitted evacuation of the wounded. Nine days later, a single UN aid convoy was admitted to Srebrenica. Conditions continued to deteriorate. On April 2, Sadako Ogata, the UN High Commissioner for Refugees, wrote to the secretary-general, "[T]hese people are desperate to escape to safety because they see no other prospect than death if they remain where they are." Ogata saw only two options: (1) immediately enhance the international presence, including UNPROFOR, turn the area into a United Nations Protected Area (UNPA), and provide lifesaving assistance; or (2) organize a large-scale evacuation of the endangered population.

Safe Areas Established

Although the special rapporteur had explicitly recommended the establishment of security zones in Bosnia and Herzegovina in November 1992, nearly six months passed before the Security Council passed Resolution 819 on April 16, 1993, established Srebrenica as a safe area—an area theoretically free from armed attack or any other hostile act. The resolution also demanded that the Federal Republic of Yugoslavia (Serbia and Montenegro) immediately stop supplying weapons to Bosnian Serb paramilitary units.[77] In May, in Resolution 824, the Security Council declared that Sarajevo, Tuzla, Zepa, Gorazde, Bihac, and Srebrenica should be treated as safe areas from which all Bosnian Serb military and paramilitary forces should withdraw.

UNPROFOR's task was not to defend a geographical area but to protect the civilian population of the area against armed attack. As Mazowiecki noted, however, the safe areas were safe only on paper. "The Security Council . . . refrained from authorizing the additional troops deemed necessary by the secretary-general to ensure full implementation of UNPROFOR's mandate."[78] The safe areas were mobbed by displaced persons, overwhelmingly women and children; people were dying daily from shelling, starvation, illness, and wounds.

By summer's end, the Serbs' shelling, sniping, and starving of civilians had resumed. In August, Resolution 836 reaffirmed the safe areas, extended the mandate of UNPROFOR to enlarge them, and authorized the use of airpower to support the mandate. But by then the Bosnian Serb leadership thought it had prevailed. Its detestable "purification" of Bosnia was nearly complete: Entire Bosnian Muslim communities had been destroyed in the merciless military campaign to establish exclusive Serbian power over the country. One by one, Muslim towns had been attacked by Serb mercenaries and regular army troops, their populations driven out, swelling the refugee population to 1.6 million. It was the largest wave of refugees in Europe since World War II.

Sarajevo's airport had been shut down for months, and the city was surrounded by Serb troops. Food was so scarce that many families lived on one meal a day. Diabetics lacked insulin; the wounded lacked antibiotics and painkillers. The international community managed little effective action. Neither the sanctions imposed on Serbia by the UN Security Council nor the arms embargoes were working. UNPROFOR was not an effective force. Cease-fires were negotiated and violated, pledges made and broken. Safe routes created for the delivery of humanitarian supplies were shut tight. Serbian mortars pounded Sarajevo's neighborhoods almost daily and fired on safe areas packed with refugees. In the last three months of 1993, there were 225 violations of the no-fly ban.

Western Non-response

The commitment, dedication, and heroic style of UN commander general Philippe Morillon and his fact-finding missions illustrated what the international community could have been doing to stop the devastation. His behavior threw the feckless performance of the UN, NATO, the EC, and the United States into embarrassing contrast, paving the way for high-ranking UN officials to call for the removal of Morillon—whom the French media were calling "General Courage"—in April 1993.

The Serb government complained bitterly that General Morillon was not "neutral and objective" in drawing distinctions between the refugees and their oppressors. In the UN Secretariat, anonymous officials alluded to Morillon's habits of exceeding instructions, violating the rule of neutrality, and attracting too much media attention. They discreetly requested his recall. No one in the Secretariat or the French government acknowledged their intentions to remove General Morillon, although France's defense minister, François Leotard, said that Morillon would probably be back in Paris by May.

The treatment of General Morillon was an exercise in cynicism, like sending NATO planes to patrol the no-fly zone and then not permitting them to enforce it. Though Boutros-Ghali insisted on his policy of neutrality, the events in Bosnia dramatized the need to rethink the theory and practice of UN peacekeeping—indeed, of UN military involvement in international conflicts and in the provision of humanitarian assistance.

It was not clear, for example, that peacekeeping forces should ever have been committed to Croatia, where their presence protected Serbian conquests, or to Bosnia, where a major war of aggression had been under way for a year and the presence of UN peacekeeping forces was repeatedly used as an excuse for inaction. The arms embargo was the clearest example of a UN action that made the situation worse. By the time Clinton was inaugurated in January 1993, the homes, families, communities, and lives of Bosnia's Muslim population had been largely destroyed.

Yet no wave of outrage swept the world at this genocide in the heart of Europe. The Serb offensive had destroyed the existing Bosnian society in less time than it took the international community to pass resolutions promising to use necessary force to deliver food and medicine. It was difficult not to conclude that the indecision and inaction of Western governments, which chose not to act effectively—either unilaterally or through the UN, NATO, the WEU, the EC, or the Contact Group—constituted passive acquiescence in Serb aggression.

It was not the first time in the twentieth century that the West had been faced with organized brutality—and not the first time Western leaders had equivocated and procrastinated and offered only the most measured and detached response. Their busy nonresponse to Serb aggression recalled the inaction of the Western leaders who confronted Adolf Hitler at the outset of World War II.

But this time the West had several activist leaders: Margaret Thatcher, Hans-Dietrich Genscher, EC president Jacques Delors, and, a bit later, Bill Clinton used the media to draw attention to the problems and raise the stakes. These leaders spoke boldly about alternatives that John Major, François Mitterrand, and George H. W. Bush had preferred not to consider. Thatcher, the former prime minister, insisted that Bosnia was a defining moment, and she was right. The world's response to Serbian genocide defined its lack of seriousness about a new world order and collective security from aggression. It illustrated the ineptitude of the elaborate international institutions constructed to deal with just such crises.

In this first test of the post–cold war period, the WEU was partially incapacitated by the EC's "dangerous lack of resolve," against which Delors had warned. NATO was partially incapacitated by French opposition to its participation. The UN was partially incapacitated by the reluctance and indifference of its member states, and by Boutros-Ghali's resistance to involvement in what he called "a rich people's war."[79] The Red Cross was partially incapacitated by the Serbs' denial of access to prison camps and by its own lack of a sense of urgency. The United States was partially incapacitated by a lack of empathy among some of the people in the administration and in Congress who were responsible for foreign policy. The UN response also was distorted by its pose of neutrality. For years, some UN officials and Security Council members clung to its contention that this war of aggression was a civil war.

The Serbs' systematic destruction of Bosnia was an ugly model for fanatical nationalists and would-be aggressors and dictators in other parts of Central and Eastern Europe. It trivialized efforts to establish collective security, and it sapped the intellectual and moral foundations of collective action anywhere.

Thatcher advised European and American leaders to issue a series of ultimatums to the government of Slobodan Milošević: cease military action and the flow of weapons into Bosnia, or else. Tell Serb forces in Bosnia to turn over their heavy weapons to an international body, or else. Permit Muslims to return to their devastated homes under international protection, or else. She advised the United States and the EC to tell Milošević that failure to do these things would result in the destruction of Serb military assets and the encouragement, by all means, of opposition groups.[80] Serbia, she noted dryly, was not a world power. It was a savage, racist regime for which there was no room in any new world order worth preserving.[81]

Yet Boutros Boutros-Ghali raised objections against these proposals for effective action, as he had more than once before. In July 1992, he had opposed implementation of the London Agreement, which called for Serbia to deliver its heavy artillery to UN forces. In September, he had opposed enforcing the no-fly zone. A few months later, out of fear of offending the Serbs, he had resisted airdropping food to towns filled with starving people.

The news that four Bosnian residents of a home for the aged had frozen to death in a single night a block from UN headquarters in Sarajevo did not help the UN's reputation for humanitarian concern and efficacy. Cyrus Vance's personal efforts to prevent U.S. officials from meeting with Bosnian president Alija Izetbegović in Washington, where he met with President Clinton in the fall of 1993,[82] did not help Vance's reputation as an evenhanded, humane mediator. Many Bosnians, Croatians, Africans, and Cambodians began to see the UN as part of the problem rather than part of the solution.

The UN Secretariat functioned like many other bureaucracies. It took action only by consensus, which was hard to build and harder to maintain. Responsibility was so widely shared and so depersonalized that ordinary moral and social disciplines disappeared. Where everyone is responsible, no one is responsible. If it is difficult to hold governments responsible for their actions, holding international institutions is even more difficult.

At the same time, the EC, the Security Council, and the member states did not challenge Boutros-Ghali's unprecedented assertions of authority. Although he was described as an activist secretary-general, he resisted all but the most limited peacekeeping in Bosnia. When the Security Council considered enforcing the no-fly zone or the delivery of humanitarian assistance, Boutros-Ghali appealed for more time to find a political solution. The result was continued Serb aggression against a background of endless negotiation. Although the UN Charter vests executive power in the Security Council, its member states accepted the secretary-general's priorities and programs as if he were the chief executive in a presidential system and the Security Council a rubber-stamp legislature.

By May 1993, the already-disastrous human situation had worsened, as Serb forces launched successive attacks on the region's remaining Muslim towns, which had grown more swollen with refugees. Despite the powerful moral argument for using American airpower to save Bosnians, all sides ventured further counterarguments on both moral and practical grounds. Britain, France, Belgium, and other countries with troops on the ground feared that air strikes would endanger their troops. Many Americans feared that air strikes would constitute an open-ended U.S. commitment that could end in a Balkan quagmire; others believed that a unilateral U.S. decision in favor of air strikes would violate international norms and endanger peacekeeping forces; still others argued that American national interests were not involved in the Balkans, and the U.S. government should not commit military forces. This left President Clinton with a critical decision: whether to commit American power, under UN command, in the pursuit of purely altruistic goals.

In his first year as secretary of state, Warren Christopher sought allied cooperation in removing the arms embargo. He sought greater use of NATO airpower for a lift-and-strike strategy (lift the embargo and strike the aggressors) to remove the Serb sieges and open the way for humanitarian assistance to the people who were trapped without food, heating fuel, or medicine. But nothing came of his efforts, and soon the administration seemed to be backpedaling from its constructive stance. "We can't do it all," Christopher said. "We have to save our power for those situations which threaten our deepest national interests. . . . [Bosnia] is a humanitarian crisis a long way from home, in the middle of another continent. [So U.S. actions] are proportionate to . . . our responsibilities."[83] In a luncheon briefing with reporters, Peter Tarnoff, the undersecretary of state for political affairs, explained that Christopher had not failed to persuade European allies to join the United States in stronger action in Bosnia; rather, he had not really tried. He had gone to Europe "to consult," not in a serious effort to rally the Europeans around U.S. recommendations to use air strikes and lift the arms embargo.[84]

This was one of several early signals of the Clinton administration's retreat from global leadership—signals that came as a surprise to many. Who could have predicted that Clinton would not approve his own "Clinton doctrine" for Bosnia? When the story of Tarnoff's briefing on Christopher's trip hit the *Washington Post*, the administration realized that such a renunciation of U.S. leadership would reduce American influence and might be taken by potential aggressors as an invitation to action.[85] The White House quickly distanced itself from Tarnoff's characterization: "That is not our foreign policy," a high-level spokesman announced. And Christopher clarified, "There is no derogation of our powers and our responsibility to lead."

But these denials did not dispel the signs of an American reluctance whose consequences were already being felt around the world. The Clinton team's intention to disengage were apparent in a series of developments: deep cuts in the defense budget; the inability of NATO defense ministers to agree on much of anything except to pass the Bosnian issue back to the UN Security Council; the concessions Clinton offered to North Korea in advance of negotiations on its nuclear intentions; the unconditional extension to China of most favored nation (MFN) status; the unconditional diplomatic recognition of the MPLA, the Popular Movement for the Liberation of Angola, government of Angola; and the unconditional release of financial assistance to the Sandinista government in Nicaragua. This same reluctance could be seen in the administration's lack of response to Iraq's provocations, and its inaction on the North American Free Trade Agreement (NAFTA), and related hemispheric matters.

In the face of such challenges, Christopher and other Clinton administration officials tended to shift ground, redefine goals, and, in the case of Bosnia, to claim impotence in the face of Serbia's escalating demands. As *New York Times* reporter John F. Burns wrote in July 1993, "For a while, Serbian political and military chiefs appeared ready to halt the seizure of territory and the raping, killing, and expulsions of Muslims that began when the Serbian military campaign began in April 1992. . . . Instead, United Nations officials say, Mr. Clinton's decision to bow to European nations like Britain and France in their reluctance to launch military strikes or lift an arms embargo against the outgunned

Bosnian government caused the fighting and the suffering of civilians to worsen rapidly."[86]

As candidate and president, Clinton expressed outrage over the Serbs' brutality. National Security Advisor Tony Lake and the U.S. ambassador to the UN, Madeleine Albright, also seemed to feel deeply about Bosnia. But Christopher had sounded a milder note in his testimony before the House Committee on Foreign Affairs. Strictly speaking, he said, "some of the acts that have been committed by various parties in Bosnia, principally by the Serbs, could constitute genocide under the 1948 convention"; he spoke of "atrocities on all sides," and said merely that the Bosnian Serbs were "most at fault of the three parties."[87] Christopher's testimony suggested that he and the administration were distancing themselves from what he described as "a multilateral problem [that] must have a multilateral response." The United States must be wise, and "being wise means acting in ways that are consistent with our national interests."[88]

But that was precisely the challenge—to know our national interests. The United States had fought two costly wars in Europe largely because American presidents had believed that our national interests were tied to the European continent. Now Christopher seemed to be suggesting that what was "at heart . . . a European problem" need be of little concern to the United States.

The notion that aggression must not be rewarded, lest it invite further aggression, had been the core concept of U.S. foreign policy at least since World War II. That was why Harry Truman signed and the Senate ratified the UN Charter, with its clear-cut prohibition on the use of force and equally clear-cut provision of self-defense in Article 51. It was the reason Truman sent U.S. troops to South Korea when North Korea attacked. It was why George H. W. Bush provided U.S. leadership for action on Kuwait. In the Bosnian conflict, however, the United States sometimes acted as though a greater national interest lay in preserving consensus with its allies rather than in discouraging aggression. And the search for consensus in the Security Council seemed to take priority over protecting a people from destruction.

Even as Christopher spoke, Karadzic was making radical new demands for more territory, more ethnic cleansing, and a Serbian state whose realization would require expelling tens of thousands more Bosnian Muslims from their homes. Karadzic said, "There will be a Serbian republic once and for good, and anybody who wants to deal with us has to take that into account." Serbian General Ratko Mladic echoed the sentiment: "We are a . . . unified people living in the land of our grandfathers." That was not true. The Bosnian Serb state was founded on nationalism and violence, consisted of bombed cities and burned villages, and included many largely Muslim towns and tens of thousands of Muslim and Croat refugees.

Many Americans—myself among them—believed that President Clinton and some of his predecessors had been right to urge air strikes; right to support the lift-and-strike option; right to refuse to sign the Vance-Owen plan because it rewarded aggression; and right to promise that no U.S. ground troops would be committed to combat in Bosnia under UN command; but wrong to announce that we would not use airpower unilaterally. By July 1993, it seemed obvious that the Clinton administration was deferring to the British, the French, and the UN on these critical issues.

As John Burns reported in the *New York Times*, senior UN officials felt powerless to mitigate the suffering; the last hope of halting the killing and slow starvation of Muslims seemed to have disappeared, they said, when Clinton decided not to commit American military forces.[89] Peter Kessler, spokesman for the UN High Commissioner on Refugees, observed that the Serbs and Croats were behaving as though they "can do anything that they want."[90] In Europe, it was said that the U.S. president had turned his attention back to domestic affairs.

But some in the Clinton administration understood that abandoning an effort in which the human stakes were so high was certain to diminish the administration's credibility, its capacity to influence events elsewhere, and, above all, its reputation for leadership. Some understood that taking this path would diminish Bill Clinton's standing in the world. Some members of the administration understood that, more than anything, Bosnians needed arms.

In July 1993, approximately 315 American soldiers were deployed with a UN peacekeeping force on the border between Macedonia and Serbia. The UNPROFOR troops were concerned that the Americans might be too heavily armed and might violate the spirit of peacekeeping. Scandinavian commanders moved quickly to teach the American soldiers to surrender on demand, fire only in self-defense, and travel in small, light, personnel carriers instead of large, intimidating vehicles. The commanders were determined that the U.S. forces would respect the UN rules of engagement, which called for strict neutrality, minimal weaponry, and nonconfrontational behavior. Unfortunately, the Serbs were not operating under the same rules—as three American peacekeepers learned when they were kidnapped and held for several days.

ARMS SUPPLIERS

One thing the Bosnian Muslims and the Croatians shared was a need for arms to defend themselves against the relentless attacks of the Serbs. In 1992, a group of three Croatians, a Pole, and two Germans were arrested for smuggling $45 million worth of arms, mainly Kalashnikov rifles and U.S. Stinger missiles, into Croatia. A second shipment included a complete Soviet anti-aerial system that was carried in by way of bogus end-user certificates from Poland. (End-user certificates name the selling and recipient countries, but may be used to send the arms to a third country.) Soon Saudi Arabia, Iraq, Iran, and Turkey were sending weapons to the Bosnian Muslims, with end-user certificates saying they were going to Africa. Once at sea, freighters said to be headed for Africa, Asia, or the Middle East changed course and headed to Bosnia. Unfortunately, many of the arms intended for Bosnia were unloaded on Croatia's Dalmatian coast. Croatia had less difficulty than Bosnia in recruiting Western military advisers and buying Western arms. Germany was ready to help, as were Italians and some American freelance military trainers. When they had arms, Croatians and Bosnians fought well, but the arms were slow in arriving.

Almost as soon as the UN peacekeeping troops arrived in Croatia in 1992 (the first place they were deployed), the Croatian government had complained that the fourteen thousand UNPROFOR troops were not acting with sufficient force to compel compliance with the negotiated truce. The Croatians promised to "liberate every inch of Croatia" with their modern army,[91] and quite soon they demonstrated they could and would do just that. The Croatians had money, foreign connections, and help in training their forces. They had financial backers abroad, mainly Germans and Croatian émigrés. Croatia's long Dalmatian coastline enabled them to receive diverse weapons, including those intended for landlocked Bosnia. A tacit agreement among some nations (presumably including the United States, as well as Iran and other Muslim states) allowed $1.3 billion of weapons to be smuggled into the country. They were also helped by the presence in Zagreb of a number of employees of a private American military consulting firm who trained Croatian officers. Croatian officers attended a course offered by private companies and headed by former U.S. Army general Harry Soyster. The course, called "Military Professional Resources," was conducted by fifteen former U.S. Army generals, colonels, and master sergeants; it prepared the Croatians for the offensive in the summer of 1995 that wrested Krajina from the Serbs. This left an impression that the United States supported the building of Croatian strength.[92]

Lieutenant-General Jean Cot, the French UN commander, who served from June 1993 to March 1994, spoke harshly of the "Pontius Pilate solution"—peacekeepers bound by UN rules of engagement that demanded that they remain neutral in the face of starvation and mass murder.[93] These rules, Cot said, rendered peacekeepers impotent, "like goats tethered to a stake"—unable to keep peace, protect unarmed victims, or prevent the spread of conflict.[94]

Why, then, did the French and British so ardently oppose lifting the arms embargo that denied Bosnians the capacity to defend themselves? Why did French defense minister François Leotard oppose NATO air strikes even in the face of Sarajevo's agonies, saying (as if there were no urgency), "With regard to any eventual action on security zones, it should be preceded by a political consultation, first among Europeans, then with Americans, with the then commander of the UN forces directly involved"? Why did the Clinton administration wait so long before mobilizing NATO and urging air strikes? NATO had already agreed to support the Security Council if it voted to enforce the ban on Serbian flights over Bosnia and Herzegovina (which would have marked the first time NATO forces defended a country outside the alliance).

Although he was personally committed to effective action, Clinton appointed two secretaries of defense who were nearly as uncomfortable with the use of force as the UN Secretariat. The first, Les Aspin, declined to provide requested heavy armor for U.S. Rangers in Somalia. The second, William Perry, said that in 1994 although "air strikes are among the options being considered now . . . I can state categorically that we will not unilaterally conduct air strikes. We may not conduct them at all." Perry said that he and General John M. Shalikashvili, chairman of the Joint Chiefs of Staff, took "very seriously the limitations of air strikes against, first of all, artillery-type targets and, second, any targets that are embedded in a civilian population."[95]

The secretary-general, too, continued to oppose the use of airpower. Why did the secretary-general oppose the use of force in Bosnia so firmly, even after the Security Council had repeatedly authorized its use? Why did he guard so jealously his power to call in air strikes, and use it so infrequently? In a confidential letter to Christopher, Boutros-Ghali offered to "spell out the reasons for [my] misgivings . . . about the U.S. plan for air strikes . . . at a time and place of NATO's choosing." His argument was simple: The "first use of airpower in the theater should be initiated by the secretary-general," he claimed, "who alone has the right to initiate the use of force."

THE DUAL-KEY PRINCIPLE

Perhaps because they did not fully understand what was at stake, the Clinton administration accepted a "dual-key" principle that seriously hampered the ability to use airpower. The dual-key, or double-veto, system required that airpower be requested by the UNPROFOR military commander and approved (or, more often, denied) by the secretary-general's special representative, Yasushi Akashi. This chain of control was added to the already extremely restrictive rules of engagement, which prevented the use of force for retaliation and provided for no use of force except to provide air cover.

In its first few months in office, the Clinton administration formulated a policy on peacekeeping that it outlined in Presidential Decision Directive 13 (PDD-13), which provided the rationale for its policies concerning command and control of U.S. forces in peacekeeping operations.

The directive and policies reflected the administration's ambivalence about the use of force and, I believe, its failure to fully understand that the heavily armed Serbs could be deterred only by much greater force. The directive provided that, while the U.S. government retained *command* over its troops in peacekeeping missions, *operational control* could be turned over to the UN or some other multilateral body.[96] This "command without control" principle was also described in the *Army Field Manual*, which discussed a dual-command structure for peacekeeping operations. A U.S. force commander would have two chains of command above him. One chain made him answerable to the U.S. commander in chief, while a second chain integrated him under the UN force commander.

Walter Slocombe, deputy undersecretary for policy, outlined the complex UNPROFOR chain of command in testimony to the House Committee on Foreign Affairs:

> [An] UNPROFOR unit commander on the ground initiates a request for close air support through his forward air controller (who is trained in NATO procedures and accompanies the unit) to the Air Operations Coordination Center located at the headquarters of UNPROFOR-Bosnia. The Coordination Center approves the request. NATO fighter-bombers are then dispatched to the location of the unit under attack. . . . Note, however, that the UN Secretary-General would have to approve the first request for close air support; i.e., the Coordination Center would relay the first request to the Secretary-General. NATO has agreed that only the very first request would have to be approved by the UN Secretary-General. . . . [T]he . . . Coordination Center would approve subsequent requests for close air support on its own authority.
>
> [In the case of] UNPROFOR troops rotating into Srebrenica or opening Tuzla airport . . . the Coordination Center can refer the request for close air support to Mr. Yasushi Akashi, the Special Representative of the Secretary-General to Bosnia-Herzegovina; the UN Secretary-General has delegated close air support clearance to him.
>
> [These] procedures . . . for providing close air support to UNPROFOR, which NATO offered last June, are quite different from the procedures to be followed in the case of conducting air strikes, pur-

suant to the NATO warnings of August 2 and 9, in response to strangulation of Sarajevo or other population centers. In the latter case, the NAC [North Atlantic Council] and UN authorities would first have to agree that the situation on the ground constitutes "strangulation," and both the NAC and UN authorities would have to agree on when and where appropriate air strikes would be conducted.[97]

The dual-key arrangement caused delay and ineffectiveness, but many months passed before it was abandoned in favor of a more effective chain of command.

It was never clear why Boutros-Ghali thought he had the right to control the use of airpower in the first place. No such right had existed previously, certainly not in U.S. experience with wars authorized and carried out by the UN. It was not described in the UN Charter. Resolution 770 stated that the Security Council could take "all necessary measures nationally or through regional agencies or arrangements" to deliver humanitarian assistance in Bosnia-Herzegovina. But Boutros-Ghali interpreted the resolution as authorizing the Security Council to take such measures *as requested by the secretary-general*. Moreover, he insisted that Resolution 816, which authorized states and regional groups to use force if necessary to enforce no-fly zones, was contingent on the secretary-general's approval, and that Resolution 836, which authorized member states acting nationally or through regional organizations to use airpower to protect UN peacekeepers, only applied *if the secretary-general approved*.

Even after the Security Council had clearly stipulated otherwise, Boutros-Ghali claimed that he alone had the authority to initiate the use of force in Bosnia. However, from 1990 through 1993—while hundreds of thousands of Bosnians suffered and died from injuries and lack of food, water, and medicine—he did not use the power he claimed; rather, he prevented its use. It was a mystery why the secretary-general so consistently opposed the Security Council in providing assistance to the besieged Bosnians, and why the governments that were providing peacekeepers—Britain, France, and the Netherlands among them—accepted his decisions. At a June 1993 NATO meeting I attended, one of Boutros-Ghali's top deputies explained to a group of high-level, mainly European officials that, in the secretary-general's view, the conflict in Bosnia should be permitted to "play itself out," after which the international community should develop plans for reconstruction. He spoke as if he believed governments that stood by passively during the slaughter of Bosnians would later offer economic aid to Serbia and Croatia.

The Bosnian government repeatedly pleaded with the United Nations and the United States to lift the arms embargo and help them get weapons with which to defend themselves against the Serbian army and Serb irregulars. Serb shells landed on the safe area of Sarajevo at the rate of a thousand a day between April 1992 and August 1995, when NATO's Operation Deliberate Force joined the action. Yet the Secretariat called for fulfillment of UNPROFOR's mandate by persuasion, not coercion, and expressed its concern that "air strikes would pose grave dangers to UNPROFOR and the humanitarian convoys and, therefore, should be initiated with the greatest restraint and, essentially, in self defense."[98]

After consulting with the UN secretary-general and the commanders of NATO and the North Atlantic Council (NAC), NATO secretary-general Manfred Worner called for stronger measures. Stressing the importance of maintaining the distinction between defensive "close air support" and offensive "air strikes," Worner interpreted Resolutions 836 and 844 as permitting the use of airpower in self-defense, in response to bombardments of safe areas, in response to armed incursions into safe areas, and in response to Serb efforts to obstruct the movement of UNPROFOR forces or humanitarian convoys. After the attack on the market in Sarajevo in February 1994, Worner suggested that the accumulation of Security Council mandates had transformed the task, making UNPROFOR more of a player in the struggle. Worner warned that attacks on the Sarajevo, Gorazde, and Bihac safe areas demonstrated that Serbs were not deterred by the demarcation of safe areas, their demilitarization, or negotiations. He added, "The recent experience in Bihac has demonstrated once again . . . the inherent shortcomings of the current 'safe area' concept at the expense of the civilian population who have found themselves in a pitiable plight."[99]

General Sir Michael Rose, designated as the next UN commander for the area, arrived in Bosnia soon after the market attack.[100] Although Sarajevo was supposed to be a safe area, attacks on the civilian population continued unabated. General Rose was able to slow or stop the shelling of Sarajevo with an impermanent cease-fire in February 1994, but he seemed almost unconcerned about the continuing attacks on Gorazde, another safe haven whose population was being decimated.

UN military observers reported that as many as seventy thousand people were trapped inside Gorazde, where up to half the houses had been destroyed. Gorazde was the last fortified town in Serb-controlled east Bosnia. The attacks were described in some detail by ABC correspondent Peter Jennings and other journalists in the area, who spotlighted the conditions in Gorazde, where approximately sixty-five thousand Muslims had gathered after being driven out of smaller towns in eastern Bosnia. The UN High Commissioner for Refugees reported seven hundred dead and two thousand wounded in the area.

Rose said flatly that the figures were exaggerated "because they [the Bosnians] want us to fight their war for them. . . . I've long since ceased to believe the first reports I hear in this country."[101] Rose sent a Canadian observer to verify the reports, but instead of supporting Rose's doubts, the observer confirmed that a major offensive was under way, with many casualties. But although Rose had the power and responsibility to call in NATO planes, and the American general Leighton Smith was ready with planes and targets, Rose chose not to accept Smith's recommendation for more robust air attacks. He told Peter Jennings, "We are not there to impose a political or military solution on any party as peacekeepers. That is not a war-fighting mission. We are not there as war fighters."[102] When Rose finally called in air strikes against the Serbs attacking Gorazde, the strikes were so mild as to be harmless.

Air strikes were a long time coming. The groundwork had been laid in August 1992, when the Security Council passed Resolution 770, authorizing member states to take "all necessary measures to facilitate the delivery of humanitarian relief to Bosnia." In October 1992, Resolution 781 imposed a ban on military flights over Bosnia-Herzegovina but did not provide for enforcement. In November, Resolution 786 described the secretary-general's plan to monitor the no-fly zone. In March 1993, Resolution 816 authorized (but did not instruct) member states to enforce the no-fly zone. In August, Resolution 836 "authorized UNPROFOR . . . to take necessary measures, including the use of force in reply to bombardments against the safe areas" and decided that member states "in coordination with the Secretary and UNPROFOR could take all necessary measures, through the use of airpower in and around the safe areas, to provide support for UNPROFOR." [103]

By the fall, typhoid was spreading in the overcrowded refugee camps, even as fresh assaults by Serbian forces were creating new Muslim and Croatian refugees. In Croatia, Serb irregulars terrorized unarmed civilians in a new campaign of ethnic cleansing in towns previously pacified under a UN-sponsored peace plan. In Kosovo, the Serbs tightened the screws on the Muslim majority, many of whom abandoned their homes and possessions to seek refuge in Albania, a country whose meager resources were already strained.

As winter came to the region, snow and ice blocked the fleeing families. A week before Christmas, NATO agreed to support the United States if it decided to enforce the ban on Serbian flights over Bosnia. The previous summer, the United States had contributed 315 U.S. personnel and equipment to the UN peacekeeping operations and had been keeping U.S. ships in the Adriatic to help enforce the blockade. [104] At the beginning of 1994, Secretary-General Boutros-Ghali hinted that the UN was ready to use airpower to achieve some relief for the Bosnians, but UN and NATO forces stood by while their planes were shot from the sky. The problem was not with the soldiers or pilots, it was with their mission, their rules of engagement, and, above all, with the UN secretary-general and his special representative, Yasushi Akashi.

Clinton's policy in Bosnia had one virtue: it did not commit lightly armed U.S. forces to a war zone under UN rules of engagement. Clinton stuck to his decision that Americans would participate as ground forces in peacekeeping in Bosnia only when a cease-fire had been agreed to by all parties to the conflict. After the traumatic experience in Mogadishu in October 1993, Clinton understood that peacekeeping could be dangerous when there was no peace to keep, but he was not yet ready to use adequate force. I wondered why not, given the failure of peace operations in Somalia, Haiti, Rwanda, and, especially, Bosnia. In Bosnia and Croatia, the UN peacekeepers had served principally to inhibit energetic resistance to the Serb attackers. They had not prevented the sieges and destruction of the safe-haven towns (Srebrenica, Tuzla, Sarajevo, and Gorazde) on the eastern side of Serbia. Sometimes they had not even tried. The UN Secretariat tried intermittently to negotiate cease-fires to prevent the complete destruction of one civilian population after another. Sometimes the threat of NATO air interventions persuaded Serbian forces to move their big guns on to another town and another target. Sometimes it had no effect at all.

Humanitarian relief was not what civilians under bombardment most desperately needed. They needed weapons and allies. There was a great deal of confusion in diplomatic and UN circles regarding appropriate measures of response to Serbian aggression. But Article 15 in the UN Charter could not be clearer: All states have an inherent, unalienable right to defend themselves against aggression and to call on others to help them. Kuwait called for help when Iraq attacked, and got it. Bosnia-Herzegovina called for help when it was attacked by Serbia and the Bosnian Serbs—but all it got was an arms embargo. The UN rules of engagement protected no one. Security against the Serbian aggressors could not be provided by peacekeepers who drew their guns only in defense of their own lives. But the secretary-general was little more to blame for the situation than the UN member states that accepted those rules. The military commanders provided by France, the United Kingdom, Belgium, and the Netherlands understood that very different tactics were required to help the Bosnians.

The United States, the UN, and NATO could have lifted the arms embargo. In fact, the state for which the embargo was created, Yugoslavia, no longer existed. But the European members of UNPROFOR were unwilling to lift it as long as they had troops in the region. The United Kingdom, France, the United States, and other allies could have granted NATO the authority to enforce no-fly zones and safe havens and protect those delivering humanitarian assistance. But those things had not happened, so no one expected much from the decision to use NATO airpower in Bosnia, especially after the defeat of a resolution to exempt Bosnia from the arms embargo. That resolution won only six of the nine votes needed to pass. Cape Verde, Djibouti, Morocco, Pakistan, the United States and Venezuela voted in favor of the resolution. Brazil,

China, France, Hungary, Japan, New Zealand, the Russian Federation, and the United Kingdom abstained. France and the United Kingdom spoke against ending the embargo.

BOUTROS-GHALI'S POWER GRAB

When the heads of state at the NATO summit decided to use airpower to open Sarajevo and other Bosnian towns under siege, the secretary-general responded by asserting that he alone had the authority to decide when the first air strike could be made in any engagement. In a meeting with the Foreign Relations Committee of the French Parliament in Paris, Boutros-Ghali claimed that the Security Council gave him this extraordinary authority, which he refused to either use or delegate to the UN commander, the French general Jean Cot. The previous commander—the Belgian general Francis Briquemont—had resigned his command the week before, complaining of inadequate resources and authority. Cot, too, eventually went public with his frustrations, telling *Le Monde* that he had repeatedly requested the authority to call in air strikes, only to be turned down by Boutros-Ghali.[105] "It is not for generals to ask me for the authorization," the secretary-general asserted haughtily. "It is for the special representative. . . . If the special representative . . . requests it, I will be the first to accede to this demand." The position of special representative (held by Yasushi Akashi) does not exist in the UN Charter; he was appointed by and responsible to the secretary-general and had no military competence.

Of course, it was not the first time Boutros Ghali had claimed such power. Yet none of the Security Council resolutions he cited—770, 816, and 836—supported his position. None of these resolutions mandated bypassing the military commanders or delegated the decision to use airpower to the secretary-general. None required his authorization. On the contrary, the UN Charter vests all decisions concerning the use of force in the Security Council and the military forces of member states. Neither the Charter nor the Security Council makes the secretary-general the supreme commander of military operations. Although Boutros-Ghali himself was unsuited by training and experience to exercise such powers, he doggedly sought to prevent NATO from using the airpower authorized by the Security Council.

His actions had terrible consequences for Bosnia, virtually ending all hope of establishing collective security or effective deterrence of aggression through the UN. He undermined the UN's capacity to act, challenged the member states' right of collective self-defense, and denied the capacity of the Security Council to decide when aggression had occurred and when a serious threat to international peace and security existed. The secretary-general claimed nothing less than a personal veto over the legitimate use of force. It was the most sweeping power grab in the history of international organizations. It should have been firmly rebuffed, but it was not.

When the heads of state of the NATO governments announced that they would use NATO's airpower to lift the sieges on Sarajevo, Tuzla, Gorazde, and Srebrenica, they seemed decisive and credible. But all too soon it became clear that few of them actually expected air strikes to take place. The pretense that significant action has been taken when nothing has really happened was a characteristic of the UN style, in Bosnia and in subsequent international crises.

American pilots were fully prepared to carry out the air strikes, but there was no agreement on their authorization to do so. The same governments that Boutros-Ghali claimed had given him a veto over air strikes could have withdrawn that power. Yet the secretary-general's military power was simply accepted by key governments—notably the United Kingdom, France, and the Netherlands. If there were no air strikes to save the starving inhabitants of Sarajevo, Tuzla, Gorazde, Srebrenica, and the other safe areas, it was because the British, French, Dutch, and American governments did not act. Instead, the leaders of the Western world tried to work out a response to Serbia's savage violence that was acceptable to Serbia. Christopher sought the approval of French foreign minister Alain Juppé, who sought the approval of Christopher and British foreign secretary Douglas Hurd. Consensus was the central value in a situation where humanitarian concerns should have dominated.

A month passed as the struggle for consensus continued. After careful consultation with Milošević and Karadzic, the secretary-general finally proposed a solution for Tuzla. He suggested that Bosnian forces should surrender the Tuzla airport and that the Serbs should be permitted to place monitors at the airport to ensure that no military use was made of it. If more violence occurred, the secretary-general would authorize his special representative to approve a request to NATO for close air support from the UNPROFOR commander, with the understanding that "close air support" did not imply punitive or preemptive air attacks. Thus, surrender of more Bosnian territory would be a condition of the delivery of humanitarian assistance. The secretary-general continued to resist the effective use of force long after it had been authorized by Security Council resolutions; he was a major obstacle to saving the Bosnians.

On February 5, 1994, Serb artillery fired into a market in the heart of Sarajevo, leaving sixty-eight dead and two hundred wounded. The ground was littered with "bodies and pieces of bodies . . . like a butcher shop." [106] By this time, Serb shells had been battering Sarajevo for twenty-two months; the market had been bombed once before, in 1992. A local journalist said, "Here in Sarajevo we have been deceived once again. Promises of planes, food, and aid have been broken. Having lied to us, the criminals have continued to destroy Zepa, Sarajevo, Mostar, Jablanica, and Gorazde. As well as monuments, they have destroyed history, hope, and goodness. It is their aim to destroy everything down to the last Bosnian." [107]

In Washington, Paris, Brussels, and London, Western leaders told one another that something must be done. This time, the secretary-general seemed to agree. In a February 7 letter, he asked NATO to launch air strikes on artillery positions around Sarajevo if they were requested by the United Nations. [108] Did this mean that NATO would function under the command of Boutros-Ghali? The U.S. Congress would support air strikes as long as NATO planes remained under NATO's control, not the UN's. The European Union (EU) issued another declaration that suggested it was ready to support the use of airpower—although it did not specify air *strikes*, and Boutros-Ghali had insisted on a distinction between the two. The new U.S. secretary of defense, William Perry, said, "I can state categorically that we will not unilaterally conduct air strikes."

No fact-finding mission was required to determine that a member state of the United Nations had been the object of savage aggression. No further action by the Security Council was required to authorize whatever steps were necessary—including air strikes—to end the strangulation of Sarajevo, Tuzla, Srebrenica, Zepa, and the other Bosnian safe areas. All that was required was that the NATO countries have the will to use the powers they had been given.

A little more than three weeks later, on February 28, American pilots in F-16 fighters operating from a NATO base in Aviano, Italy, under the command of Admiral Jeremy M. Boorda, senior NATO commander for southern Europe, twice warned Serb planes in positions around Gorazde to leave. When four Serb planes persisted in violating the no-fly zone, the Americans shot them down. This was the first NATO military engagement in its forty-five-year history. "If it was a test, I think we passed," Admiral Boorda said.[109] Yet the fact that the American planes came from a NATO base made U.S. officials nervous, and they took "unusual steps to assure the Russian government that the incident was not intended as a hostile act toward Moscow," according to a *New York Times* report. "Thomas Pickering, the American ambassador to Moscow, was dispatched to convey the message personally to a representative of President Boris N. Yeltsin."[110]

IF CRIMES GO UNPUNISHED

By 1994, more and more Americans were convinced that compelling moral and strategic reasons existed for caring about Bosnia. By then, more than 100,000 Bosnians had been killed in the war, and more than 1.6 million people had been driven from their homes and villages. Many Americans believed that the United States could not remain indifferent to such pain without brutalizing ourselves; that we could not deny empathy to the Bosnians without dehumanizing them and us. It seemed ever clearer that our indifference to their suffering would encourage their tormentors, and others. As Sigmund Freud argued in *Civilization and Its Discontents*, crimes that go unpunished only incite other crimes and incite others to commit crimes. Commentators referred to the lessons of Hitler's seizure of the Rhineland, Mussolini's seizure of Ethiopia, and Neville Chamberlain's capitulation at Munich, which showed that ignoring, appeasing, and even rewarding aggression encourages more aggression.

The strategic argument for caring about Bosnia was as clear as the moral argument. The pattern of aggression and expansion in which

Serbia had already engaged—in Slovenia, Croatia, Bosnia, and Macedonia, and toward Vojvodina, Kosovo, and Albania—made it clear that Serb leaders were operating from a large appetite and an ambitious strategic plan that was especially dangerous at a time of great instability in Central and Eastern Europe. Restricted to Serbia, Milošević was a tyrant; permitted to expand, he would prepare the way for other tyrants.

Bosnia's fate had special significance for another reason: it was a test of the ability of the United States, the UN, NATO, the EU, and the CSCE to cope with an aggressor. By establishing diplomatic relations with Bosnia and admitting it to the UN, the United States and the EU had endowed it with the legitimacy and rights of states in the contemporary world. If they could not make collective security work for Bosnia, there was no reason to think they could make it work anywhere else.

Bosnia was an especially important test for the United States, the UN, and the EU. When the Security Council imposed an arms embargo that deprived Bosnia of the right to self-defense, it assumed responsibility for that defense. Again and again, the Security Council adopted resolutions promising the delivery of humanitarian assistance, the establishment of safe zones, and the protection of Bosnian towns against strangulation by sieges and annihilation from the air. The credibility of the major powers and of collective security itself were at stake.

Bosnia was also a test for the American government, which had assumed a special responsibility for the collective security structures constructed after World War II. We had planned them, vouched for them, and come to believe our own assurances. But the elaborate architecture of collective security was not proving up to the challenge. An international failure in Bosnia would be our failure as well.

Finally, Bosnia had become a test of the Russian-American relationship, in which the Bush and Clinton administrations had invested much. Russia's unexpected demand for a major role in a Bosnian settlement suddenly gave the problem a new great-powers dimension. But effective action proved elusive.

In the meantime, the UN peacekeepers suffered continued attacks, demonstrating how Boutros-Ghali's determination to control the use of force by UN troops had hamstrung operations. Take the following case, by no means unique: At 6:30 PM one evening in the fall of 1994, a Serb attack pinned down French peacekeepers near Bihac. At 7:00 PM, the peacekeepers called the UN commander for Bosnia, General Michael Rose, at his headquarters with a request for air support. At 8:30 PM, the request was transmitted to General Jean Cot, UN commander for the former Yugoslavia. At 10:40 PM, Cot transmitted the request to the secretary-general's special representative, Yasushi Akashi. A spokesman for the UN Secretariat claimed that the authorization for air strikes was granted at 11:35 PM, but that weather conditions made the strikes impossible. According to a NATO spokesman, the UN authorization never came—despite that French forces were in clear danger and NATO planes were overhead.

After several such episodes, including one a week later in which a French soldier was killed, French prime minister Edouard Balladur and defense minister François Leotard visited French forces in Bosnia, where Balladur demanded that procedures be promptly and definitively revised so that reinforcements could be provided when they were needed. "But," said foreign minister Alain Juppé, "the problem is not only delay. It is also will. One does not get the impression that UN representatives on the ground—those who represent the secretary-general of the United Nations—have a firm determination to use force each time that it is necessary."

Bihac was by no means an isolated problem. On February 22, 1994, Swedish peacekeeping troops near Tuzla came under heavy mortar attack and requested NATO air cover. General Rose relayed the request to General Cot, who was away. Cot's chief of staff presented the request to Akashi, who turned it down, saying the encounter was not serious enough to warrant a NATO air strike. Five Swedish peacekeepers were wounded. General Cot went public with the complaint that he had repeatedly asked for air strikes to protect UN forces, only to have his requests turned down by the UN representative on the ground. He made it clear that the situation of Bosnian towns under siege did not improve after NATO decided to undertake air strikes because the secretary-general's cumbersome rules of engagement and chain of command prevented the effective use of airpower. The French ambassador to the UN demanded that the power to call for air support be vested in military commanders. In Washington, Secretary of Defense William Perry agreed, describing the decision-making process as "torturous" and insisting that there must be a more streamlined command authority. Perry noted that Security Council resolutions permitted UN military commanders to request help directly from NATO headquarters and stated his expectation that, in the future, they would do so.

It was past time for the governments participating in UNPROFOR to assume responsibility for the security of their forces and the success of their operations. They needed to recognize that the secretary-general had neither the military expertise nor the authority to exercise command and control. The UN Charter vests authority over military operations in the Security Council, and the Security Council had already authorized the use of force in Bosnia-Herzegovina in three resolutions:

- Resolution 770 provided that all necessary measures could be taken nationally or through regional agencies or arrangements to deliver humanitarian assistance in Bosnia-Herzegovina. That meant that NATO could use necessary force to deliver food and medicine to civilian populations under siege. No further authorization was required.

- Resolution 816 authorized states and regional groups to use necessary means to enforce no-fly zones. No further authorization was required.

- Resolution 819 authorized member states acting nationally or through regional organizations or arrangements to use airpower to protect UN peacekeepers. That meant that NATO, for example—or France or the United States—was authorized to provide air strikes to protect peacekeepers when they came under attack. No further authorization was required.

Sometimes it seemed that Boutros-Ghali understood this. In a meeting with the Committee on Foreign Relations of the French Chamber of Deputies on January 11, 1994, he was asked: "If tonight or tomorrow NATO formally decides to undertake air strikes, what will be your attitude? Will you support that demand?" He waffled, replying, "I don't have this power. It is for the Security Council to accept or refuse, to authorize or not to authorize . . . It can give a mandate to NATO, or not give a mandate. . . . It is for them to decide."

Yet Boutros-Ghali continued his inept efforts to micromanage military matters. In mid-winter of 1994, for example, he replaced a French unit that had been in the Bihac area for months with Bangladeshi troops who were new to the region, inadequately armed and trained, and not supplied with winter clothing or footwear. Several Bangladeshi soldiers were immediately captured. (Three hundred and forty-nine UN peacekeepers were being held hostage throughout Bosnia at that time.) Serb officers promised to cease harassing Bangladeshi forces only if NATO ceased its air strikes against Serb forces. The captured Bangladeshi soldiers were bound, gagged, and forced to remain on an airfield without food or water for many hours. One soldier died of bronchitis, asthma, and exposure. More hostages were taken, including a Jordanian major who also became ill and died. Because the area was under siege, it was not possible to quickly provide warm clothes. Those who fell ill were denied medical treatment. Three UNPROFOR hostages were tied on the runway of the Banja Luka airport for eight hours in November 1994 after a NATO air strike targeted a Croatian airfield. Several UN observers in Croatia were denied food for twenty-four hours at a stretch.[111]

If Boutros-Ghali recognized that the authorization lay not with him but with the Security Council, why did he so often insist otherwise? The explanation, I believe, lies in the secretary-general's ceaseless efforts to increase his power. These efforts had become a huge obstacle to effective peacekeeping and war making. His redefinition of the peacekeeping mission and the UN rules of engagement sent lightly armed forces into war zones with no reliable arrangements for reinforcement or defense. This is what happened to French, Swedish, and Bangladeshi troops in Bosnia.

It was the U.S. government's acceptance of such rules of engagement for NATO operations that allowed Bosnian Serb forces to shoot down an American F-16 and its pilot, Captain Scott O'Grady, on June 2, 1995. O'Grady was ultimately saved by his own initiative, stamina, and good luck, and by the determined efforts of his rescuers.[112] In the days after the F-16 was shot down, the U.S. government neither retaliated nor expressed much outrage over this deliberate targeting of an unprotected U.S. plane on a routine, nonviolent mission.[113] The Clinton administration gave no one, including the offending Serbs, any reason to fear American displeasure. The Serbs had been permitted to install the SAM missile batteries that brought down O'Grady's plane in the area patrolled by American planes. Then U.S. planes were sent up without the protection of readily available, highly effective electronic equipment that provides notice to fighter pilots that they are being targeted by missiles.

It was past time for the members of the Security Council to accept responsibility—not just for providing troops, but also for ensuring adequate weapons, realistic rules of engagement, and competent military commanders.

The essential elements of the Bosnian conflict were changing. On August 28, 1995, the Bosnian Serbs carried out a third brutal attack on Sarajevo. Abandoning its usual neutral posture, the Security Council declared that beyond a reasonable doubt the Bosnian Serbs were responsible. The secretary-general, who habitually opposed the use of force regardless of the provocation, flatly condemned the attack. UN peacekeepers quietly left Gorazde to ensure that they would not be taken hostage.

The NATO attacks continued and intensified. From September 10–20, 1995, thirty-four hundred sorties were flown. Communication and transportation resources and military stores were targeted. At the same time, the now-adequately-armed Croatian and Bosnian forces began to sweep Serb forces out of land they had captured. Both Croats and Muslims wanted to keep fighting, but Washington insisted that the war should end.

After the second wave of bombing hit Sarajevo in February 1994, Christopher signaled to our associates that we would act only in conjunction with them, asserting "there are atrocities on all sides." A new joint action plan was organized by the United States, Russia, France, the United Kingdom, and Spain as a containment plan; it called for sealing Bosnia's borders and establishing Muslim safe areas. Clinton himself became active in the search for a settlement and was prepared to settle for a three-way partition with terms dictated by Serbs and Croats.

The United States undertook the effort to negotiate peace and a unified state but was unable to do so. In fact, a war seemed likely between the Muslims and the Croats. For a while, Muslims permitted the Serb forces

(JNA) to assume a number of policing functions. Then cooperation be-tween Croats and Muslims broke down almost entirely, and the Croats, while still formally allied with the Muslims, effectively became allies of the Serbs; they were complicit in some of their most brutal attacks, in-cluding the siege of Srebrenica, and participated in creating more refu-gees and more misery for the Muslims. Croat militia members murdered dozens of Muslim civilians in Ahmici and other villages. Milošević and Croatian president Franjo Tudjman plotted settlements that would leave less and less land for Muslims and press them into smaller and smaller areas.

In March 1994, under pressure from Washington, Bosnian Muslims and Croats stopped fighting and agreed to form a federation. Coopera-tion between these two republics of former Yugoslavia got under way after Tudjman transmitted a message that Iran was ready to ship arms to Bosnia. The Clinton administration quietly acquiesced, without inform-ing its allies or Congress. Arms shipments began in April 1994, and from that point on, Bosnian forces received a steady supply. The U.S. govern-ment delivered no arms itself, but covert support was given by Islamic nations—including Iran. The arms were accompanied by small but in-creasing numbers of mujahideen.

The Bosnians undertook offensives that opened roads to Tuzla and defeated Serbs in Bihac, which was under heavy shelling. Soon the newly armed Bosnian forces began to win battles and undertake more military initiatives. Croatia and Bosnia won battle after battle, and by 1995 con-trolled most of Krajina.

Tension over the continued lack of a UN response to Serbian aggres-sion had been building, with Americans calling for more vigorous use of airpower, implementation of the Security Council resolutions, and more determined use of force to punish and deter the attacks. Congress had made repeated calls to lift the arms embargo and permit Bosnia to de-fend itself.

As 1994 drew to a close, it seemed clear that it would not be easy to heal the rift between the U.S. government and its NATO allies on the matter of Bosnia. The NATO military operation, which had been under way since April, revealed differences that were broader, deeper, more un-pleasant, and more important than anticipated. Many Americans had been surprised by how passively the British and French greeted the carnage in Bosnia, and by Boutros-Ghali's opposition to the use of force to stop the brutal aggression. These differences would not be easily overcome; nor would the differences between the United Nations and its critics in Congress.

In November 1994, against European objections, the Clinton administration announced that it would no longer enforce the arms embargo. That same month, Republicans won control of Congress; Majority Leader Robert Dole and House Speaker Newt Gingrich announced their support for the lift-and-strike policy, and noted that a congressional majority supported their position.

The following month, former president Jimmy Carter negotiated a four-month cease-fire. At its end, the Bosnian and Croatian governments undertook more military offensives after the Serbs attacked safe areas and took several hundred UNPROFOR troops hostage. UN negotiators struck a deal with the Serbs to release the hostages in exchange for a promise that there would be no more NATO air strikes against Serb forces.

Each of the principals had his own priorities and goals. Radovan Karadzic, leader of the Bosnian Serbs, understood that his forces and resources were dangerously thin and that he was likely to be betrayed by Milošević, whom he accused of treason. "You have turned your back on the Serbs," he charged, having concluded that Milošević would do nothing to help them. He was right. Milošević was ready to abandon the Bosnian Serbs as part of his plan to achieve the lifting of the sanctions on Serbia imposed since 1992. The Croatians were preoccupied with extending their control over Krajina, an area that Clinton's national security advisor, Anthony Lake, had targeted as a center of a future state of Bosnia-Herzegovina, which he was eager to cobble together out of the Muslim-Croat federation and the Bosnian Serb ministate.

Richard Holbrooke, the assistant secretary of state for Europe, understood that the task of dealing with Milošević was his, and he relied principally on American airpower and tough talk. Milošević was furious with the Bosnian Serbs and warned them that he was ready to cut them off. Finally, and very reluctantly, the Bosnian Serbs agreed that Milošević could negotiate for them. NATO began heavy bombing on August 31,

1995, flying thirty-four sorties against the Bosnian Serbs over a period of two weeks.

Milošević asked the Contact Group to stop the bombing. In return, Holbrooke asked Milošević to accept a division of Bosnia-Herzegovina into two parts, with the Muslim-Croat Federation taking 51 percent and the Bosnian Serbs 49 percent. When Bosnian Serb general Ratko Mladic and Karadzic rejected the idea, Holbrooke walked out. Milošević told Mladic and Karadzic that NATO would destroy their forces if they continued to hold out against Holbrooke's proposal. Then, just as it appeared that the conference was about to break up, leaving Croatia and Bosnia in a stronger position, Washington applied more pressure, this time leaving Mladic and Karadzic to face the prospect of fighting alone.

In November 1995, when the United States and the other Contact Group members finally brought the warring factions together for peace talks, the Croats' top priority was eastern Slovenia. The Bosnian Serbs were determined to split Sarajevo. Silajdzic refused to meet with them and told Milošević he wanted Gorazde as a symbol of Bosnian presence on the Serbian border. Milošević capitulated. There would be three states: one Serb, one Croat, and one Muslim.

The four-month-long truce helped, but it did not prevent repeated attacks on Bosnian safe areas. The next month, another NATO bombing led to Serb withdrawal from Gorazde and marked the end of the total vulnerability of the Bosnian Muslims.

In Bosnia, clashes continued between the Americans and the secretary-general's special representative. One incident concerned Gorazde. NATO had declared a deadline of April 23 for the Bosnian Serbs to pull back their troops, but when the Serbs failed to comply, Akashi blocked the promised NATO air strikes. The United States complained that Akashi allowed Bosnian Serb tanks through to Sarajevo in clear violation of an understanding that no tanks would be permitted in the city.

CLINTON'S DISTASTE FOR
U.S. UNILATERAL FORCE

The long-awaited NATO air strikes on Serb positions around Gorazde, which began on April 10, 1994, demonstrated that the White House was ready to use limited airpower to achieve limited objectives. But President Clinton's comments on the strikes sounded more like a disclaimer than a statement of purpose: "This is a clear expression of the will of NATO and the will of the United Nations," he said, as if the United States had no voice or responsibility in these decisions. "We have said we would act if we were requested to do so. We have now done so and we will do so again if we are requested."[114] One can only wonder why Clinton chose these words. Did he regard the American pilots' mission in Gorazde as more legitimate if it was specifically authorized by the UN? Was he trying to suggest that attacks by American pilots in American planes should be seen not as Americans but as representatives of a multinational force acting on behalf of a multinational body? Clinton's language was a clear reminder that many of his key advisers had a longstanding distaste for unilateral American use of force, and put their faith instead in a policy of active global multilateralism.

Among the most prominent statements of this policy was the 1992 essay "Military Action: When to Use It and How to Ensure Its Effectiveness" by William Perry, soon to become Clinton's secretary of defense. In the essay, which developed themes from an earlier article he coauthored with Ashton Carter (then assistant secretary of defense for policy) and John Steinbruner of the Brookings Institution, Perry advocated renouncing the use of American force in favor of a policy of "global engagement" through an international police force. Perry's recommendations were straightforward: all nations should reduce their military forces to those required for defense of their own territory, except for the United States and a few other major states, who would maintain some additional forces to supplement multinational forces as needed. Any new international aggression would then be deterred by those multinational forces, rather than any individual state. Though a Pentagon spokesman denied that Perry would follow its prescriptions as secretary of defense, their relevance to his performance was obvious.

Perry's proposals resembled those of other members of the Clinton team, including Morton Halperin (then at the National Security Council) who recommended that U.S. armed forces be drastically reduced and the unilateral use of force renounced. Like Perry, he looked to multinational forces to meet the occasional challenges created by outlaw states.

Halperin emphasized that the United States should act in the world through the UN, and he expected that national defense would become the defense of world peace. Peace, he anticipated, would be protected mainly by the moral force of a united world community. "The threat of military force should be sufficient to obviate the need to use it if the right military and political conditions are met," Perry wrote.[115]

The vision of investing less in national defense and more in international peacekeeping was emphasized in the pronouncements of various Clinton officials. In a June 15, 1995, speech before the Philadelphia Bar Association, Madeleine Albright emphasized (and, I think, exaggerated) the diverse tasks that could be performed by a UN peacekeeping operation, which, she said, could "separate adversaries, maintain cease-fires; facilitate the delivery of humanitarian relief; enable refugees and displaced persons to find homes; demobilize combatants and create conditions under which political reconciliation may occur and free elections may be held. It can help to nurture new democracies; lower the global tide of refugees; reduce the likelihood of unwelcome interventions by regional powers; prevent small wars from growing into larger conflict."[116] Spurred on by this vision, American participation in peacekeeping had spread rapidly into all the areas where the Clinton administration found cause to use force to achieve a goal.

In the meantime, the costs of these missions were mounting. On March 8, 1995, the General Accounting Office (GAO) reported to the House of Representatives that United States had deployed the following troops:

- 26,000 troops to the peacekeeping mission in Somalia

- 14,000 troops to monitor repression of the population in southern Iraq

- 11,700 troops to enforce the arms embargo in former Yugoslavia

- 20,000 troops on what the Clinton team called "returning Haiti to democracy"

- 2,000 troops to enforce the no-fly zone in Bosnia-Herzegovina

- 1,000 troops to help provide humanitarian assistance in Bosnia

- 1,500 troops for the security of safe havens for the population in northern Iraq

"[A]s the number, size, and scope of peace operations have increased dramatically in the past several years," the GAO report observed, "the nature and extent of U.S. participation have changed markedly. Recently, the United States has used much larger numbers of combat and support forces to respond to events in a number of locations." The title of the report: "PEACE OPERATIONS: Heavy Use of Key Capabilities May Affect Response to Regional Conflicts."[117]

More Busy Nonresponse

For all the expenditures the U.S. government was making around the world, however, the international response to Bosnia was still anemic compared to the challenges it faced. It takes a strong stomach to watch a town ˙encircled; its population—swollen with refugees—bombed, strafed, and picked off by snipers; its crowded hospitals targeted; its water and electricity cut; and its inhabitants denied the food and medicine waiting for them just beyond the big guns. Yet, time and again, this is exactly what the UN and NATO forces managed to do: stand by and wait while the Serbs surrounded Bosnian towns.

As embodied in 1993's PDD-13, the Clinton administration's policy committed the United States to broader participation in global peacekeeping. The discussions regarding this participation raised many questions, because in Bosnia, for example, peacekeepers had repeatedly inhibited efforts to protect Bosnians rather than inhibiting Serbian attackers.

Only determined action could have stopped the spread of violence in the former Yugoslavia and restored credibility to the UN. It was clear that continuation of the feckless policies applied thus far in Bosnia would negate the idea of collective security for another generation. But NATO's belated decision to resist Serb violations of UN-declared "safe areas," including Sarajevo, provoked a violent Serb reaction.

For months, the UN forces had tolerated Serbs blocking the delivery of food and medicine and reclaiming their heavy weapons from UN custody. But eventually the situation of UN peacekeepers and the civilian Bosnian population had deteriorated so badly that action was required. On May 25–26, 1995, NATO planes dropped bombs on Serb ammunition dumps near Pale. Although the targets had no military importance, the bombing enraged Bosnian Serb leaders, who responded with rocket attacks on downtown Tuzla that killed seventy civilians. Serb shelling of Sarajevo was stepped up, and serious hostage taking began.

In June 1995, a group of Bosnian Serbs—dressed in stolen French uniforms and the blue berets of UN peacekeepers—infiltrated UN lines, seized a bridge in Sarajevo, and stole a half dozen UN tanks, two dozen armored personnel carriers, and assorted other vehicles and supplies. By week's end, the Serbs held four hundred hostages, mainly British and French UN peacekeepers. Many of the hostages were chained in exposed positions as human shields to prevent another NATO raid.[118] The foreign ministers of France, Germany, Great Britain, the United States, and Russia, who happened to be meeting, called this an "outrageous act." The French proposed an international rapid reaction force to reinforce the peacekeepers. The British dispatched another fifteen hundred troops and put additional troops on alert. Clinton spoke of sending American ground forces to help evacuate or reposition UN troops. Defense Secretary Perry said that U.S. ground forces might be sent only as "part of a NATO operation in order to extract UN forces that are in danger." Such an operation, he said, would have to be under NATO command and authorized by Congress.

A few days earlier, on May 26, NATO secretary-general Willy Claes had made clear that air strikes would come only at the request of the UN and for the purpose of helping peacekeepers. "NATO has no intention to take sides in the conflict," he said, "but will continue to act within the framework of the UN Security Council resolutions and the ongoing efforts of the international community to achieve a negotiated solution." It was another ludicrous invocation of "NATO neutrality." Claes also said that NATO's actions were precipitated by "persistent and flagrant violations of the safe areas by the Bosnian Serbs . . . and represent a threat to the viability of the UN mission and a challenge to the will and the credibility of the international community."

Finally, after forty-eight hours of delay, the Security Council passed a resolution, authorizing NATO to take "necessary measures" to enforce the no-fly zone, "subject to close coordination" with the secretary-general and the UN peacekeeping forces. With the UN resolution in hand, the permission of the president of Croatia, and a resolution passed by the sixteen ambassadors to NATO, Claes instructed General Bertrand I. Lapresle, the military commander of UN forces in Bosnia, to take appropriate military action. After consultation with the secretary-general's special representative, NATO authorized a limited attack on the Serb airstrip in Croatia, taking care not to destroy Serbian planes, hangars, or vehicles. This was not exactly decisive force. Admiral Leighton W. Smith, NATO commander in Southern Europe, said of the raid: "If I wanted to put that airfield out of commission, and to make sure nothing ever took off from it again, we would have taken out all the aircraft. . . . We would have hit their ammunition dumps and we would have taken out all the buildings anywhere around that airfield."

The halfhearted attack only emboldened the Serbs. Concluding that NATO was incapable of acting forcefully, Serb forces resumed their attacks, bombarding Bihac and the surrounding villages again with tanks and a helicopter gunship. Surface-to-air missiles were fired at British planes, and no significant NATO response was forthcoming. Bihac was soon completely surrounded. French foreign minister Alain Juppé said the events in Bosnia raised serious doubts about whether NATO could ensure European security in the post–cold war world. "Never has NATO appeared so little capable of maintaining security on the old Continent. Never have events in Bosnia shown it in so bad a light."

Spring and summer of 1995

By the spring of 1995, three Muslim towns that had been declared safe zones in 1993 remained outside the control of the Bosnian Serbs: Srebrenica, Zepa, and Gorazde. They had been shelled and starved and were full of refugees from the surrounding towns. Each had given up its weapons as the price for becoming a safe area under UN protection. They were "protected" by a skeleton crew of two dozen Dutch soldiers, who were themselves surrounded by well-armed Serb troops.

The events that followed, in the spring and summer of 1995, have been investigated by the Dutch and French parliaments, various journalists, and scholars seeking to understand what led to the mass murder of some eight thousand men and boys in Srebrenica, and the rape and brutalization of many of the girls and women. The details of the massacre of Srebrenica are so horrific that in 2002, after reviewing a report on the incident, the entire Dutch cabinet resigned in shame.[119] "Someone must take responsibility" for the thousands killed in the massacre, one official said, "and no one else was willing to do so."

On July 6, 1995, the Serbs moved on Srebrenica, mounting a fierce attack with tanks and artillery. The Dutchbat commander made a series of requests for close air support, first in Sarajevo, then in Srebrenica. Supposedly, the commander was told by the UN that NATO planes would soon arrive to conduct air strikes. But NATO planes were never called in. Later investigations indicated that Akashi, who reported directly to Boutros-Ghali, did not inform him or other top UN officials of the requests.

On the afternoon of July 6, approximately twenty thousand persons—mainly women, children, and old people—converged on the Dutchbat headquarters in Srebrenica, demanding air strikes. But there were no calls and no air strikes. The commander cut a hole in the fence, and four to five thousand people came through, under the illusion that it was safer inside the compound than outside it. Another fourteen to fifteen thousand refugees remained outside the compound. Thousands more unarmed refugees sought to make their way to Tuzla. On July 8, the Dutch peacekeepers abandoned three posts under direct fire and again requested air strikes. On July 10, close air support was finally requested.

On July 11, the Serbs took control of Srebrenica. Mladic insisted that Muslim boys and men between the ages of seventeen and sixty must be disarmed and then questioned one by one. He said they would be well treated. In fact, virtually all of them were slaughtered. On August 8, the massacres were described in a *Newsday* article. On August 10, Ambassador Madeleine Albright showed her UN colleagues aerial photographs

of Bosnian Serbs killing hundreds of men and boys held captive in a soccer stadium.

Akashi later insisted that he had no advance intelligence of the
planned attack, but he had many warnings. The Bosnian government
had urgently and repeatedly provided him with accurate information
about what was happening, including the forced departure of thousands
of Bosnian males from their homes and villages. Later investigations
confirmed that UN and U.S. intelligence had detected signs of an upcoming Serb offensive, but not of Serb intentions to annihilate thousands
of Bosnian males. The Bosnian government, better informed, had inter-
cepted Serb radio communications describing plans for mass murder,
but no one else was taking these reports seriously. (That the killings were
planned in advance has since been confirmed by the testimony of Bosnian Serb officers at The Hague.[120])

UN officials said they did not regard Bosnian government intelligence as reliable, although they later acknowledged that Bosnian officials
had accurately reported the systematic slaughter of Bosnian men and
boys. UN and U.S. government officials later acknowledged that they had
too little confidence in information from the Bosnian government.[121]

A year later, *Newsday*'s Roy Gutman compiled the following timeline
of events:

JUNE 4 French general Bernard Janvier, supreme UN military commander for former Yugoslavia, meets with
Bosnian Serb general Ratko Mladic to discuss the release of UN hostages and an end to NATO air strikes

JUNE 7 Serbs release 111 peacekeeper hostages

JUNE 9 Special representative Yasushi Akashi announces that
the UN will abide by "strictly peacekeeping principles" (i.e., no use of force)

JUNE 13 Serbs release 118 more hostages

JUNE 17 Serbs release the remaining UN hostages

JULY 6 Serbs attack Srebrenica

JULY 11   Serbs capture Srebrenica. They drive men and boys out of the town, then slaughter them

JULY 12–18   Serbs kill approximately 7,000 to 9,000 men and boys from Srebrenica in cold blood

AUGUST 8   *Newsday* reveals the massacres

AUGUST 10   U.S. ambassador Madeleine Albright shares CIA photographs with UN colleagues as proof of the mass executions[122]

The Dutch UNPROFOR contingent that was supposed to be protecting the refugees was very lightly armed. The Serbian forces, on the other hand, were ready for war—with tanks, armored vehicles, heavy artillery, communications, and intelligence. For months, the Bosnian Serbs had blocked all deliveries of food, fuel, and spare parts to the Dutch troops. These conditions led the Dutchbat force commander to conclude, "My battalion is no longer willing, able, or in the position to consider itself impartial due to the policy of the Bosnian Serb government and the BSA [Bosnian Serb Army]." The commander spoke repeatedly with General Mladic, appealing for help. Meanwhile, progressively more desperate refugees were gathering in Srebrenica: Four thousand were already there, and their numbers were swelling rapidly. In a letter, Boris Yeltsin placed blame firmly on the aggressors:

> This long-lasting and severe situation is no longer acceptable for the soldiers. Therefore, it is my strongest opinion that this Bosnian Serb government should be blamed for it in the full extent, as well as for the consequences in the future.

Apparently no one received Yeltin's message until much later.[123]

French General Bernard Janvier, supreme UN military commander for former Yugoslavia, opposed the use of NATO airpower even before he negotiated the release of UN forces held hostage by the Serbs. By most accounts, it was Janvier who struck the deal with Serb commander Ratko Mladic that NATO would halt air attacks and the Serbs would release the hostages. As U.S. ambassador to Bosnia John Menzies later confirmed, they agreed over the opposition of the United Kingdom's General Rupert Smith, who favored using more force to try to save the Bosnians. General Smith argued that analysis of Serb behavior indicated that their "intention is to finish the war this year and take every risk to achieve it. They will destroy the eastern enclaves this year." Javier believed that the Serbs' desire for international recognition would prevent them from going to such extremes.

According to the *Newsday* accounts, Javier told UN headquarters that Bosnia's forces were adequate to defend Srebrenica (which they were not) and that NATO airpower was not needed (which it was). Akashi, always opposed to the use of force, informed the Serbs that UNPROFOR would strictly observe peacekeeping principles (that is, refraining from the use of force). Another, divergent account of the deal asserted that it was an agreement not between UNPROFOR forces and Bosnian Serbs, but between UN officials and Bosnian Serbs. Richard Holbrooke commented, "To this day, Washington has never been sure of what was actually agreed to, but after the hostages were released, the intensity of Bosnian Serb military effort increased dramatically, with no further UN or NATO air strikes."[124]

After the release of the UN hostages, word circulated of a secret deal between Javier and Mladic, under which UNPROFOR commanders agreed never to ask for NATO airpower again in Bosnia. This agreement was publicly affirmed by Milošević and the Bosnian Serbs, but denied by French and UN officials. In fact, however, no further NATO air strikes were carried out. Boutros-Ghali removed from General Smith the authority to call for NATO air strikes, saying that he would "personally make all further decisions . . . from New York."[125] There were no further calls for air strikes.

The *Washington Post* later ran a story stating that twelve thousand Bosnian men had set out on foot from Srebrenica at dawn on July 11 in an attempt to escape death at the hands of the Serbs. Fewer than half survived the march. The article described five or six massacre sites at which large numbers of Bosnian men and boys were forced to dig trenches. They were then shot and buried. One Bosnian man, who had observed the process, from a hiding place said that the slaughter and burial were systematic and utterly brutal, and that General Mladic was personally in charge of the entire operation. This eyewitness said that "the Dutch peacekeepers made little attempt to defend the civilian population. They were worried about their own hostages." The account continued: "Refugees later described how many of the peacekeepers were forced at gunpoint to strip to their underwear by Bosnian Serb soldiers, who then strutted around in UN uniforms themselves."[126]

A very few men and boys managed to survive by pretending to be dead, hiding among the corpses. No precise number of those killed is available; the common estimate is approximately nine thousand. Later investigations by a French parliamentary committee and by the Dutch government confirmed that the men and boys of Srebrenica were murdered in cold blood over a period of several days. They were undefended, with no assistance from the UNPROFOR peacekeepers, who seemed less interested in protecting Bosnians in a UN safe haven than maintaining satisfactory relations with the Serbs and protecting their own security.

The Serb capture and mistreatment of UN peacekeepers, and the slaughter of the undefended male population of Srebrenica in July 1995, were deeply humiliating for UN forces. It was the most shocking mass murder in Europe since World War II, and the widespread international revulsion that followed finally mobilized NATO troops to action. French president Jacques Chirac (who had approached the British and been turned down) called President Bill Clinton to propose jointly planned responses in the future. The two agreed to abandon the UN's "proportionate response" and dual-key strategies if the Bosnian Serbs continued to attack Muslim towns. Henceforth, NATO would be authorized to launch air strikes on the request of the UNPROFOR force commander without seeking approval from the UN secretary-general. NATO carried out thirty-four hundred sorties against the Bosnian Serbs in less than two weeks

The Strain on U.S.-European Relations

The Bosnian conflict challenged the Euro-American relationship, because Americans were generally more ready than Europeans to use force to defend the Bosnians. More Americans than Europeans saw Bosnia as the victim of Serbian aggression, ethnic cleansing, and conquest, and wanted to help—although they did not want their country to become involved in a ground war. Many Americans disapproved of the UN's posture of neutrality between aggressors and victims, and did not believe it was morally permissible for a UN member state to be denied the right of self-defense through an arms embargo. The disagreements primarily concerned the arms embargo, the use of airpower, and the right of peacekeepers to fend off Serb assaults. American impatience grew with the UN's inefficient policies, cumbersome chains of command and control, and unrealistic rules of engagement.

Tension over the lack of response to continued Serbian aggression had been building, with Americans in and out of government calling for more vigorous implementation of the Security Council resolutions and more determined use of force to punish and deter armed attacks. A bipartisan coalition in Congress called for lifting the arms embargo to permit Bosnia to defend itself. But Secretary-General Boutros-Ghali had established his control over NATO air strikes, and he used that control to block the use of air strikes against the Serbs.

In an effort to impose a settlement, the Contact Group issued an ultimatum: if either side did not accept the latest proposed plan, disincentives (including military force) would be used. This was a take-it-or-leave-it approach. General Michael Rose called Admiral Leighton Smith in Naples and asked him to put NATO aircraft overhead, which Smith did within five minutes.[127] Akashi denied having turned down a NATO request for permission for an air strike, saying that "NATO will take the lead on this, because the violation of the no-fly zone and the direct attack on its planes are direct challenges to the alliance." But permission to use airpower was never granted.

Of the growing differences between the Americans and the Europeans, Ambrose Evans-Pritchard wrote in the London *Daily Telegraph*: "Something has finally snapped in the relations between the United States and Britain. The irritation that has been festering over Balkan policy for three years has reached the point of irreparable rift."[128] NATO was finished, Pritchard opined, and with it the intimate British-American relationship that had existed since World War II.

There was no enforcement of Security Council resolutions, no enforcement of no-fly zones or of safe havens, little delivery of humanitarian assistance, no deterrence of attacks, no protection for peacekeepers or refugees, and no enforcement of NATO ultimatums. Each deadline set by NATO and UNPROFOR was missed, each safe haven violated, each effort to alter the UN mode of operation rebuffed, and each decision to alter it ignored. Humiliation of UN forces was tolerated and thus intensified, and attacks on civilians spread. NATO was permitted to attack only a runway and three missile sites. This extremely weak response was defended as necessary to protect UN peacekeeping forces. To U.S. complaints, French, British, and UN officials essentially replied that the United States had no standing because it refused to contribute troops to UNPROFOR.

In the fall of 1995, the Clinton administration reluctantly announced that it would no longer enforce the arms embargo. At that point, an open rift developed between the United States and the British and French, who critiqued the American withdrawal as unhelpful and inappropriate behavior for a country that had no forces on the ground. The WEU pledged to maintain the arms embargo, and Willy Claes announced that NATO would continue to enforce the embargo because it had been mandated by the UN.

Tensions multiplied after the Security Council authorized NATO to attack Serb targets in Croatia. UN officials debated and delayed, and finally permitted fifty NATO planes to attack three Serb missile sites in Croatia in a pinprick effort to slow Serb aggression against Bihac. Things got worse after NATO failed to endorse a U.S. plan to save Bihac. Two days later, NATO joined UN officials in announcing that nothing would be done to mitigate the Serb assault on Bihac.

The Move to Dayton

The war had taken a hideous toll on the people of Bosnia. By the end of 1994, at least two hundred thousand people had died and roughly half of Bosnia's four and a quarter million inhabitants had been driven from their homes.[129] Possessions had been seized and destroyed, property confiscated, and all deeds and records that established ownership burned. Many males between the ages of sixteen and sixty had been killed. The devastation virtually guaranteed that for many there would be no home to return to, yet many still longed to return home.

The changed balance of power on the ground gave Milošević an interest in negotiating to end the fighting. Serb supremacy in the war had rested on military strength. The Yugoslav National Army (JNA) had been one of the strongest and best equipped on the European continent. It was not exclusively Serb, but when Yugoslavia split into pieces, the largest part of the army remained with Serbia. At the beginning of the war, Croatia and Bosnia lacked the men and arms to defend themselves, and the 1991 UN arms embargo had made it nearly impossible for them to acquire arms and military training. But Croatia had money and connections, and once the Croatians found ways to buy weapons and hire military advisers, they were able to defend themselves and help Bosnia gain access to arms. It was because of this that the Serb monopoly on power was ended.

Meanwhile, the United States, France, and NATO began to use their overwhelming advantage in airpower and tightened the economic sanctions against Serbia. Milošević had important incentives to end the war. The split between Milošević and his Bosnian Serb allies, Karadzic and Mladic, had become public. Milošević wrested from them an agreement that he alone would represent all Serbs in the negotiations. Holbrooke had categorically refused to meet with Karadzic and Mladic, who had been indicted as war criminals by the international tribunal for the former Yugoslavia in July 1995.

Preliminary negotiations included a reluctant agreement from the Bosnian Serbs that Bosnia-Herzegovina would consist of two entities: 51 percent of the territory would go to the Muslim-Croat Federation and 49 percent to the Bosnian Serbs. Russian president Boris Yeltsin helped by writing a letter to the Serb leadership guaranteeing that he would send Russian troops into areas from which the Serbs withdrew, so they could not be quickly occupied by Muslim forces. NATO officials helped with assurances that the NATO commanders would control the use of weapons around Sarajevo. Bosnia-Herzegovina would have a collective, rotating presidency. The Bosnian representative, Alija Izebegović, would take the first eight-month term, followed by the Serb, Momcilo Krajisnik, and then the Croat, Kresimir Zubak.

The Serbs wanted to divide Sarajevo, but the Muslims refused to discuss it. The Americans pressured the Croats and Muslims to accept a cease-fire and create a viable Muslim-Croat federation. Finally, the three parties agreed to the basic provisions for the creation of Bosnia-Herzegovina and also maintain the October 5, 1995 cease-fire.

The Bosnian president, Izetbegovéc, clung tenaciously to the right of Muslim refugees to return to their homes, thus securing their hold on eastern Bosnia. For a time, the conflict between the Muslims and the Croats became almost as bitter and as murderous as the earlier conflict between Muslims and Serbs.

Then the Muslim foreign minister, Haris Selajdzic, announced that he wanted Gorazde, the last Muslim safe area in eastern Bosnia. Twice the negotiations deadlocked and twice Milošević compromised, enabling an agreement to be reached.

The Dayton Peace Accords

Finally, on November 1, 1995, the United States and the other Contact Group members—France, Germany, Russia, and the United Kingdom—brought the warring factions together for formal peace talks, at Wright-Patterson Air Force Base in Dayton, Ohio.

The three strong-willed, difficult presidents—the Croat Franjo Tudjman, the Serb Slobodan Milošević, and the Bosnian Muslim Alija Izetbegovéc—ultimately accepted that the talks would take place under rules set by the Americans, and that Ambassador Holbrooke would lead the Contact Group, assisted by General Wesley Clark. Clinton's secretary of state, Warren Christopher, welcomed the three presidents to Dayton; the negotiations would take place under his skillful guidance, drawing on Holbrooke's tactical skills, knowledge of the area's politics, and understanding of Milošević.

The U.S. goal at Dayton was to turn the sixty-day cease-fire into a permanent peace. The Dayton Peace Accords ended the war and established Bosnia-Herzegovina as an internationally recognized sovereign state.

In the discussion of ending the war in Bosnia, the French insisted on a "new constitution, re-creation of normal conditions of life, and a secure environment." The French also indicated their intention to introduce civilian authority into every level of the military command system, emphasizing that civilian authorities would report not to the military commander and NATO but to the EU and the UN. Finally, the issues were settled.

The Dayton Peace Accords provided both more and less than the participants expected. The agreement, initialed on November 21, 1995, did not solve all the problems, but it resolved enough of them to provide a pause, and for all parties to feel they had achieved minimum goals. In addition to a cease-fire, an agreement on arms reductions, and boundaries, the accords created a new state that included two multiethnic entities, established boundaries between them, and committed the parties to reversing ethnic cleansing, arresting war criminals, and helping refugees return to their homes. The agreement provided a basis for building democratic, multiethnic institutions for southeastern Europe, with closer ties to the rest of Europe. The Organization for Security and Cooperation in Europe (OSCE) would supervise elections, and Carl Bildt, the Swedish prime minister and UN special envoy for the Balkans, would oversee implementation of the political provisions of the accords.

The accords embodied the parties' understanding that force was a necessary component of implementing an agreement in the Balkans, allocating power among two entities and three ethnic groups. The three national groups that had been at war for years would survive. Each group would keep its army (by this time all three armies had weapons, although the Muslims lacked heavy weapons). The agreement would be overseen and enforced by a sixty-thousand-person NATO-led implementation force (IFOR), to which the United States contributed twenty thousand troops. IFOR would be headed by an American, Admiral Leighton Smith, who made clear the new rules of engagement: "The senior soldier on the scene has the right and responsibility to protect himself and those under his command or leadership. If he feels threatened, he will use whatever force is necessary to neutralize that threat."[130]

A NATO-led stabilization force (SFOR), thirty thousand strong, would also be deployed throughout the country to assist in the continuing implementation of the accords and to protect the unified state. SFOR would be overwhelmingly made up of NATO troops but would also include Russians, Ukrainians, Moroccans, and other Eastern Europeans. It was estimated that about two thousand mujahideen from Iran, Afghanistan, and other Muslim states were in Bosnia. The rapid departure of these foreign Islamic fighters was specifically agreed to in the accords.

The accords were signed on December 14, 1995, in Paris. The official transfer of authority in Bosnia, from the UN peacekeeper mission to the NATO peace enforcement operation, was scheduled for December 19. Meanwhile, the U.S.-Bosnia negotiating team worked to resolve the last details. Holbrooke believed that four key issues still needed to be addressed for the implementation of the accords:

- Corruption must be eliminated; transparency and accountability must be established.

- War criminals must be arrested.

- Refugees must be allowed to return to their homes, and the pace of return must be quickened. Srebrenica must be repopulated. The perpetrators of the Srebrenican massacre must be jailed.

- Freedom of the press must be established.

Additional, unwritten commitments related to the Dayton Accords rested on each party's sometimes unspoken understanding of the stakes. These commitments included the U.S. promise that the UN arms embargo would be lifted promptly and that the United States would assume responsibility for arming and training Bosnian forces to enable them to achieve parity with the Serb forces operating in Bosnia. And the entire agreement was reinforced by the unwritten commitment that NATO would respond to Serb aggression with overwhelming (not proportionate) force.[131]

Throughout the negotiations, Richard Holbrooke provided brilliant leadership on the American side. When his principal assistant, Robert Frasure, and two colleagues were killed in an accident on the treacherous mountain road leading to Sarajevo, Holbrooke, devastated but determined, took charge of the negotiations.

However, many provisions of the accords have not been carried out, and several problems remain unresolved: the capture of war criminals, the status of Kosovo, and respect for the rights of Kosovars. This phase of war in the Balkans had ended, but no one was greatly surprised when it resumed three years later in Kosovo.

The Verdict on Srebrenica

War crimes trials in The Hague have provided details about the mass murder at Srebrenica, making it clear that it was a well-planned, deliberate killing operation. In November 2003, two senior Bosnian Serb officers gave the war crimes tribunal detailed accounts of the orders they received from General Mladic for the murder and burial of more than seven thousand unarmed Bosnian boys and men.[132]

A French parliamentary commission, established at the urging of the humanitarian organization *Médecins Sans Frontières* (Doctors Without Borders) to investigate the French role at Srebrenica, published its report on November 29, 2001. The responsibility for the disaster, it concluded, was shared by France and its Western allies for failing to prevent the massacre after the fall of the "Muslim enclave" of eastern Bosnia in July 1995. The report concluded that the tragedy "was ultimately caused by the absence of political will in France, the United Kingdom, the United States, and [among] the Bosnians, themselves"[133]—a list that should also include a failure of political will by the United Nations.

Four of the ten deputies on the commission did not agree with all the conclusions, especially the conclusion that France had made a deal with Ratko Mladic to secure the release of UNPROFOR troops. General Javier testified about the disagreements among himself, Akashi, and General Rupert Smith. All witnesses saw General Mladic as the person responsible for the Srebrenica massacres, and many witnesses also held him personally responsible for the siege and the repeated attacks on the unprotected citizens of Sarajevo.[134]

Tadeusz Mazowiecki, who served as the UN's special rapporteur on human rights for Yugoslavia from August 1992 through July 1995, resigned after the fall of Srebrenica and Zepa in August 1995 in protest against the UN's hypocritical claim to be defending Bosnia when in fact it had abandoned it. The Geneva-based UN Human Rights Commission expressed support for Mazowiecki's "moral and courageous stand and his resignation in protest of the perpetuation of gross violations in Bosnia and Herzegovina." Secretary-General Boutros-Ghali had no comment.

Mazowiecki was finishing his eighteenth report on human rights violations in former Yugoslavia when he learned of the massacres. The establishment of safe zones where civilians would be protected by UN peacekeepers had been his idea—in fact, his first recommendation. He had asked for sufficient troops to protect the safe areas. The UN Secretariat had requested thirty-four thousand troops to defend the safe areas, but the Security Council had supplied only seventy-six hundred for the mission.[135]

"Speaking of protecting human rights is meaningless in the context of the lack of consistency and courage on the part of the international community and its leaders," Mazowiecki wrote. He never hesitated to accuse all sides of the conflict of criminal acts, but he concluded that the Bosnian Serbs were guilty of perhaps 80 percent of the human rights violations and war crimes. He saw the fall of Zepa and Srebrenica as a critical moment that called into question the whole international order—much like the failure of the League of Nations to confront Mussolini's invasion of Abyssinia, from which the League never recovered. NATO's very credibility, Mazowiecki believed, was now at stake.

Boutros Boutros-Ghali, meanwhile, chose this moment of defeat and humiliation to pay a visit to Africa. "Why," asked Michael Ignatieff, in an interview for the *New Yorker*, "why not return [to Bosnia]?" "Because if I do," said the secretary-general, "all the African countries will tell the world that while there is genocide in Africa—a million people have died in Rwanda—the secretary-general pays attention only to a village in Europe."[136]

But Srebrenica was no village, and this was not the first time Boutros-Ghali had made clear his view that Africa's problems should have priority over Europe's. In July 1992, Boutros-Ghali had advised the Security Council to ignore Lord Peter Carrington's efforts to end the fighting in the Balkans and focus instead on Somalia. In September 1992, he opposed enforcement of the no-fly zone. In the spring of 1993, he reiterated the extraordinary notion that UN peacekeepers should remain neutral in a conflict in which civilians were being starved and shelled. In April 1993, he requested the recall of the French general Philippe Morillon, who with great personal courage had led a UN convoy of food and medicines into Srebrenica, where civilians—even then—were starving. And in the fall of 1994, Boutros-Ghali requested the recall of a second French general, Jean Cot, the commander of UN forces for former Yugoslavia, who had offended the secretary-general by pressing too hard for prompt and effective air cover for his forces.

In his interview with Ignatieff, Boutros-Ghali commented that humanitarian efforts often fail. He recalled other instances of failure and genocide. "Everywhere we work, we are struggling against the culture of death," he said.[137] But he did not struggle hard enough. Instead, he claimed control over the use of force by the UN and then deprived victims of help in their struggle against aggression, redefined peacekeeping so that it was of no use to anyone, wrote rules of engagement that did not permit UN forces to protect civilians under their care, and shackled NATO.

THE OLD EUROPE AND THE NEW

In the complicated diplomacy that finally silenced the guns of Sarajevo and produced the Dayton Peace Accords and the new state of Bosnia-Herzegovina, the outlines of a new Europe could be discerned. It was not the Europe of the cold war, though NATO played a role. It was not the Europe of Brussels, though the European Union representative, David Owen, continued his negotiations. It was the historic Europe of nation-states—*L'Europe des patries*, Charles de Gaulle had called it—that stretched from the Atlantic to the Urals.

Lenin was no help in understanding the politics of historic Europe; with his innate contempt for nation-states, he would never have understood why, for example, Boris Yeltsin moved four hundred troops to Sarajevo, where the Serbs greeted them as saviors. More relevant than Lenin is the history of the Ottoman Empire, whose fourteenth-century conquest of Serbia still moves Russians to sympathy, Greeks to rage, and Serbs to mobilization.

During the cold war, ideology and bloc politics had replaced ethnicity and history as organizing principles of European politics. With the end of the cold war, ideology gave way again to ethnicity and history.

Many observers believed the Old Europe was dead, a casualty of the two world wars, bolshevism, Nazism, fascism, and the European Union. But as Yugoslavia disintegrated, old patterns of trust and mistrust, affinity and hostility reemerged, and European powers reached across centuries to find "natural allies" from earlier times: Serbs and Russians; Germans and Croats; Bosnians and Bulgarians.

Nowhere were the old patterns more clear than in Russia, where fanatics like Vladimir Zhirinovsky, the leader of the Liberal Democratic Party of Russia, threatened Russian air strikes if NATO attacked Serbs, and liberals like former economics minister Yegor Gaidar and the political class recalled the "old Slavic ties" that linked Russia and Serbia.

Radio Moscow reminded listeners that it was a Russian tradition to support the Serbs, "those ethnic Slav people who are orthodox Christians." This sentimental recollection of historic Russo-Serbian-Orthodox solidarity not only reinforced the "Slavic brotherhood," it reminded Russia that it should have a voice and a role in any Balkan settlement. And so, rather suddenly after the cease-fire was signed, Yeltsin moved Russian troops from Croatia to Sarajevo and threatened a veto in the UN Security Council.

Suddenly, Russia was back: a superpower playing an independent role in the first European crisis since the end of the cold war; imposing compliance, of a sort, in the Serbs around Sarajevo and forestalling the NATO air strikes.

Russia was not the only country that was profoundly affected by the Balkan crisis. It stimulated Germany to make its first wholly independent foreign policy initiative since World War II. When it backed Croatia with diplomatic recognition and economic help, Germany's EU colleagues were shocked and offended by its unexpected interest in traditional national goals. And Germany's concern with Croatia aroused anxiety in France, where politicians began to worry that if Germany's attention turned east rather than west, it might drift away from the many ties and institutions designed to anchor it in democratic Western Europe.

So France, confronted with the ghost of the Old Europe, worked harder to strengthen its national and multilateral diplomacy. As the

Balkan crisis deepened, socialist president François Mitterrand and his neo-Gaullist prime minister, Edouard Balladur, worked to extend France's alliances and influence in the east as well as in NATO and Brussels. De Gaulle had seen a Franco-Russian entente as the natural protection for France against a resurgent German nationalism, and Serbia as the southern anchor of a Paris-Moscow arrangement.

How did the Americans fit into the new European politics, where alliances were based less on shared values than on shared history? The American legal-moral tradition in foreign affairs was very different from the historical European balance-of-power tradition. This was one reason that the Clinton administration had a difficult time finding common ground on Bosnia and Kosovo with European allies, and an equally difficult time explaining its actions and intentions to the American people (and perhaps to itself) as the situation developed. The savage war against Bosnia illuminated some important differences in the political sensibilities and reflexes of the Old Continent and the United States—and foreshadowed changes in that relationship that are still emerging.

5.

KOSOVO

The events in Kosovo, especially in 1998–99, offered further proof of the dangers of wishful thinking and the underestimation of bad actors—in this case, the same bad actor who had been at work in Bosnia. Once again, UN threats were unpersuasive; once again, NATO was divided. In this case, however, the United States was able to coordinate action with NATO and bring the ethnic cleansing to an end—perhaps the Clinton administration's only real foreign policy victory, although it was a long time coming. And the UN has proved useful, after the fact, in peacekeeping and conflict resolution.

BACKGROUND OF THE CRISIS

Soon after the death of Josip Tito on May 4, 1980, rumors spread among the Serbian population of Yugoslavia that Albanians in the southern province of Kosovo were engaged in a plot to eliminate Serbs from the area, and even that they were planning genocide. On September 24, 1986, a "memorandum" from the Serbian Academy of Sciences was published in the mass-circulation paper *Vercernje Novosti*, bitterly attacking Tito's 1974 constitution. That constitution had made Kosovo and Vojvodina autonomous provinces with seats in the federal presidency, had given them a voice and vote equal to those of Yugoslavia's six republics, and had established autonomy in most areas. The destruction of this constitution became a principal goal of Serb nationalists, and Slobodan Milošević was their chief spokesman.

By the time U.S. ambassador Warren Zimmerman arrived in Belgrade in March 1989, Slobodan Milošević had already made his reputation as an "ambitious and ruthless" (in Zimmerman's words) nationalist leader.[1] He used Communist organizational tactics to launch his campaign of repression in Kosovo and extend his personal power in Serbia and Yugoslavia. Wielding populist rhetoric, Soviet-style ruthlessness, and the Communist Party network, Milošević packed meetings and purged opponents in his campaign to create a new post-Communist Yugoslavia. He did not hesitate to use force to make life increasingly difficult for the Kosovars.

Milošević's intentions were already evident in 1989. Soon after being elected president in Serbia, he revoked the Statute of Autonomy for Kosovo. That revocation was the prelude to a campaign of discrimination and ethnic cleansing against the Albanian Muslims who constituted 90 percent of Kosovo's population—a campaign that ultimately drove Yugoslavia to war and the NATO governments to countermeasures.

It was hard for Americans to comprehend how aggressive and persistent Milošević was in the eight years following Tito's death, just as it has always seemed difficult for us to grasp the scope and ambitions of any coercive ideology or personality throughout history. Milošević pressed Serbia's claim to sovereignty over Yugoslavia and Kosovo, and managed to persuade many Europeans and Americans of its justice, even though Serbia had no historic or constitutional right to such sovereignty. He was the "leader" who drove Slovenia, Croatia, and Bosnia-Herzegovina into wars that left millions homeless and tens of thousands dead, the man whose megalomaniacal drive for power reduced the Yugoslav federal state to a fraction of its former size. Slovenia, Croatia, and Bosnia-Herzegovina eventually acquired weapons and won their independence, but not until Croatia and Bosnia had endured brutal ethnic cleansing and war had spread to Kosovo.

From the beginning, Milošević ruled by force and responded only to force, rebuffing the efforts of seasoned diplomats like Cyrus Vance and the United Kingdom's Peter Carrington and David Owen to find peaceful settlements to the conflicts. Milošević was not interested in agreements that gave minorities equal rights—or any rights. He was not interested in negotiating agreements or implementing agreements that had been negotiated. Peace and stability were not his priorities.

As chairman of the then Central Committee of the League of Communists of the Serb Party, Milošević launched his plan to gain control of the federal presidency and parliament in early 1989, through a series of carefully drafted constitutional "amendments," imposed arbitrarily by fiat, that progressively restricted and ultimately eliminated Kosovo's autonomy. The amendments deprived Kosovo and Vojvodina of autonomy and gave Milošević a near majority in the federal presidency. Milošević was determined to undermine Kosovar Albanians' political, economic, and social rights, and he used these amendments to seize control over Kosovo's police, courts, and civil defense; its social, economic, educational, and administrative policy; and even over the choice of an official language. He eliminated Albanian language instruction, segregated Albanian schools, and abolished the Academy of Sciences in Kosovo. Ethnic Albanians were fired from state jobs and replaced by Serbs. The names of streets were changed from Albanian to Serbian.

The Albanian majority sought to defend its rights under the 1974 constitution, launching demonstrations against Milošević's amendments; he responded by declaring a state of emergency, enforcing it with large numbers of Serbian security police.

In March 1990, Milošević spearheaded a program designed to increase the power of Serbs in Kosovo at the expense of Albanians by manipulating property rights and sales, helping Serbs acquire houses and work in Kosovo, encouraging a low birth rate among Albanians, and replacing Albanians with Serbs in desirable jobs, including civil service, education, and the professions.

On July 2, 1990, Albanian members of the Kosovo parliament passed a resolution affirming that Kosovo was "an equal and independent entity within the framework of the Yugoslav federation." This was simply a restatement of its status under the 1974 constitution, but Serb authorities reacted by dissolving the assembly. Two months later, the Kosovar delegates declared Kosovo a "republic" and held elections and a referendum that pronounced overwhelmingly in favor of independence. Out of these activities emerged the Democratic League of Kosovo (LDK), with

Ibrahim Rugova, a pacifist and Sorbonne-educated professor of literature and history, as its leader.

Yet the LDK was unable to stem the tide of nationalism Milošević had inspired. The attraction of violent nationalism in some regions of southeastern Europe illustrates the danger Milošević and ethnic extremism posed to the peace of the region—and the danger of violent ideologies to stability throughout regions of the world. It was not dominoes we needed to be concerned about in the Balkans; it was the contagion of mass murder. The only known antidote to such a crisis is the imposition of law and civilization by direct intervention. That was why, in 1999, NATO finally went to war.

THE 1990S: EVOLUTION OF THE CRISIS

By the time he was elected president of Serbia in January 1990, Milošević had made his priorities clear. He was a Serb nationalist, not a Yugoslav nationalist, and he was determined to "make Serbia whole" and to strip Kosovo and Vojvodina of their autonomy, their votes in the federal presidency, their rights under the federal constitution, their officials, and their constitutions. By abolishing the autonomy of these provinces and taking their votes, Serbia would directly control three of the eight votes in the federal presidency. With Montenegro, whose support he could count on, Milošević would control four votes—clearing his way to pursue his campaign against the Albanians without political opposition.

Milošević was able to pursue his designs on Kosovo throughout the early 1990s largely unchallenged—in large part because of the international community's hands-off attitude toward the region. In 1991 the Council of Ministers of the European Community appointed France's former minister of justice Robert Badinter to the Arbitration Commission of the peace conference on the former Yugoslavia. When the newly renamed Badinter Arbitration Commission took on the subject of violence and legal challenges in Yugoslavia and its republics, it concluded that the international community should recognize Yugoslavia and its republics but ignore the autonomous regions. The status of Kosovo and anything that occurred within its borders, such as human rights violations, were considered to be internal issues.[2]

The Commission's position was not unique; rather, it was the traditional position held for years, and echoed at the Munich and London conferences on Yugoslavia. At both conferences, concern was expressed about the bleak situation in Kosovo, and "the Serbian leadership" was urged to "respect minority rights" and "refrain from further repression in Kosovo,"[3] but that was as far as it went. The only other mention of Kosovo during that time in international forums was during the Organization for Security and Cooperation in Europe (OSCE) Helsinki summit in July 1992, when the Declaration on the Yugoslav Crisis urged "the authorities in Belgrade to refrain from further repression" of Kosovar Albanians. Once again, however, the issue was treated as an internal Yugoslav matter.[4] And once again, words did not dissuade Milošević's plans.

The United States took a distinctly different position. In December 1992, in the so-called Christmas warning, President George H. W. Bush declared, "In the event of conflict in Kosovo caused by Serbian action, the United States will be prepared to employ military force against the Serbians in Kosovo and in Serbia proper." He was making the point that escalation of violence in Kosovo would be viewed as more than an internal problem.[5] This warning was restated at a news conference in February 1993 by the incoming Clinton administration's newly appointed secretary of state, Warren Christopher, who reiterated Bush's position: "We remain prepared to respond against the Serbs in the event of a conflict in Kosovo caused by Serbian action."[6]

But the strong words of the new administration were not followed by strong actions. Instead, the worsening Bosnian crisis monopolized American and European top officials' attention, claimed their resources and energy. In the interest of halting the widely publicized mass murders in Bosnia, the Western leaders dropped Kosovo from their immediate agenda.[7] Anxious to treat Milošević as a peace broker, they withheld criticism of his treatment of Kosovo, giving the impression that NATO and the Contact Group had no interest in the troubled region.

Meanwhile, by the mid-1990s, Milošević's treatment of Kosovar Albanians had degenerated into widespread state-sponsored violence. In 1994, the Council for the Defense of Human Rights and Freedoms in Kosovo recorded "2,157 physical assaults by police, 3,553 raids on private dwellings, and 2,963 arbitrary arrests."[8] Milošević was preoccupied with the so-called Serbian Question in Yugoslavia (his name for the decline of the Serbian population in the province of Kosovo). His campaign of terror drove thousands of Kosovar Albanians to flee to Italy between November 1994 and mid-January 1995 alone.[9]

After unleashing war in Slovenia, Croatia, and Bosnia, and imposing unspeakable misery on the people of these republics, Milošević agreed to the 1995 Dayton Peace Accords under heavy American and European pressure; he partially implemented them, only because they were backed by substantial force. But the negotiations at Dayton neglected to deal with Kosovo; the Western powers never made it an issue, and Milošević was hardly likely to allow democratic self-government for the region or to honor the human rights of the Kosovar Albanians, on his own recognizance. And so, in the late 1990s, he escalated his campaign of violence and ethnic cleansing.

The Serbian government justified its aggression against ethnic Albanians by insisting that its actions were a reaction to increasing violence against the Serbs, but until 1995, Kosovar Albanians had offered only nonviolent resistance to the government in Belgrade. Even after the outbreak of violence in 1998, most Kosovars followed their pacifist leader, the poet and intellectual Ibrahim Rugova.[10] Rugova believed that the only way to restore autonomy for Kosovo was through diplomatic pressure from the west. However, as Serb aggression increased, support for the Kosovo Liberation Army (KLA) also increased. The first violent incidents took place In October 1997, when Serb police assaulted a peaceful protest of two thousand students in Pristina, the capital of Kosovo. Later that year, the KLA mounted what the Belgrade news agency Beta called "a series of terrorist actions," attacking a group of Serbian police.[11] In fact, the KLA was a small group of poorly armed and poorly organized rebels. Not until the summer of 1998 did the KLA grow into a force of about one thousand soldiers, capable of resisting Serb security forces and large enough to gain international attention.[12]

By 1998, less than three years after the signing of the Dayton Accords, the refugee crisis in Kosovo had grown so extreme that it threatened to destabilize neighboring countries. The crisis, which lasted from March 1998 through March 1999, was the culmination of Milošević's decade of oppression.

Reports of appalling violence came regularly from the region. On March 5, 1998, Serb forces brutally killed Kosovo Liberation Army (KLA) regional commander Adem Jashari, who had dared to stand up to Serb aggression, along with more than fifty members of his family. Confronted with such atrocities, the Western powers wrestled over the appropriate response: Americans favored a strong response to Serb aggression, but the Europeans resisted military action, raising the same tired arguments that had been made against the use of force in Bosnia.

The result of the lack of any serious international interest in Kosovo, coupled with the increased Serb abuses of ethnic Albanians, undermined Rugova's peaceful resistance attempts and stimulated greater violence by the KLA as they fought off the Serbs. By the spring of 1998, the KLA had begun guerilla warfare against the Serb security forces garnering attention. Responding to pressure from the Contact Group, the Council of Europe held debates on the situation in Kosovo. In January 1998, Council of Europe Resolution 1146 expressed concern about the "deterioration of the political situation in the Federal Republic of Yugoslavia" and the "serious implications for the stability of the Balkan region." The Council "condemn[ed] the continued repression of the ethnic Albanian population in Kosovo." [13]

As in 1992, this expression of concern from Europe or from any nation did not prevent or deter Milošević from continuing his military campaign against the ethnic Albanian population. Having lost Bosnia, he merely redirected his focus of nationalist passions on Kosovo, which became central to his hold on power.

Government propaganda, a favorite Milošević tool, and the Serb media inflamed the situation, and the Serb security forces took ever more violent action. In February 1998, Yugoslav forces (claiming to rid Kosovo of terrorists) attacked the Drenica region using helicopters and armored vehicles. According to Amnesty International, however, "Between 28 February and 1 March the Serbian police killed 26 ethnic Albanians in the villages of Likosane and Cirez. There was evidence that many of these were unlawfully killed." [14] A subsequent Amnesty International report indicated that the deteriorating security situation in Drenica had led to "hundreds of civilian deaths, many apparently a result of deliberate or indiscriminate attacks" and that the "attacks on civilians have been part of the reason why more than fifty thousand people have fled their homes."[15] These attacks further inflamed an already volatile situation.

Although targeting civilians had long been a policy of the Milošević government, the brutality of the massacre at Drenica upped the level of violence and captured the attention of foreign policy officials in the U.S. government and elsewhere. At a meeting in Rome with Italian foreign minister Lamberto Dini on March 7, 1998, Secretary of State Madeleine Albright addressed the "explosive and worrisome situation in Kosovo."[16] She had dealt with Milošević before—after five years of work on Bosnia as the U.S. ambassador to the UN, she believed that his actions were moral outrages and feared that Milošević could seriously destabilize the region. Moreover, her personal experience of escaping Nazi-occupied Europe left her ideologically opposed to appeasing dictators, for she had direct experience of the deadly consequences when ambitious leaders resort to coercion and violence to achieve their goals unchecked. She knew that Milošević understood only decisive action, and she remained committed to her word that the United States was "not going to stand by and watch the Serbian authorities do in Kosovo what they can no longer get away with doing in Bosnia."[17] Albright was determined to take serious measures to punish the Belgrade government.

But the division among the allies persisted after years of war in the Balkans, and they had only deepened despite the Dayton Peace Accord. One who shared many of Albright's views was General Wesley Clark, who had been NATO's Supreme Allied Commander in Europe (SACEUR) since 1997. Clark's appointment was controversial within the military; his views were more hawkish than those of many in the Pentagon—including Clinton's second-term secretary of defense, William Cohen, and the service chiefs, who were reluctant to engage militarily in Kosovo. But Clark had experience with Bosnia: having dealt with Milošević in Bosnia and Croatia, and attended the Dayton negotiations, he was aware of Milošević's penchant for violence.

When Clark attended an inspection of U.S. troops stationed in Macedonia in March 1998, Macedonian president Kiro Gilgorov[18] warned him that there would be trouble in Kosovo. Milošević, he said, "likes to use military force. And, though he might say he would negotiate, he does this to complicate situations, so he can seek advantages for himself. In the end, he respects only the threat of military force."[19] Clark's own experience in Bosnia told him that Gilgorov was right.

President Clinton agreed with Secretary Albright that something would have to be done about Kosovo, and he encouraged her personally to take charge of the situation. She later wrote: "I concluded that we should not be content to follow the consensus on Kosovo, which was going nowhere. The NSC [National Security Council] and Pentagon did not desire to become involved in another war in the Balkans. We warned Milošević repeatedly not to launch a war. . . . On March 19, we met with the President to review our options. Several of us felt that if we did not confront Milošević now we would have to confront him later."[20] Clinton agreed, saying, "In dealing with aggressors in the Balkans, hesitation is a license to kill."[21]

On their own initiative, Albright and Clark met with the Contact Group on Yugoslavia in London on March 9, 1998.[22] The Contact Group gave Milošević ten days to deescalate the conflict, demanding that he agree to political negotiations with ethnic Albanians in Kosovo led by Rugova. The group demanded that he remove the Serb special police from Kosovo and that the Yugoslav government grant full access to the International Committee of the Red Cross and the United Nations Commissioner for Human Rights. Albright emphasized the need for "action, not rhetoric" as "the only effective way to deal with this kind of violence."[23]

Diplomatic Attempts to Check Milošević

Albright's work through the Contact Group resulted in a UN resolution on March 31, 1998, imposing an arms embargo on Yugoslavia.[24] Security Council Resolution 1160 stated that "failure to make constructive progress toward the peaceful resolution of the situation in Kosovo would lead to the consideration of additional measures" by the UN.[25] The resolution did not specify what these measures might be; the very fact of mentioning enforcement, however, signaled that the Security Council no longer considered Kosovo a purely internal Yugoslavian matter. Yugoslavia's ambassador to the UN, Vladislav Jovanovic, denounced the resolution as "an unprecedented interference in internal affairs."[26] The

Russian and Chinese delegates joined in, stating that interference in a country's internal matters "might have wider negative implications."[27]

But the denouncements and outrage in New York and the UN resolution had little effect on the situation in Kosovo. Violence pressed forward. Masked Serb forces entered villages and terrorized civilians— killing some, forcing others to leave, separating families, and confiscating property. Many in the targeted villages were marched to train stations and packed onto trains headed toward the Macedonian border.[28] In response to pressure from Western Europe and the United States, Milošević held a referendum in April 1998 in which the Serbian people voted against foreigners meddling in their affairs. Empowered by the 95 percent vote of confidence, he stepped up military operations and ethnic cleansing throughout Kosovo.[29]

The United States responded by pushing for new sanctions, but again the Europeans were not willing to take a tough stand against Serbia. Finally, the United States said it was "prepared to abandon the Contact Group . . . if the group balks at imposing new sanctions on the Belgrade government when it meets in Rome."[30] The United Kingdom backed the U.S. position, and in the spring of 1998 the Contact Group agreed to freeze Serb and Yugoslav assets and warned that it might block all foreign investment if Milošević did not agree to mediation and to talks with Rugova.[31] A month later, the Contact Group imposed an investment ban on Yugoslavia, and the U.S. government sent Richard Holbrooke to Belgrade for the first round of talks.[32]

To show his personal determination to stop the violence, Clinton visited Kosovo, met with Rugova, and pledged U.S. aid. He asked Rugova to continue negotiations with Milošević, and promised that the disaster of Bosnia "should not be repeated and will not be repeated."[33] His support, and Holbrooke's negotiating skill, led to a U.S.-brokered meeting between Rugova and Milošević on May 27, 1998. But despite this achievement, the refugees continued to pour out of Kosovo. By the end of May, an estimated three hundred people had been killed since the start of Yugoslav operation in February. In addition, approximately twelve thousand refugees had fled into Albania.[34] Milošević refused to stop the violence, and Rugova broke off the talks.

At the NATO ministerial meeting in Luxemburg on May 28, chaired by British foreign secretary Robin Cook, the Balkans dominated the agenda. Clark and the German general Klaus Naumann, the head of NATO's Military Committee, reported to the group on the situation. The Germans and the British agreed with the United States that something had to be done about Kosovo. Klaus Kinkel, German foreign minister, said that "a clear red line must be drawn."[35] Cook pledged that a "modern Europe will not tolerate the full might of an army being used against civilian centers."[36] In June, after another month of violence, Britain's new prime minister, Tony Blair, declared: "[T]he only question that matters is whether you are prepared to use force. And we have to be."[37] NATO would have to get tough with Milošević. Among all the NATO members, however, only the United States, the United Kingdom, and perhaps Germany were prepared to use airpower at this point, and only Great Britain was prepared to use ground troops.

The Ineffective Threat of the Use of Force

In June 1998, columns of Serbian tanks entered Kosovo; they turned Kosovar villages into dust and rubble, and killed more than 250 people.[38] Another five to ten thousand refugees poured into Albania, causing a humanitarian crisis in that nation. Albanian premier Fatos Nano warned that the refugee crisis threatened to destabilize the region.[39] In July, the Yugoslav army and Serb police launched a major offensive against the KLA, comparable to the ethnic cleansing carried out earlier in Bosnia. More than two thousand civilians were killed, two hundred villages destroyed, and three hundred thousand civilians displaced.

Faced with this new aggression, the Clinton administration began to consider war options. Convinced that putting ground troops into combat in Yugoslavia would be unpopular among Americans, Clinton flatly ruled out this option. Pentagon officials also had serious concerns about sending American troops into the hostile regions of Serbia.[40] At the same time, they did not believe that the war could be won from the air. As is often the case, some in the U.S. government believed that a political rather than a military solution was required. The administration did not lack congressional support to send troops into Kosovo, but no one wanted to do it. Speaking about the use of force in general, Benjamin

Gilman (R-NY), chairman of the House International Relations Committee, observed that "such solutions do not eliminate the underlying problem. . . . They promise to drag on indefinitely, at high cost to our own nation."[41] However, the atrocities committed by the Serb forces in the summer of 1998 made war against Serbia seem inevitable.

Although Clark initially lacked Pentagon backing for the use of force, he had the support of General Naumann and NATO's Military Committee, which consisted of the top military officers of all member countries and was responsible for advising the North Atlantic Council (NAC) on military matters. On June 11, the NATO defense ministers met in Brussels to consider the use of air strikes and NATO's role in solving the conflict. The United States and Britain were ready to back up their diplomatic efforts with force, as was NATO secretary-general Javier Solana, who had supported Clark's views from the beginning. "Milošević has gone beyond the limits of tolerable behavior," Solana acknowledged, "and . . . we are showing that we are willing to back up international diplomacy with military means."[42]

In May 1998, the United States sent Christopher Hill, ambassador to Macedonia, to join Holbrooke in negotiations with Milošević in Belgrade. Milošević agreed to end the violence in Kosovo, promised that "no repressive action will be taken against the civilian population,"[43] and agreed to meet with Rugova. But soon this new agreement, too, was tossed away. Milošević knew that the NATO members did not all agree on the use of force, and he dug in his heels.

On June 16, 1998, NATO staged an air exercise called Operation Determined Falcon to show Milošević what he was facing. About one hundred NATO aircraft took off from their bases, flew through Albania to the Serbian border, and then flew east in Albanian and Macedonian airspace.[44] The Kosovars rejoiced at the sound of the engines, and the Serbs heard them, too.

The Europeans, notably the French, remained determined to find a diplomatic solution. France called for an international plan to restore autonomy to Kosovo and hinted that it would be willing to support military intervention. But it still maintained that "a mandate from the UN Security Council would be needed before NATO could go into action in

Kosovo."[45] French foreign minister Hubert Vedrine said, "[The] use of force could provoke exactly the opposite of the desired result."[46] Russia and China remained large obstacles to any use of force; they opposed taking an aggressive stand against Yugoslavia or "any action that would penalize Yugoslavia."[47]

The differences between American and European attitudes on how to deal with the humanitarian crisis renewed the strain on U.S.-European relations that had developed in the early 1990s over Bosnia. The European leaders' indecision made it impossible for the United States to negotiate effectively with Milošević. The Germans complained that "there seems to be very little willingness [on the part of the Americans] to treat the Europeans on an equal footing."[48] But the European diplomats were passive, and it fell to the United States to lead the initiative to find a solution in Kosovo. Holbrooke knew that the Europeans had accomplished little since the days of Bosnia, and he recognized that they were "not going to have a common security policy for the foreseeable future. We have done our best to keep them involved," he said.[49] Holbrooke believed that his successful Bosnian strategy—calling for Serbian compliance with the UN and other confidence-building measures—would work again in Kosovo, and Albright agreed.

Still, the violence continued. On September 23, 1998, the Security Council passed Resolution 1199, demanding "that all parties, groups and individuals immediately cease hostilities and maintain cease-fire in Kosovo."[50] The resolution demanded that Milošević comply with the earlier demands of the Contact Group for the withdrawal of forces from Kosovo, allow international verifiers, and resume negotiations. China did not support the resolution and abstained from voting. Unexpectedly, Russia came on board. The Russian representative, Sergei Lavrov, said, "The resolution was in line with Russia's principles because no measures of force and no sanctions at this stage are being introduced by the Security Council."[51] The Russian vote marked a diplomatic turnaround in the balance of forces on Serbia's war in Kosovo. With Russia siding with the United States, the threat of force became more real. Milošević understood that a NATO war against Yugoslavia was a possibility, but he did not think it would be prolonged or costly.

Clinton Rallies the International Community

In clear defiance of the west, three days after UN Resolution 1199 was passed, Yugoslav forces renewed their bombing runs. When the village of Gornje Obrinje was attacked on September 26, 1998, it decided to fight rather than surrender. The village was largely destroyed. Most homes were burned. Cattle were left dying in the street. The Serbs stripped men of weapons, then killed them. The descriptions of the killings—of men and boys being beaten to death or shot at close range, of entire families found dead in the forest—resembled accounts of earlier Serb attacks on Croats and Bosnians. Once again there were beatings in the police station and on the street; men humiliated and forced to sing Serb songs.[52] The Yugoslav army and Serb police latest move largely drove the KLA into the mountains.

With no end in sight to the violence, President Clinton sought to rally international support for the use of force. Placing calls to French president Jacques Chirac and German chancellor Helmut Kohl, he got their agreement in principle for an air campaign. Banimino Andretta, Italy's acting defense minister, said that his country's forces were ready to join their NATO allies. Russia joined the group in backing Holbrooke's diplomatic efforts, but Russian foreign minister Igor Ivanov restated Moscow's opposition to any military strikes. Secretary Albright replied that the United States would not be deterred from acting because of Russian opposition.

On September 24, 1998, pressure from the United States led NATO to adopt an Activation Warning Order (ACTWARN), signaling a warning status for the deployment of air forces. The North Atlantic Council's approval of a limited air option and phased air operation (code named Operation Allied Force) moved NATO one step closer to a war against Serbia. ACTWARN was not binding, but on October 12 the NATO countries took one step closer to action, specifying which forces they would contribute for an impending strike by approving activation orders (ACTORDs) authorizing preparations for a limited bombing campaign. With more than four hundred aircraft standing by ready for a possible air campaign, NATO authorities voted to authorize strikes if security forces did not withdraw from Kosovo within ninety-six hours.

To give Holbrooke the maximum leverage during his ongoing negotiations, the activation warning orders were issued before his meeting with Milošević in Belgrade. First, Holbrooke demanded that Milošević withdraw an estimated four thousand Serb troops from Kosovo. Second, he demanded that Milošević allow up to two thousand civilian observers or verification cease-fire monitors under the auspices of the OSCE to ensure that the troops were withdrawn. Finally, he demanded that Milošević set a timetable for negotiations and adhere to it. Milošević agreed all too quickly, raising swift doubts about his sincerity in the minds of U.S. officials. So Solana and Clark flew to Belgrade to sign an agreement with Yugoslav military officials that would permit NATO reconnaissance planes to fly over Kosovo to verify the withdrawal. To everyone's amazement, by the end of October, large numbers of Yugoslav troops withdrew, and the Kosovo Verification Mission monitors were deployed. Holbrooke was cautiously hopeful that the agreement could end the conflict.

But the Serbian pullout was by no means complete. Large numbers of Serb police forces stayed in Kosovo, and violence continued. On October 24, 1998, the Security Council passed Resolution 1203, which reinforced previous resolutions and stressed the need to address the current "humanitarian situation and to avert the impending humanitarian catastrophe." [53] On November 17, the Security Council passed Resolution 1207, calling on Yugoslav authorities to comply with the requests of the International Criminal Tribunal for the Former Yugoslavia (ICTY) and arrest certain persons. [54] Milošević ignored both resolutions.

At the beginning of 1999, the Contact Group was still looking for ways to avoid fighting in Kosovo, although France had reluctantly lined up behind the American threat of NATO bombing. French defense minister Alain Richard said, "We will share our part of the responsibility as the head of Europe and the Alliance." But France articulated caveats, including that did not demand a Security Council mandate or threaten a veto; it emphasized its opposition to Kosovar independence and insisted that its troops would only get involved once U.S. troops were involved as well.

Before proceeding with its air strikes, the Contact Group—which by now expanded to include Italy and Germany—decided to make one last attempt at a negotiated settlement, scheduling a meeting at a chateau in Rambouillet, France, near Paris, in February 1999.

The Road to Rambouillet

In January, before the negotiations at Rambouillet, the Serbian forces intensified their activities in Kosovo. On January 15, all adult males in the village of Racak were hunted down and killed. As Milošević braced for the coming NATO attack, he only intensified the terror campaign in Kosovo. The massacre in the village of Racak was a preview of atrocities to come.

The Serb offensive that followed, called Operation Horseshoe, was a massive campaign to kill or expel all Kosovar Albanians. NATO observers thought the Serbs were preparing for a spring offensive that would target KLA strongholds. As Clark said later, "We never expected the Serbs would push ahead with the wholesale deportation of the ethnic Albanian populations."[55] According to Clark, the Serb forces numbered about fifteen thousand at this time. They included the regular Serbian police, the blue-uniformed Serbian Interior Ministry troops, the local police, and paramilitary troops commanded by indicted war criminals such as Zeljko Raznatovic, known as Arkan.[56] Their tactics—sealing a town, killing all military-age males, and sending the rest of the population packing—were seen as evidence of a planned genocide directed from the highest levels of government.[57] In January and February, Yugoslav officials reportedly collected "key documents and records from different villages in central or western Kosovo for 'safekeeping,' " and "valuable religious icons, paintings, and historical manuscripts were removed from museums and libraries and trucked north toward Belgrade."[58] Belgrade was preparing a full-on campaign to create an ethnically pure Kosovo.

The cease-fire Holbrooke had negotiated with Milošević in October 1998 did not end the fighting because it did not secure the withdrawal of Serbian forces from Kosovo. Instead, it spent the NATO consensus and support for forceful air intervention to secure nothing more than a cease-fire that Milošević never intended to honor. This approach was taken in spite of the lessons presumably learned in Bosnia. The presence of Serb forces was the primary cause of continuing violence in Kosovo, and we should have demanded their removal. Peace would have to be imposed, not brokered.

Meanwhile, the United States and the Contact Group met in London and set a final deadline of February 19 for the Yugoslav government and the Kosovar Albanians to accept a negotiated settlement. "[T]he Contact Group has made it unmistakably clear that the consequences of failure to comply will be swift and serious," Albright declared.[59]

The peace negotiations at Rambouillet began on February 6, chaired by the British and French foreign ministers, Robert Cook and Hubert Vedrine. The French and British invited the OSCE, the EU, and the Russians, as well as U.S. Ambassador Chris Hill. Conspicuously absent, however, were NATO's commander, General Clark, or any other high-ranking U.S. military officers. Kosovar Albanian leaders from both KLA and LDK were invited, as were the representatives of the Federal Republic of Yugoslavia. The threat of air strikes did compel the Serbs to attend. But Milošević himself expected the talks to fail, and he declined to attend, sending Serb president Milan Milutinovic as his deputy.

Milošević took NATO's absence to suggest that the United States and the Contact Group would not follow through on their threat of air strikes. Three factors caused him to question NATO's credibility: that NATO did not compel him to attend the peace talks, as it had for the Dayton negotiations (Milošević claimed that he might be exposed to arrest if he went to Rambouillet); that the OSCE still had a thousand unarmed civilian monitors in Kosovo, whose safety would be endangered by bombing; and President Clinton's perceived reluctance to commit ground troops. (Clinton had pledged to decide by February 1 whether he would send U.S. ground troops to Kosovo; on February 4 he said only that he was "committed to *considering*" [emphasis added] authorizing the troops.)

The Rambouillet conference moved forward largely as a result of the determined efforts of Secretary of State Madeleine Albright, who arrived after the meetings began and proceeded to personally manage the extremely complicated relations within the Contact Group and among Russia, France, the United Kingdom, the United States, the OSCE, Serbia, the Kosovars, Congress, and the media.

The Contact Group presented its demands at Rambouillet as non-negotiable: (1) the KLA must disarm; (2) Serb forces must withdraw from Kosovo, under the supervision of thirty thousand NATO troops; (3) Yugoslavia must restore Kosovo's autonomy and independent institutions. The issue of the future status of Kosovo would be considered in three years.

U.S. diplomats found the Kosovo Albanians to be much less helpful than they had hoped. But the Albanians were not given crucial draft texts of political and security agreements, and therefore they never had the opportunity to express their concerns. Conference organizers permitted twenty-nine-year-old Hashim Thaci, a former Kosovar Albanian student leader who helped found the underground movement that became the KLA, to be seated as head of the Kosovar delegation, but after twelve days they handed him an annex to the agreement that called on the KLA to disarm and disband.

After two weeks, there was progress, but at least one major sticking point: the Albanians refused to sign the agreement unless it guaranteed them a referendum on independence within three years. Albright was disappointed by the Kosovar Albanians' lack of cooperation; Albanian delegates blamed hardliners inside the KLA for this failure. Veton Surroi, a moderate in the sixteen-member Albanian delegation, blamed Adam Demaci, a former influential political adviser to the KLA, for wrecking the consensus.

Still, by the time the Rambouillet meetings broke off in late February—with plans to begin again in March—the Kosovar Albanian delegation had committed to sign the agreement when the talks resumed. The sequence was supposed to unfold as follows:

1. Kosovar accepts the three-year interim self-government offer

2. The agreement is imposed on Belgrade, using NATO air strikes, if necessary

3. NATO ground troops are introduced to enforce the political settlement

The eighty-two-page Rambouillet agreement spelled out terms that met most of the Kosovars' goals, but the Contact Group had ruled out their key goal: independence. Moreover, in an effort to accommodate the Serbs, the negotiators agreed to the disarming of the KLA without a parallel disarming and withdrawal of the Serb army, and they failed to impose meaningful limits on the number of Serb soldiers and police who could remain in Kosovo. They also offered to secure limits on the size of the peacekeeping force that would be deployed in Kosovo. Without even attending the meeting, then, Milošević had gained most of what he wanted.

The Rambouillet summit was doomed from the start. One major error was to grant the French demand that General Clark be barred from the conference, despite the Kosovars' request to meet with him. An effective NATO ground presence was essential to establishing a credible threat, and air strikes were critical to sustaining Albanian morale. Clark should have been present. The Contact Group's refusal to accept the KLA as a permanent feature of the geopolitical landscape was yet another substantial policy error. And there were other errors: the war crimes provisions of the agreement was downgraded, and the United States missed an opportunity to repeat its Christmas warning.

Perhaps the greatest flaws in the agreement were its concessions to Milošević. It permitted Serbia to maintain a large and intimidating force in Kosovo: four thousand military and police forces, twenty-five hundred Ministry of the Interior special police, and fifteen hundred army troops. It offered no provision for resolving the situation after the three-year interim—for example, through a referendum, a major demand the KLA eventually demanded. And it outlined no mechanism for enforcement.

Notwithstanding these accommodations, however, the Rambouillet accord was doomed to failure. Though the talks did resume, and the Albanian faction did sign the agreement on March 18, the Serbs walked away, and further negotiations were deemed futile.

THE ARRIVAL OF WAR

The extent of the devastation wreaked in Kosovo in the late 1990s can best be communicated through numbers: more than ten thousand people were killed and 1.3 million ethnic Albanians were driven from their homes. The failure of Rambouillet left only one option: war.

When the NATO governments finally decided the time had come to use force against Serbia, the United States and NATO agreed to assist Kosovo, as they had finally provided arms and assistance to Croatia and Bosnia, enabling them to defend themselves against Milošević. There was still resistance to the idea of sending U.S. forces to Kosovo to help protect the Kosovars; some argued that such an action was contrary to the will of Yugoslavia, the nation of which Kosovo was a part, and that to send troops into a province in a sovereign state would constitute an invasion. But at Rambouillet the Contact Group had announced that such an action might not be considered an invasion if its purpose was to protect Kosovar civilians from Serb armies. Whether such an action amounts to an intervention, or "coercive peacekeeping," it was nothing new: the Bush and Clinton administrations had undertaken much the same kind of action—without the consent of the nations in question—in Somalia and Haiti.

By now it was clear that Milošević could not be controlled by diplomatic means. He had ignored his October 1998 cease-fire agreement with NATO, boycotted Rambouillet, and ignored the resulting accords signed by the Kosovars. By 1999, Milošević's nationalist campaign had long since been transformed into a shocking reign of terror directed at Kosovar civilians, including children.

President Clinton's goals for Kosovo were humanitarian. They included the return of refugees—the goal of all interventions in which unarmed civilians are driven from their homes. Clinton's team had tried to negotiate a settlement and a cease-fire and identify a space for peacekeeping.[60] NATO's war in Kosovo came three years after the signing of the Dayton Accords and after nearly a year of unfettered Serb violence against the Kosovar Albanians. By the time NATO bombs started falling on Serbia, nearly 90 percent of the Kosovar Albanians had been displaced.

Before NATO could go to war, approval was required under the U.S. Constitution from the U.S. Congress. Congress was deeply divided on Kosovo, both between and within the parties. But the situation on the ground was rapidly deteriorating; Kosovars fleeing to Albania were being killed en route by Serb irregulars. Finally, on March 23, 1999, the Senate voted to approve the NATO bombing. (The House, however, refused to do so when it voted one month later.)

Many analysts have characterized the inability of the United States and the European countries to prevent Milošević's violence as a failure of diplomacy. I disagree. What these analysts fail to understand was that the problem was not a diplomatic one. The problem was Slobodan Milošević himself—a power-hungry dictator who responded not to diplomatic commitments but only to force, a man whose main goal was to fashion, at any cost, an ethnically pure Kosovo inhabited only by Serbs.

Despite their lengthy experience with Milošević in Bosnia, where cease-fires had been repeatedly ignored, the Europeans and Americans took a remarkably long time to realize that Milošević would never willingly negotiate a solution to the Kosovo dispute. The Europeans' faith in dialogue, and their aversion to force, made them especially ineffective in dealing with Milošević. Ultimately, the destruction and mass murder in Kosovo were stopped only by the application of American force and NATO airpower. And this came to pass only because key officials in the Clinton administration—especially General Clark, Secretary Albright, and Clinton himself—grasped that diplomacy had to be backed by force in dealing with such a bloodthirsty dictator.

Force and More Force

In March 1999, for the first time in its fifty-year history, NATO went to war. The purpose of the action was not to defeat the Serbs, to conquer the Yugoslav armed forces, to cause a regime change, or to win independence for Kosovo. Rather, it aimed to persuade Milošević to stop the violence.

On March 24, NATO aircraft started the bombing campaign against Yugoslavia.[61] General Clark's first priority was to incapacitate the Serb air defense and military communication systems. NATO used precision bombs that targeted the Serbian military and avoided civilian targets. After four weeks of bombing, however, Milošević remained defiant. In fact, he decided to speed up the killing and expulsion of ethnic Albanians. According to Human Rights Watch, between the beginning of the NATO air raids and their cessation in June 1999, the Serbs undertook a "systematic campaign to terrorize, kill, and expel the ethnic Albanians," an initiative "organized by the highest levels of the Serbian and Yugoslav governments."[62] On March 25, the day after NATO strikes began, Serb forces burned two hundred to six hundred homes in the town of Djakovica.[63] During the NATO air campaign, approximately 863,000 sought refuge outside Kosovo. An additional 590,000 were internally displaced. Altogether, over 90 percent of the Kosovar Albanian population was displaced.[64] The *Washington Post* called the Serbian action "the most ambitiously ruthless military campaign in Europe in half a century."[65]

On April 3, the Serbian people showed their support of Milošević and their defiance of the West by holding a rock concert in Belgrade.[66] Within a month, however, the economic effects of the war began to be felt, though most people still did not have to go far for electricity or water.

The turning point came when the daily lives of the Serbian people were seriously affected. On March 27, day four of the air strike campaign, the second stage of the air war began. Now, in addition to bridges and highways, NATO began bombing Serbia's military-industrial infrastructure, its electricity plants, and its oil pipelines and refineries, thereby destroying the Serbian economy. According to the Independent International Commission on Kosovo, NATO knocked out 70 percent of Serbia's electricity production capacity and 80 percent of its oil refinery capacity. By the end of May, there were "increasing signs that Belgrade is feeling the heat," according to State Department spokesman James Rubin.[67] The morale of the people began to crumble, and antiwar demonstrations and protests replaced pro-war rock concerts throughout Serbia. In the Serbian towns of Cacak and Krusevac, protesters reportedly shouted, "The dead do not need Kosovo!"[68] Between five hundred and a thousand Serb soldiers deserted in Kosovo.[69]

NATO threatened to use ground troops if Milošević did not surrender. After Russian prime minister Victor Chernomyrdin held an unsuccessful meeting with Milošević in Belgrade on April 24, the NATO allies and the U.S. administration began seriously considering the use of ground troops. Britain's chief of defense, General Charles Guthrie, threatened that Yugoslavia's "war machine is going to be weakened, again, day after day, week after week. . . . And we will be able to choose our time" for entering with ground troops.[70] The increasing unpopularity of the war at home, the inability of the Serb government to control its military forces, and NATO's threat of broadening its campaign convinced Milošević that his chances of retaining power would be greater if he gave in to NATO's demands and ended the war.

Turning Points

Before the first bomb was dropped on March 24, Serb military forces and paramilitary gangs had swarmed into Kosovo's capital, Pristina, and the surrounding cities, driving Kosovars out of their homes; burning houses, stores, factories, and restaurants; raping, beating, and killing. Once the bombing started, foreign journalists, especially Americans, were subjected to hostile treatment—shoved, slapped, threatened, arrested, interrogated, and expelled, their cameras and other equipment smashed.

After the foreign observers and journalists were driven out, it became harder to get detailed information about events inside Kosovo. As the war progressed, however, it became clear that the Serbs had launched a full-scale ethnic cleansing campaign, moving from village to village; torching houses; seizing and sometimes executing Albanian teachers, journalists, and political leaders; and beginning the ominous separating out of boys and men that had preceded the massacres in Croatia and Bosnia-Herzegovina. It is said that this deadly work was carried out by the followers of the war criminal Arkan, among the most brutal ethnic cleansers in Croatia and Bosnia.

Because the myth of "NATO aggression" was already being circulated by Milošević apologists, it is important to remember that the buildup of Serb forces was already taking place before NATO reluctantly started bombing. Milošević's intentions were clear: to use his growing forces to drive the 1.8 million Albanians (Muslims) out of Kosovo before war even began. He had already driven 400,000 of them from their homes, destroyed more than 500 villages and 22,000 houses, and killed more than 2,000 persons. One of the original objectives of the air war—protecting Albanians from Serb aggression—proved unattainable. Wave after wave of refugees arrived daily in neighboring countries, with fresh tales of abuses and killing by Serb forces. Eventually, General Clark acknowledged that bombing could not stop the harsh mistreatment of the victims.

Not every aspect of the NATO campaign was an immediate success. In early June, when NATO offered the Kosovo rebels their first NATO air support, in an unsuccessful bid to break a Serbian siege along the Albanian border and to show the Serbs (and perhaps themselves) that they were able to fight, the experience showed that the Serb military remained a military force of considerable strength. U.S. secretary of defense William Cohen commented that the Serbs had not yet surrendered. "At this point, not one single Serb soldier has withdrawn from Kosovo,"[71] he conceded. NATO forces had not definitively defeated the Serbs, and American pilots were still flying B2 bombers on thirty-hour missions from Missouri to Kosovo.

Yet by this time the die, after nearly three months of bombing, was cast. After several discussions between Milošević and the Russian envoy Victor Chernomyrdin, U.S. deputy secretary of state Strobe Talbott met with the Russians in Rome to discuss how to end the war. Many of the outstanding issues were resolved in talks among Chernomyrdin, Clinton, and Vice President Al Gore. They planned the withdrawal of Serb troops and an occupation with NATO in the central role. At the Group of Eight (G-8) meeting in Cologne in June, Russia ended its opposition to the UN peace plan, and the principles of the settlement were adopted at the G-8 meeting on June 2. The proposal required the following:

- Immediate and verifiable end to violence and repression in Kosovo

- Demilitarization of the KLA forces

- Withdrawal from Kosovo of Serb military, police, and paramilitary forces with a phased detailed timetable and a buffer zone behind which Serb troops were to withdraw

- Deployment of an international civil and security presence that would include NATO forces under the auspices of the UN

- Establishment of an interim administration in Kosovo, to be determined by the UN Security Council

- Safe return of all refugees and displaced persons

- Access for humanitarian organizations

- Establishment of a political process aimed at substantial self-government for Kosovo

- Comprehensive approach to the economic development and stabilization of the crisis[72]

Although Russian foreign minister Ivanov said that his country's participation in peacekeeping was not certain, all the G-8 countries signed the agreement.

Three days before the war ended, after eleven weeks of air strikes, the Serbs finally withdrew from the Kosovar village of Kacanik, stripping it as they left. KLA soldiers welcomed the NATO forces. They quickly found a mass grave containing the bodies of twenty-eight men, and several areas that had been sacked and burned by forces led by Serbia's justice minister, General Dragoljub Jancovic.

On June 3, Finnish president Martti Ahtisaari, Talbott, and Chernomyrdin went to Serbia to present the peace plan to Milošević. It was a take-it-or-leave-it proposal. If Milošević left the meeting without agreeing to the proposal, they told him, NATO's pounding of Yugoslavia would continue, and the terms of any future agreement would be worse.[73] They also presented Milošević with a military technical agreement that detailed the phased pullout of Federal Republic of Yugoslavia (FRY) forces in twelve days.[74] According to the technical agreement, the withdrawal of the Serbian military from sections of Kosovo would be synchronized with the entry of the Kosovo Force (KFOR). With no other option, Milošević agreed; the Serb parliament formally approved the peace plan the same day. The use of force had prevailed.

The NATO bombing would continue until three conditions were satisfied in the following twenty-four hours:

1. The Serbs had to begin their withdrawal from Kosovo;

2. NATO would then begin a twenty-four-hour pause in the bombing; and

3. The Security Council had to pass a resolution endorsing the deal.

On June 10, 1999, Serbian troops began pulling out of Kosovo. British general Mike Jackson notified NATO secretary-general Javier Solana that the pullout had begun, and Solana directed Clark to call off the air strikes. President Clinton was on the telephone with Tony Blair when Javier Solana called to inform him that the Serbs were moving out of Pristina and Kosovo. Milošević claimed a victory for the "best army in the world,"[75] declaring, "We never gave up Kosovo." The Albanians waited for the Serbs to leave. An estimated 840,000 refugees began to filter back into Kosovo.[76]

After the Serbs' withdrawal, the United States, the United Kingdom, and NATO devised a partition of Serbia. Within four days of the end of the war, the Russians unexpectedly rolled into Pristina, throwing the delicate plans and timetable into disorder. Serb women rushed forward to embrace the Russians, who saw their success in arriving first in the capital of Kosovo as a great victory. Russia's foreign minister, Igor Ivanov, called the early arrival "an unfortunate mistake," and NATO generals feared that their Russian counterparts might have ordered the advance without the approval of the Russian civilian leadership.

ENDING THE WAR

Fourteen of the fifteen members of the Security Council voted for Resolution 1244, which was adopted on June 10, 1999. The resolution set forth a plan that gave the UN sweeping responsibilities in Kosovo: policing the province, overseeing human rights abuses, organizing elections, and overseeing the return of more than a million refugees and displaced persons. The resolution gave NATO sweeping powers, including the task of "deterring renewed hostilities." Secretary-General Kofi Annan was to appoint a special representative to ensure close cooperation with the NATO-led force.

The ending of the war pointed up some of the frictions among UN members over the NATO action in Kosovo. China withheld approval of Resolution 1244, complaining that NATO's attacks had violated the charter of the UN, though it did not veto the agreement. Russian Security Council representative Sergei Lavrov, one of the resolution's cosponsors, said his country "sternly condemned the NATO aggression against Yu-goslavia." The resolution left ambiguous the future role of Russian troops, who were outside the NATO chain of command. Cuba's represen-tative accused the NATO allies of "genocide" in Serbia. As Annan's direc-tor of communications, Shashi Tharoor, said, "the war was a NATO war, the peace will be a United Nations peace."

A series of further Security Council resolutions followed. Resolu-tion 1244, passed on June 10, governed the details of ending the war, though it made no specific reference to NATO's plan to deploy fifty thousand peacekeepers in Kosovo.[77] When those troops began to move in for the occupation that followed, they found Kosovo riddled with evi-dence of Milošević's rule of terror. The Serbs had raped, beaten, and killed thousands of Albanians, and some 800,000 Kosovars had been dis-placed. The UN High Commissioner for Refugees (UNHCR) estimated that Kosovars had fled to other countries in the following numbers: Albania, 444,000; Macedonia, 247,000; Montenegro, 69,600; and Bosnia-Herzegovina, 21,700. Peace was much needed, although there was little peace to keep.

The cost of a year of negotiations was high. According to the prosecu-tor of the International Criminal Tribunal for the former Yugoslavia Carla Del Ponte's report to the UN Security Council, from January 1998 until the end of the NATO campaign the Serbs killed more than eleven thousand people in Kosovo.[78] Del Ponte's office reported 2,108 bodies ex-humed from 195 of 529 known mass graves, and perhaps 6,000 bodies in all the mass graves.[79] Europe had not seen atrocities of such magnitude in the years since the Second World War. The fact that genocide was possible in Europe at the dawn of the twenty-first century seemed incredible, but after the experience in Bosnia, it is amazing that no one reacted sooner to Milošević's ethnic cleansing of Kosovo. Some have said that Clinton made a mistake in announcing that he would not send American ground troops to Kosovo—that it only emboldened Milošević, who believed he could outlast an air campaign. But Clinton ultimately provided enough force and leadership to prevail, and the NATO air campaign was definitely a success. Approximately 750,000 refugees have returned to their homes, "a remarkable testament to the success of U.S. and NATO policy."[80]

AFTER THE WAR

There is no doubt that life in Kosovo improved considerably after the NATO bombing, which brought an end to Milošević's campaign of violence. After the defeat, and in the face of economic deterioration in Serbia, Milošević lost popular support and was overthrown in the parliamentary election of 2000. Vojislav Kostunica was elected as president, and reformist Zoran Djindjic served as Serbia's prime minister until his murder—by a Belgrade-based criminal gang, according to the Serbian government—in 2003.

In the years since, Belgrade has made greater efforts to cooperate with Kosovo authorities and the West. In June 2001, it handed Milošević over to the International War Crimes Tribunal in The Hague, and the Yugoslav government has been relatively more responsive to Western demands for democratization.[81] Strains in relations between Belgrade and Pristina persist, however, as the question of Kosovo's ultimate status remains unresolved. The Serbian government expects Kosovo to come under Serbian administration once the UN interim administration ends. The Kosovar Albanians, on the other hand, expect Kosovo to become independent. In March 2002, when the leaders of Podgorica and Belgrade signed an agreement to form a union of Serbia and Montenegro, the issue of Kosovo's status returned to the fore.[82] The mention of Kosovo as part of Serbia in the preamble of the draft Serbian constitution renewed mutual suspicions and sparked again demands for an independent Kosovo.[83] The Serbian government was not inclined to allow Kosovo to separate, but the agreement allowing Montenegrins to decide freely the future status of their union with Serbia offered hope that Belgrade and Pristina might reach a similar peaceful agreement for greater Kosovo autonomy.

The political framework established by Security Council Resolution 1244 is important to Kosovo's improved situation. The resolution adopted the proposals from the G-8 meeting in Cologne, which still classify Kosovo as constitutionally part of Serbia but under an international protectorate. It is administered under the civil authority of the UN Interim Administrative Mission in Kosovo (UNMIK), established on June

10, 1999.[84] Pending final settlement of the matter, UNMIK was responsible for promoting the establishment of substantial autonomy and self-government in Kosovo; for performing basic civilian administrative functions; for organizing and overseeing the development of provisional institutions for democratic and autonomous self-government, including holding elections; transferring administrative duties to these institutions as they are established; for facilitating a political process to determine Kosovo's future status; for supporting the reconstruction of key infrastructures; for maintaining law and order; for protecting human rights; and for ensuring the safe and unimpeded return of refugees and displaced persons to their homes in Kosovo.

Security in the region was administered by an international security force, KFOR, which includes forces from Canada, Finland, France, Germany, Greece, Italy, Lithuania, Norway, Poland, Portugal, Sweden, Turkey, the United Kingdom, and the United States.[85] The problem of how and whether to incorporate Russian troops was solved when the Russians marched over from Macedonia and took over the Pristina airport.[86] The incident ended with the incorporation of Russian troops into NATO forces and the assignment of 2,850 Russian troops to the U.S., French, and German Multinational Brigade Kosovo Sectors. The Russians also retained responsibility for security at Pristina Airport, with another 750 troops.[87]

But life in Kosovo is by no means back to normal. The economies of both Kosovo and Serbia were destroyed. And although the genocide was halted, the simmering hatred between the two sides, and the extremism and discrimination cultivated over a decade of violence, cannot be easily eliminated.

Despite these problems, the prolonged UN presence has contributed positively to achieving long-term peace and stability in the region. UN efforts at reconciliation have encouraged the Serb and ethnic Albanian communities to cooperate. On July 23, 2000, representatives from both communities signed the Airlie Declaration at the Airlie House in Warrenton, Virginia. They agreed to work together toward "building a peaceful accommodation, despite great pains and sorrows suffered in past conflicts."[88] They agreed to cooperate in the development of democracy, free media, civil society, and the return of the displaced people and to create conditions for greater and safer participation of the Serbian community in local elections.

This cooperation, however, was more easily achieved on paper than in reality. Despite efforts by the UN to involve the Serbian community in the democratic process, the Serbs boycotted municipal elections on October 28, 2000, and elections for the Kosovo Legislative Authority on November 17, 2001.[89] Still, the UN reported that the elections went smoothly, with voter turnout of 64.3 percent of Kosovo's 1.25 million eligible voters, and without any major incidents.[90] Kosovars elected Ibrahim Rugova's moderate party, the Democratic League of Kosovo (LDK), with 46 percent of the vote, although the LDK did not get enough votes to form a majority government.

A much longer-term problem is the huge damage that was done to Kosovo during the 1990s. Although thousands of Kosovar refugees were eager to return from their temporary refuge in Macedonia, Albania, and Montenegro, the landscape they found when they arrived was devastated, full of destroyed villages and uninhabitable houses. Hashim Thaci, the leader of KLA, pledged that Kosovo would be a democratic nation and encouraged people to think that "Kosovo [would] have respect for human rights, a free media, democratic institutions, and free and democratic elections."[91] As reported by the *New York Times*, Thaci told a news conference that "the Yugoslavs should finally come to the realization that the future of Kosovo will be decided in Kosovo proper."

In retrospect, most Americans and much of the NATO leadership underestimated the strength and determination of Milošević and the Serb leadership. *Winning Ugly*, the excellent study of the Kosovo conflict by Ivo Daalder and Michael O'Hanlon, provides ample documentation of the pervasive, consistent underestimation of Serb strength. For example, they quote Secretary Albright on *The NewsHour with Jim Lehrer* on March 24, 1999, the night the bombing began: "I don't see this as a long-term operation. I think this is something that is achievable in a relatively short period of time."[92] This expectation, the authors say, was "widely shared in Europe among civilian and political leaders alike."[93]

Still, however they may have underestimated the timing, the Western powers scored an important moral victory in Kosovo. Faced with a violent aggressor who refused to heed warnings, NATO took action—armed not with a mandate from the UN Security Council, but by consensus of its member nations, including the United States Congress. Even with congressional approval, some still feel that NATO should not have acted without a specific Security Council mandate. At the time, some members of the Contact Group were uncertain about the legal basis of their action. But the Yugoslavs had refused to comply with numerous demands from the Security Council under Chapter VII of the UN Charter. This non-compliance underlay the North Atlantic Council's decision to act; they expected another Security Council resolution would be passed in the near future. In the meantime, however, Slobodan Milošević was posing an ever-increasing threat to international peace and security. Time was of the essence. And NATO seized the day.

6.

CONCLUSION: AFGHANISTAN AND IRAQ

The doctrine of non-intervention, to be a legitimate principle of morality, must be accepted by all governments. The despots must consent to be bound by it as well as the free States. Unless they do, the profession of it by free countries comes but to this miserable issue, that the wrong side may help the wrong, but the right must not help the right. Intervention to enforce non-intervention is always rightful, always moral, if not always prudent.[1]

—JOHN STUART MILL

When terrorists attacked the United States on September 11, 2001, we were given ample reason to reassess the direction of our foreign policy during the preceding years. With disturbing clarity we could see that small groups, which were essentially anarchical and homeless in nature, had the will and the ever greater means to attack a superpower—more so, even, than the nations who harbored or sponsored them. Our foreign policy had become embroiled in peacekeeping and nation building under the interventionist efforts of a world collective of nations; we had set our national security second in priority, misleading ourselves that creating a world of peaceful nations would ensure the protection of our own national security interests.

These trends in our foreign policy played out in two very different military actions after the September 11 attacks: one in Afghanistan, one in Iraq. One military action I could support without reservation, the other I did not support.

Throughout my career, I have been careful not to criticize any sitting president, and I was not inclined to change my position in that regard when President George W. Bush sent troops across the border of Iraq. In fact, when asked, I even agreed to defend his actions. I believed then as I believe now that President Bush had the legal right to invade Iraq, if not entirely for the reasons his administration claimed. However, I also believe that he had neither the obligation nor the need to expand his military offensive into Iraq after sending troops into Afghanistan.

What struck me then, and has remained with me ever since, is that the aggressors of this moment in time may be the small number of masterminds who direct the network of terror known as al Qaeda, but that they have their precursors in the bad actors we have encountered at other times and places in our history: the likes of Joseph Stalin in the former Soviet Union, or Slobodan Milošević in the former Yugoslavia. Aggressors are a constant in history. They seek to impose their will on governed masses, which are denied any voice in their own destinies and any recourse to justice. The rule of law, the sovereignty of states, and basic human rights become collateral damage before such ambitions.

When great men in our history—Woodrow Wilson, Franklin Delano Roosevelt, and George Herbert Walker Bush most recently—have promoted the concept of a new world where peace is a priority shared by all nations, they have been guided by a certain vision of common purpose, of the importance of building a democratic world order. The resurgence of destructive aggressors may challenge the basic beliefs of great leaders, but it does not prove them fools. The United Nations, and the League of Nations that preceded it, have played a crucial role in the dialogue among nations, particularly in dealing with world aggressors. The founders of the UN, however, never intended that its role would extend to exercising sovereign rights reserved to its member states. In the years since the end of the cold war, efforts within the UN to usurp this power have contributed to undermining the peace and the stability of nations.

As peacekeeping and nation-building efforts advanced over the fifteen years before the 2001 attacks, the role of the UN was expanded de facto. At the same time our foreign policy was losing focus; our attention to national security was subsumed by a desire to promote democracy, as if democracy alone could imbue chaotic societies and unstable governments with a respect for what we respected: the rule of law, basic human rights, and a peaceful world order. As the emergence of al Qaeda demonstrates, peace and stability among nations are not the priority of all groups who seek power in the world.

As America watched the horrific images of the World Trade Center's twin towers crumbling, the hull of the Pentagon burning, and a crater smoldering in a desolate field in Pennsylvania, it had never been more apparent how deeply the world had been affected by our efforts at peacekeeping and nation building—and how uncertain, and dangerous, the results of those efforts could be.

THE SOVIET INVASION OF AFGHANISTAN

When the Soviet Union invaded Afghanistan on December 25, 1979, it devastated much of the country and kidnapped many Afghans. Suddenly, the United States was forced to formulate a policy with regard to this unfamiliar, distant region. Crafting policy in Afghanistan turned out to be far more complicated than most Americans realized, and the foreign policy decisions made in the following years set off a series of events from which we must learn if we are to understand the crossroad where our nation today stands.

Afghans constitute a complex society with multiple tribes and clans who have fought one another for centuries. They are a tough people with a history of defeating whoever goes to war against them, including the Soviets—a fact of which they remain justly proud today.

When the Soviets invaded, Afghanistan had experienced six years of political tumult after King Zahir Shah had been overthrown in 1973 by his cousin. The king's anemic monarchy had been unable to assimilate with the strong forces of the tribal clans, and the unintended consequence was that the nation became a kind of vacuum, weakened by vying factions, ripe for foreign invasion. Sensing an opportunity to establish a presence in Asia, with a close proximity to the Middle East, the Soviets moved into Afghanistan, imposing their military and social justice by force.

At the time of the invasion, we were in the midst of the cold war, at the height of the nuclear arms race, and the Soviet threat was our most urgent task. At this precarious moment in history, Americans became involved on the periphery of the Soviet war in Afghanistan. We helped the Afghans, or at least we tried to. We were not always certain which warlords we should help or trust, and we supported the Afghan resistance, led by the mujahideen, to defeat the Soviets. It is clear today that we made some mistakes, but we did our best, and eventually, with our help and Afghan ferocity, the Soviets withdrew from Afghanistan in February 1989 after ten years of bloody battles.

At that time, I was serving in the U.S. government as a member of Ronald Reagan's cabinet, his National Security Council, and an inner circle called the NSPG (National Security Planning Group). During our discussions on these subjects, we gave little consideration to whether it was prudent for the United States to leave Afghanistan after the Soviets withdrew. The conflict was done, and it seemed appropriate to leave, so we did. Talks of maintaining a U.S. presence, of extending an occupation, or even of nation building were not seriously contemplated. This was not due to a lack of compassion. It was because we assumed the people of Afghanistan, like people of all countries, would be eager to take control of their own affairs and govern themselves. This belief is shared by most Americans; that is fundamentally why the United States has never developed a colonial empire.

Not long after the U.S. personnel left, fighting spread among the various warlords, ethnic groups, and factions in Afghanistan. The wars among these warlords and ethnic groups were bitter, characterized by personal violence among families, clans, and groups. The long struggles further fractured an already fragile society, until the Taliban emerged as the strongest and the most violent. Dogmatic and harsh, the intolerant Taliban moved ruthlessly to eliminate opponents and consolidate power. Their near-universal repression was far more onerous than anyone foresaw. Soon after seizing power, the Taliban began systematically subjugating other clans, and girls and women of every clan. Afghan women, who had previously enjoyed and participated in all facets of social and professional life, were swiftly sent home, ordered to resign from the hospitals and schools they once staffed and to leave the professional roles to men. Young girls were summarily turned away from their schools. Those who refused to follow Taliban rule were harshly punished. The Taliban's punishments were frequent, ruthless, and carried out in public places: public floggings of women and men, amputations, executions, and stoning became frighteningly commonplace under Taliban rule. Many Afghans were deeply shocked by the new regime, and many of them went into exile. Large Afghan refugee communities formed in Europe and the United States, where many exiles detailed the cruelty and ruthlessness of the Taliban. Their ominous words and warnings should have been taken seriously by any government who sought to engage with Afghanistan.

While the Taliban controlled much of Afghanistan, they were also providing refuge, education, and training to al Qaeda while the terrorist group prepared, trained, and plotted jihad against the West, culminating (so far) in their 2001 attacks on the United States. When the United States dismissed the possibility of occupying Afghanistan after the Soviets withdrew, or of participating in a nation-building effort there, we had no foreboding or expectation that anarchy and totalitarian terror might be a consequence. Even in retrospect, it is hard to imagine how anyone might have divined such devolution of circumstances.

In fact, looking back at history and the decisions that were made, I believe that if we had understood the likely consequences, the United States would not have withdrawn from Afghanistan when the Soviets left. Instead, we might have preserved a presence, in order to help support a moderate faction cope with the challenge of Taliban extremists. In hindsight, we might also have recognized that at least one of the warring groups in Afghanistan would seek to settle the issues of government by force, if only because that is one turn of history we might always count on—the resurgence of coercive powers seeking to dominate others. Instead, the United States withdrew, for a noble but perhaps shortsighted reason: because Americans have no inclination to occupy other countries.

Seventeen years after the end of the Soviet-Afghan war (and three U.S. administrations later), President George W. Bush took to the airwaves to announce that the first military response to the 9/11 attacks had begun. "On my orders," he announced, "the United States military has begun strikes against al Qaeda terrorist training camps and military installations of the Taliban regime in Afghanistan."[2] The Bush administration immediately declared its reasons to the people it governed and to the world. In explaining the launch of Operation Enduring Freedom, the U.S. rightly argued that the 9/11 attacks were part of a series of strikes on the United States that had begun in 1993. This campaign of attacks had now intensified, and that escalation warranted an escalated response. Moreover, the United States and United Kingdom had the strength of evidence that more attacks were impending if they did not take action. Action was required in the name of national security, for threats from the Taliban and al Qaeda had already been carried out and more were imminent. This marks one critical distinction between the Bush Doctrine as it was applied to Afghanistan, and as it was later expressed in justifying the invasion of Iraq.

Another distinct difference between the Bush administration's military operations in Afghanistan and in Iraq was found in the reactions in the world community and the UN. When Operation Enduring Freedom began, world opinion rallied alongside the United States and its stated objectives in Afghanistan. Notably, Secretary-General Kofi Annan refrained from criticizing the United States or Britain during or after their military strikes in Afghanistan. When he was asked why he had not done more to fight terrorism, and if he thought the UN had been sidelined during Operation Enduring Freedom, his response suggested a view of the United Nations that goes beyond its traditional role:

> Countries have to cooperate as the Security Council has indicated in refusing shelter for terrorists, in denying them the use of financial resources, and making sure there is no logistic support. And I believe the actions that the Security Council and the General Assembly have taken provide a solid basis for international action and international cooperation around the globe. . . . The military action on which we are fo-

cused for the moment in Afghanistan is quite frankly a very small part of the fight against global terrorism. The Council in its resolution indicated that the perpetrators must be brought to justice. And the Council also indicated that all means must be used to prevent attacks of that kind. So when we talk of the fight against terrorism, I would disagree with you that the United Nations is sidelined. In fact, on the key issues, the initiatives and the foundation are being laid by the United Nations.[3]

Today, with the help of our allies, including the Dutch, the Canadians, and the British with NATO in the lead, the United States continues to engage the Taliban in battle in Afghanistan. We work closely with freely elected Afghan President Hamid Karzai's administration to establish the rule of law critical for the hope of peace and democracy for the people of Afghanistan. As President Karzai said in September 2006, "Afghanistan is a country that is emerging out of so many years of war and destruction and occupation by terrorism and misery."[4] And a future peaceful and stable Afghanistan will serve not only the region but our national security as well.

Had we known then what we know now, would it have been possible to have prevented today's conflict in Afghanistan? That question is impossible to answer in retrospect, but after 9/11 we cannot claim ignorance when making foreign policy decisions. We know enough now about the possibilities of anarchy, violence, conquest, indoctrination, and civil war. In Afghanistan, we saw how those elements can interact and create a very real threat on our own national security. In the war we fought in that region to protect our national security, we also chose to follow a more prudent course by supporting more moderate and legitimate forces against the aggressive forces that have proven their intentions as rulers. We have come to see that we should—indeed must—recognize the need to help others cope with these threats in their homelands, as they gain experience in governing themselves and in sustaining stable institutions.

Today, we need the sort of wisdom Tocqueville had in mind when he wrote about the varieties of experience needed to make policy in remote places: "Experience, instruction, habit, and all the homely species of practical wisdom that are required to make sensible rules about everyday life are required for those who would govern and those who would advise those who would govern."

America, among all the nations in the world, has such breadth of experience woven into its very cultural fabric. As Harry Truman, one of my favorite presidents, observed in his memoirs:

> Our populace, unlike that of any other great nation, is made up of strains from every population around the world. And when we became the most powerful nation in the world, we tried to put into effect the ideals of all races and nationalities. All of which we have written into our Constitution and Declaration of Independence.

If our country and culture are informed by people of all kinds, as Truman suggested, then it is no wonder that we assumed that we were safe in projecting our ideals and concerns to all races and nationalities. As a melting pot of nationalities and races, we assumed that our national character could be a reliable guide to the expectations of others around the world. We made what I think were democratic assumptions about goals of other people and places, even those that did not necessarily share our ideals and concerns. We assumed that the goals and values of peoples in other regions were similar to ours, and we expected their approach to achieving those goals to be familiar to us.

What September 11 taught us—or should have—is that our assumptions, though reasonable, were not universally true. It had come time to admit our error. There are people committed to, and indeed driven by, goals and values that run violently counter to our own.

Perhaps we had simply grown accustomed to dealing with modern Europeans who live comfortably in industrial democracies. Perhaps, as we worked to develop a collective world organization, we assumed that such ideals as human rights and the rule of law, which had evolved in Western civilization, were universally appealing. For whatever reason, even as we won the cold war without firing a gun or a missile, we lulled ourselves into a false sense of security, and so began to believe that world stability and security could be self-sustaining, within or among nations.

Yet that belief is not consistent with the recurring lesson of history—

which is that our world is periodically changed by violent aggressors, who bring about chaos and tragedy in their efforts to advance their own objectives.

THE AMERICAN INVASION OF IRAQ

On a personal note, I have dedicated much of my professional life to reconciling what I consider the twin goals of American foreign policy, and that is why President George W. Bush's decision to go to war has troubled me deeply.

These twin goals of our foreign policy are, first, ensuring our security and, second, promoting democracy and human rights. An appropriate balance between the two must exist, and that balance must be determined within the unique circumstances of any situation. Yet, for democracy to take hold in a given region, it must be preceded by institutions that are receptive and willing to support democracy—because democracy requires security as a prerequisite. That is why, throughout history, if the single force of political stability in a region is removed without critical institutions in place to fill the resulting vacuum of power, the security of societies and their budding institutions will be precarious at best.

Unfortunately, what we face in Iraq today is a vacuum of power, a lack of stable institutions needed to govern, and the problem that the promise of democracy for which our nation stands may be lost in the essential scramble for safety and stability in the streets. This is one of the reasons I am uneasy about the war we have made here—for we have helped to create the chaos that has overtaken the country, and we may have slowed rather than promoted the pace of democratic reform.

My role in the Iraq war began at its outset, in March 2003, when I was recalled to active duty by the Bush administration to serve as the head of the U.S. delegation to the United Nations Human Rights Commission[5] in Geneva. At the time, historic forces were already in play. U.S. and coalition forces were destined for Baghdad. The Bush administration was justifying the legitimacy of its use of military force in Iraq, citing the right of preemptive self-defense. Meanwhile, credible rumors were circulating in diplomatic circles that a resolution to condemn the Iraq invasion was in the works—an international response that was quite different from what we had received when we counterattacked our enemies in Afghanistan. Just two days after the Iraq invasion, on March 22, the Arab League voted in favor of a cease-fire resolution while meeting in Cairo. The resolution carried no consequences and had no real momentum, yet it marked a watershed in the growing international opposition to George W. Bush's decision to invade Iraq—and the United States could not afford for the movement to gain momentum.

At the request of the Bush administration, I had gone to work with an able team, including Allan Gerson as legal counsel. We began by swiftly delivering letters to the Arab League members and the African Union, designed to counter their budding efforts to condemn the U.S. entry into war. Spearheaded by Syria and Cuba, the opposition was joined by Zimbabwe, South Africa, Malaysia, Sudan, Libya, Burkina Faso, and even Russia. Then, on March 26, that group convened a special session in Geneva to introduce a resolution entitled "Human Rights and Humanitarian Consequences of the Military Action Against Iraq." A careful read of the resolution's text revealed that humanitarian concerns were secondary to its real purpose. This was a declaration to condemn the coalition action in Iraq as "clearly in violation of the principles of international law and the United Nations Charter" and to call for "an immediate end to the unilateral military action against Iraq."[6]

Had such a resolution passed, it would effectively have been a triumph for Saddam and a potential diplomatic disaster for the United States. Unlike in the UN Security Council, the United States does not enjoy absolute veto power at the UN Human Rights Commission. Suddenly, Geneva had surpassed the UN General Assembly and the UN Security Council as the international diplomatic theater for the efforts to challenge Bush's invasion of Iraq. Reasoning behind the debate in Geneva was predictably chaotic, as the debate that the first President Bush had brought before the Security Council was now reexpressed, by his son's administration, in humanitarian and human-rights terms. Some member states argued that justifying force as preemptive action could open the door to unlawful reprisals from Israel or other nations under other circumstances. Other states were reluctant, with good reason, to support the use of military force as an intervention to promote democracy.

Ultimately, the member states were struggling against allowing the

Bush Doctrine to set a precedent that condoned the use of military force preemptively as a sufficient (and independent) cause for entering into war. Preemptive action, as Bush's then national security advisor, Condoleezza Rice, had stated, "must be treated with great caution. It does not give a green light—to the United States or any other nation—to act first without exhausting other means, including diplomacy."[7] Rice's statement was and is correct, but it did not reassure a very worried international community, nor was preemptive self-defense a legal argument to use to block a resolution calling for a cease-fire.

I had grave reservations when George W. Bush made the decision to invade Iraq, and I was privately critical of the Bush administration's argument for the use of military force for preemptive self-defense. But I was not critical of the Bush administration's lawful purpose and could confidently affirm our legal rights when I was tasked to block the resolution whose purpose and substance belonged before the UN Security Council. That is why, when I agreed to represent the United States at the Geneva Human Rights summit, I did so—on the condition that I could abandon the Bush Doctrine of preemptive self-defense. In my opinion, the analysis that had led the administration to the Bush Doctrine was deficient. Despite this, however, a valid argument could be made under international law that President Bush was within legal bounds when he used military force to enter Iraq in 2003. What was at stake, in Geneva, to my mind, were the two concepts of the rule of law and of sovereign rights, and both deserved forceful advocacy in the chaotic efforts of the opposition to confuse the basic issues at work. I could go to work in Geneva representing President Bush's right to use force in Iraq, based on these personal and professional moral convictions—even though I did not agree with the president's choices.

What I argued in Geneva with my team of advocates was that the 2003 act of force on Iraq was not *going* to war. It was, rather, the continuation of the 1991 Gulf War, and thus wholly permissible under the rule of law. UN resolution 687, which contains the terms of the cease-fire with Iraq, and which was negotiated during the interruption of the first Gulf War, had clearly been violated. Indeed, Iraq had never fulfilled the terms of the 1991 cease-fire agreement. Resolution 687 was as valid in 2003 as it was in April 1991 and as it is today. The legal authority to use force to address Iraq's material breaches was and remains clear, and is a matter of record. So the case I presented to the international community in Geneva in March 2003 at the bequest of the Bush administration was an argument based on the rule of law, not an argument on behalf of the Bush administration's assertion of its right to preemptive action in self-defense. This distinction was not lost on the world community.

On March 27, 2003, a week after the invasion of Iraq, a vote for a "humanitarian" resolution on Iraq was called in Geneva, proposing to discuss purported human rights violations connected with the war—and thus, by implication, to question the legitimacy of the war. Its outcome was uncertain, and tensions were high. Standing firmly with the United States were the European Union, Germany, and the chairman of WEOG (the Western European and Others Group). In Africa, the Cameroons and Uganda broke ranks with the African Union to vote against the resolution; Senegal, Togo, and the Democratic Republic of Congo abstained. Swaziland and Sierra Leone decided to absent themselves from the day's business, as did Ukraine. Thailand, despite pressure from the Asian bloc, voted with the United States and her allies. The final tally was twenty-five against the resolution, eighteen in favor, and seven abstentions.[8] The U.S.'s legal right to use force in Iraq, a nation in breach of the 1991 ceasefire agreement it had signed, was affirmed, and in that way reason—and justice—did prevail in the hallways of Geneva.

THE LEGALITY OF WAR QUESTION

As of this writing, nearly three thousand U.S. service men and women have given the ultimate sacrifice in America's latest war. While voices from both sides of the aisle, and from around the globe, continue to debate contentiously most aspects of Iraq's current and future affairs, we have a responsibility to ensure that history is recorded accurately and honestly. We must unflinchingly acknowledge our mistakes as well as celebrate our successes—and must learn from both. Only then can we set a wise and prudent course in future foreign policy decisions.

When weapons of mass destruction, which lay prominently within the Bush administration's decision for preemptive military action in self defense, were not located by Americans or other weapon inspectors—as the U.S. government had seemed certain they would be—criticism of the United States and its credibility naturally escalated. These credible criticisms fed less credible, in fact quite incredible, attacks on America's character, policies, and actions. One result is the reemergence of a question that should have been put to rest long ago in Geneva, where the opposition's argument was routed by an appeal to the rule of law: Was the U.S. invasion of Iraq legal?

It is very important to note that it was former secretary-general Kofi Annan who resurrected this question in the public sphere. During a September 2004 interview with the BBC, Annan bluntly called the 2003 invasion of Iraq "illegal" and alleged that it was not in conformity with the UN Charter.[9] Not surprisingly, he made no reference to the defeat of the Geneva debate, and the Human Rights Commission's resolution in favor of the rule of law and of sovereign rights. Annan's charge against the United States and its allies was a glaring mistake; the invasion was not illegal, at all, even if it was not supported by member states—or by every American. For Annan to express such an inaccuracy on the world stage would not have been all that noteworthy, except that his effort to reignite controversy involved this unprecedented act: a secretary-general of the United Nations had overridden, in word if not in deed, the rule of law. To my knowledge, in the fury of debate over the war, no one effectively covered this historic travesty.

Recently, I have been asked by several foreign journalists whether I think the United States will regain its reputation and its credibility in the wake of the secretary-general's charges and the failure to locate the weapons of mass destruction that were at the heart of the Bush administration's case for war in 2003. My answer is yes. I am confident that America will survive these attacks. But such assaults are never welcome, and overcoming them is never simple.

To begin, however, we must bring some focus back onto the UN and the role of the secretary-general in the ongoing controversy about the American role in Iraq today—and in perpetuating the damage to the reputation and credibility of the United States. Not only was it shocking and atypical for a sitting secretary-general to lodge such an unwarranted charge against a permanent member of the Security Council, it was not consistent with UN culture. As the United States must weather the storm that surrounded its choice to invade Iraq without UN support in 2003, so must the UN culture renew its commitment to the rule of law—and to the will to enforce the legal order of nations.

For example, instead of retracting his accusation against the United States and coalition forces, Annan continues to reiterate it—unchallenged. Among other things, his condemnation was notably selective, for Annan was fully aware that the Soviet Union's march into Afghanistan and France's efforts to assist Iraq in acquiring and constructing a workable nuclear weapon in violation of the Nuclear Proliferation Treaty, both constituted *true* violations of international law. Yet former secretary-generals refrained from accusing either of these permanent member states of illegal behavior. Nor did Annan charge or condemn the Clinton administration or NATO nations with illegal behavior when NATO bombed Milošević's Serbs in Kosovo without Security Council authorization. It is true that Annan complained during Kosovo, but his complaints never rose to the serious charge of breaching the rule of law by engaging in illegal behavior.

Perhaps most ironic of all, for a decade the nation of Iraq repeatedly violated the series of Security Council resolutions that followed the 1991 cease-fire—yet Annan did not rebuke Iraq for illegal behavior. And yet, by condemning the U.S. action in Iraq in 2003 as "illegal," he provided an excuse for many member nations to refuse to back any effort to force Iraq to abide by the terms of the 1991 cease-fire and, in doing so, to absolve themselves of their crucial responsibility as member states to take action against true crimes against international law and to aid in the restoration of peace for the people of Iraq.

A secretary-general acts as the UN's chief diplomat. His comments and actions carry global impact, with potential long-term implications. So it is imperative for all secretary-generals to act responsibly, to speak honestly, and to work within the UN Charter—that is, to affirm the rule of law among nations as a first priority. The more a secretary-general can live up to the gravitas of his role, the more warranted will be efforts such as those by George H. W. Bush to demonstrate respect for the UN and its Charter, as he did with the first Iraq war. In the process, the UN culture itself may shift toward enforcing stability and prosperity based on the rule of law—not the whim or greed of individuals. Yet, always, the will to oppose forces hostile to the rule of law must remain vigilant, as a review of even the few years between the two Iraq wars attests.

On the Road to War

By the time the United States was preparing for war in 2003, the UN oil-for-food program was in its sixth year of operation. It was established by the Security Council in December 1996 to provide temporarily for the humanitarian needs of the Iraqi people during economic sanctions, which had been imposed after Iraq's invasion into Kuwait. The oil-for-food program allowed Iraq to sell oil under UN auspices in order to use the proceeds to purchase food and medicine. Shortly after the 2003 invasion, the program was dismantled.

On February 3, 2003, U.S. secretary of state Colin Powell addressed the UN Security Council to review Iraq's disarmament obligations under Security Council Resolution 1441, which passed by unanimous vote in November 2002.

By then, Saddam had violated some seventeen previous resolutions demanding his verifiable disarmament. Resolution 1441 was passed after U.S. and European inspectors agreed that Iraq had not fully complied with its obligations under previous resolutions, and that it was in "material breach" of Resolution 687. Moreover, Resolution 1441 listed *nine binding steps* Iraq must take to give inspectors "unrestricted access" to sites where weapons might be hidden or face "serious consequences." As of that February day when Secretary Powell addressed the Security Council, no steps had been taken to abide by that resolution.

Resolution 1441 was critical, as Secretary Powell commented:

> The purpose of that resolution was to disarm Iraq of its weapons of mass destruction . . . Iraq already having been found guilty of material breach of its obligations, stretching back over sixteen previous resolutions and twelve years!
>
> Resolution 1441 did not deal with an innocent party, but a regime this Council has repeatedly convicted over the years. It gave Iraq one last chance to come into compliance or face serious consequences.[10]

Secretary Powell reminded the Security Council that Resolution 1441 had placed the burden on Iraq to comply and disarm—not on the inspectors to find weapons Iraq had worked to conceal. He reminded the council of the assessment made by Dr. Hans Blix, the chief UN weapons inspector, whose January 27 report stated that "Iraq appears not to have come to a genuine acceptance, not even today, of the disarmament which was demanded of it." [11] Powell reinforced Blix's findings using the IAEA director's December report to the Council, in which Dr. Mohammad El Baradei demonstrated that Iraq's declaration "did not provide any new information relevant to questions that have been outstanding since 1998." [12]

Secretary Powell's case also reviewed the many conversations among Saddam and his closest collaborators about their efforts to hide weapons. Describing the Iraqi scientists' reports, he recapped three key points:

1. Saddam Hussein has used horrific weapons on another country (Iran) and on his own people. . . . No other country has had more battlefield experience with chemical weapons since World War I.

2. Saddam Hussein has never accounted for vast amounts of chemical weaponry now unaccounted for .

3. Iraq's record on chemical weapons is replete with lies. [13]

For Saddam Hussein, the UN threats were not persuasive; their challenges were merely verbal, after all, not military. In the meantime, his noncompliance was advancing his own agenda and his own self-interest. Saddam himself could see that he had been involved in behavior counter to his own agreements with the world community for over a decade with no real consequence. What was permissible on the international stage was uncharted territory well beyond the rule of international law.

KOFI ANNAN EXPANDS HIS UN ROLE

Annan's charge that the U.S. invasion was "illegal" was only the latest in a series of similar remarks he had made from the beginning of the Iraq debate. Like his immediate predecessor, Boutros Boutros-Ghali, during the Bosnia crisis, Annan made unprecedented assertions of his authority and power to define the legitimacy and "legality" of a member state's decision to use force. Before the invasion began, for example, Annan told a press conference in The Hague: "Security Council members now face a momentous choice. If they fail to agree on a common position [concerning Iraq], and action is taken without the authority of the Security Council, the legitimacy and support for any such action will be seriously impaired. If the U.S. and others were to go outside the council and take military action it would not be in conformity with the Charter." [14]

Then, less than two weeks after the invasion, on April 2, Annan told the Arab network Al-Jazeera that he had "raised questions about the legitimacy" on the use of force by U.S. and coalition forces, "and whether it was in conformity with the Charter." [15] Once again, Annan claimed a unique right to decide how the issue of Iraq should be dealt with, and by whom.

Some five months later, speaking to the General Assembly on September 23, 2003, Annan took another step, presuming to define the limits of member states' "inherent right of self-defense" as defined in Article 51 of the UN Charter. Article 51 states that "Nothing in the present Charter shall impair the inherent right of individual or collective self-defense if an armed attack occurs against a Member of the United Nations, until the Security Council has taken measures necessary to maintain international peace and security." Annan, in his September speech, unilaterally added to Article 51 an entirely new assertion. "Until now," he said, "it has been understood that when states go beyond that and decide to use force to deal with broader threats to international peace and security, they need *the unique legitimacy provided by the United Nations*." [16] This is a troubling formulation, one that raises more serious questions than it answers. By whom has Annan's claim been so understood or authorized? Certainly not by the Security Council. For that matter, who decided that the Security Council was the source of the "unique legitimacy" provided by the United Nations? Certainly not the sovereign member states who compose the United Nations. Annan's effort to expand the power of the UN itself, and with it his own role internationally, garnered little attention as international opposition to the Bush administration was gaining ground.

No answers were forthcoming when Annan informed the General Assembly that he intended to challenge the belief of some states that they have the right and obligation to use force preemptively without Security Council agreement. His remedy was to establish a high-level panel of eminent personalities (of his own choosing) to reflect on the problems of the United Nations and to alter its composition and procedures. That in itself was a shocking expansion attempt of the secretary-general's powers as defined in the UN Charter. Article 99 states that the secretary-general *may bring* to the attention of the Security Council any matter which *in his opinion* may threaten the maintenance of international peace and security.[17] The UN Charter, however, gives the secretary-general no authority to *act* unilaterally on his personal opinions. Secretary-generals are meant to play a constructive role, not a destructive one. They work at the service of the Security Council, not the other way around. Any attempt to usurp the authority of the Security Council is dangerous, jeopardizes the rule of law, and does not conform within the UN Charter. And yet Annan was using the circumstances created by Saddam's flagrant disregard for the 1991 cease-fire as a pretext to further his own expansionist goals—with the preservation of the rule of law a distant second priority.

As U.S. and coalition forces began to battle insurgents on the ground, and to rebuild the Iraqi nation torn apart by the rule of Saddam and the war that ousted him, Annan continued to advance his vocal march forward to discredit the invasion. In a meeting with Arab journalists on March 8, 2004, Annan claimed, "I myself indicated that a war would not be in conformity with the Charter and the credibility of any such action would be widely questioned and the legitimacy would be widely questioned. And that is what has happened."[18] What the war was not in conformity with, of course, was Annan's own self-interested vision of the UN and its Secretariat, and his hopes to expand their international power.

No matter what any individual secretary-general may desire, however, the United Nations itself does not, and cannot, provide unique legitimacy to any action, whether peaceful or forceful. Only its member states may do so, because their decisions and actions derive presumably from the governed. Here the ideals of constitutional democracy and the rule of law are imbedded in the UN from the world leaders who created it. That is why the UN cannot presume a supranational role, but must instead remain a collective of sovereign nations. That is why Secretary-General Annan's opportunism in the wake of the Iraq war cannot, in the end, succeed. Even as the United States works to repair its international image, in some part due to Annan's expansionist behavior, the UN itself must also turn to a new secretary-general to help it survive the corruption and disarray in its own house, which are part of Annan's legacy. Until the UN reaffirms that its legitimacy is found in the sovereign rights of its member nations and in the rule of law, its effectiveness will be compromised.

This is borne out by many instances in recent history, in which, if a country had chosen not to exercise its sovereign rights until it had divulged a mission to protect or rescue its citizens, catastrophic consequences might have resulted. One such instance, which I recounted previously in this book and bears repeating, was Grenada, where more than six hundred American medical students were held at gunpoint by a group who held the island's entire population hostage with a shoot-on-sight curfew. This outlaw group had already killed five members of Grenada's cabinet in cold blood. The hostages in Grenada survived because U.S. troops, in a surprise landing, rescued them. Had the Reagan administration waited for the permission of the Security Council—or the "unique legitimacy" Secretary-General Annan claims can only be granted by the United Nations—before a rescue was attempted, many people, Grenadans and Americans alike, would have needlessly died.

The U.S. & Coalition Response

Along these lines, before President George W. Bush launched the invasion into Iraq, he did review Iraq's record. Like his father during the Gulf War, he explicitly argued that the United States has never taken the position that the use of force is legitimate only if it is specifically authorized by the Security Council. Nor could any serious nation have taken such a position in the decades since the UN was established. Nothing in the UN Charter can challenge sovereign rights. Even if an American were to disagree with the Bush Doctrine, as I did, no one can disagree with this fundamental analysis of the right of a nation to defend itself—or, under appropriate circumstances, to enforce a cease-fire by military means. The U.S. action in Iraq had been authorized by the president and the U. S. Congress. As Vice President Cheney commented, the United States requires no "permission slip" from the UN to use its military.

Moreover, as in the first Gulf War, the United States did not fight this war alone. Thirty countries joined the effort, among them the United Kingdom, Spain, Australia, Bulgaria, and Poland—all of whom rejected Annan's charge that the action was illegal. Though he was facing a tight reelection race where his support for the Iraq war was a crucial issue, Australia's prime minister, John Howard, declared: "The legal advice that we had, and I tabled it at the time, was that the action was entirely valid in international law terms." It was also valid in terms of U.S. law. The use of military force in Iraq in 2003 might have been ill-advised, in other words, but it was legitimate.

The U.S. permanent representative to the United Nations at the time, Ambassador John Danforth, took Annan to task for his statements and defended the sovereign rights of the United States and other member states:

> We don't agree with the Secretary-General on this point. I personally have the very highest regard for the Secretary-General, but when you consider there were sixteen Security Council resolutions and Resolution 1441 held that the then government of Iraq was not in compliance with the previous resolutions, which is clearly the case, and promised that there would be serious consequences if they were not in compliance, then the question for those countries that were part of the coalition was: Well, do all these UN resolutions mean nothing? Does the Security Council mean nothing? Is it totally ineffectual? When it says there are going to be consequences, that's really meaningless? So it seems to me that it would undercut the rule of law had there been no action.[19]

Ambassador Danforth's position was far from singular. Secretary of State Powell expressed strong disapproval of Annan's assertion, calling it "not a very useful statement to make at this point." Powell maintained that "that the U.S. Constitution gives the United States government the right to act in its own self-defense without UN approval. What we did was totally consistent with international law."

In addition, there was an overlooked question of precedent. As Richard Holbrooke, President Clinton's UN ambassador, reminded us:

> Three times President Clinton did what many Democrats are now saying Bush can't do. He did it in Bosnia in '95, in Iraq with Desert Fox in December of '98, and in Kosovo in '99. In the Balkans case he had no Security Council Authority. In the case of Iraq, in December '98, the UN was starting its meeting when they got word that the bombing had begun, and President Clinton simply said "We are bombing under UN authority because Iraq is in material breach." Since the end of the Gulf War in 1991, American and British warplanes have repeatedly bombed military targets in and around the no-fly zones in Iraq and all along their actions have had no [specific] UN mandate. They have been tolerated by the Security Council, presumably because most of its members maybe understood the need.[20]

Lord Goldsmith, the United Kingdom's attorney general, added to criticism of Secretary-General Annan's charge, arguing that the authority to use force derived from the combined effects of resolutions 678, 687, and 1441, all of which where adopted under Chapter VII of the UN Charter, which specifically allows the use of force to restore international peace and security. In his view, the combined effect of these resolutions meant that "military action against Iraq was legal without a second resolution."

William Taft, the senior legal adviser at the Department of State, concurred with Britain, though for somewhat different reasons. Taft emphasized prudential concerns, saying that "we are now acting because the risks of inaction would be greater," and asserting "the necessity of using force to protect against further harm." Taft also argued that "the inherent right of self-defense embodied in the UN Charter must include the right to take preemptive action; otherwise the original purpose is frustrated. We cannot wait for a first strike under such circumstances."

It is worth remembering that the U.S. government never made its argument to invade Iraq in 2003 *solely* from the need to eliminate weapons of mass destruction. From the beginning it also asserted the imperative to secure a "regime change," and that imperative finds justification in the long history of moral reasoning. It has been nearly a thousand years since John of Salisbury wrote a treatise justifying the murder of tyrants. Legal and moral theory has frequently considered the use of military force simply in order to remove a tyrant who threatens and provokes his own people and the norms of the community. Saddam Hussein was a textbook example of such a tyrant.

The moral challenge implicit in international affairs has been expanded with the evolution of the "right to intervene." Bernard Kouchner, appointed by François Mitterand to the post of Secretary for Human Rights for France, enunciated a right of "humanitarian intervention" to provide recourse to persons trapped in a state which seriously abused their human rights. The concept of the "right to intervene" is embodied in Security Council Resolution 688, which was adopted in April 1991 after Saddam Hussein had driven thousands of Kurds and other minorities into freezing weather without shelter or means of self-defense weeks after signing the cease-fire agreement. Eventually, this concept formed another strong pillar in the Bush administration's argument to invade Iraq.

Professor Myers McDougal of Yale University, a leading expert on international law, wrote that Article 2(4) of the UN Charter prohibits violence, but stressed that this prohibition must be seen in the context of the whole Charter and as complementary to Article 51. The member states who accepted the UN Charter, he emphasized, did so in the expectation that member states would cooperate in maintaining world peace, even if such action were to involve violence. Other legal scholars have noted Article 51's explicit statement that the UN Charter was not intended to "impair the inherent right of individual or collective self-defense if an armed attack occurs against a member of the United Nations, until the Security Council has taken measures *necessary* to maintain international peace and security."

As I consider today the questions surrounding the legality of the invasion in Iraq, our Declaration of Independence comes to mind, especially the following paragraph:

We hold these truths to be self-evident, that all men are created equal, that they are endowed by their creator with certain unalienable rights,

that among these are life, liberty, and the pursuit of happiness. *To preserve these rights governments are instituted among men deriving their
just powers from the consent of the governed.*

This historic document, which held democracy up before an un-
knowing world as a viable form of government, is also a brilliant show-
case for a doctrine of political legitimacy in political philosophy. It
reminds us always that the very function of government is to preserve the
rights of the governed—and that legitimate government policy, foreign
and domestic, requires the consent of the governed. Without the consent
of the governed, policy cannot be legitimate.

Within our doctrine of legitimacy, therefore, the United States and
our allies were not obliged to seek the consent of the Security Council before force was used in Iraq. Our doctrine of legitimacy is based not on the
consent of the United Nations, but on the consent of *the governed.* In theory, and in fact, the rule of law protects a government based on consent.
In launching the war against Iraq, the United States had that consent.
Thus, whatever other debates may persist about the war, the contention
that it was "illegal" is itself illegitimate.

UNINTENDED CONSEQUENCES OF INACTION

When one argues that there are circumstances when war must be made
to keep the peace, one should also consider the ramifications of refraining from war. When the decision is made not to use force in the interest of
keeping the peace, or appearing neutral, unintended consequences may
emerge—among them violence, bloodshed, and even war. History offers
too many examples of such cases.

When Kofi Annan was in charge of peacekeeping in the UN-declared
"safe area" of Srebrenica in 1995, the UN, UNPROFOR (the United Nations Protection Force), and NATO stood by as some nine to eleven thousand Bosnian men and boys fell victim to mass murder. The men of
Srebrenica sought refuge at a UN post, but the local command failed to
protect them, and they were bused out by Serb militias to their death.

In Rwanda, where eight hundred thousand people were slaughtered
in approximately one hundred days in 1994, neither Annan's United

Nations Peacekeeping Organization nor the permanent members of the Security Council (the United States, the United Kingdom, China, Russia, and France) made a serious effort to take urgent action to prevent or stop the genocide.

In Darfur, since fighting initially broke out between government forces and rebels in March 2003, scores of people have been massacred, and more than a million civilians have fled their homes, pouring into neighboring countries and creating "one of the worst humanitarian crises in the world."[21] The catastrophe in Darfur, a scorched-earth campaign of ethnic cleansing according to Jan Egeland, chief of the UN's Humanitarian Affairs and Emergency Relief Agency, remains a specter before the United Nations and the world community, one that demands resolution in a meaningful and lasting way.

When seventy-five Cuban doctors, teachers, journalists, and librarians were arrested and harshly imprisoned in the summer of 2003, no help was offered by the United Nations. The UN's Commission on Human Rights did not even mention their arbitrary arrest or harsh treatment.

And the negligence continues. During the conflict between Israel, Lebanon, and Hezbollah in the summer of 2006, four unarmed and defenseless UN observers tragically died when caught in the cross fire. The peacekeepers, part of the United Nations Interim Force in Lebanon (UNIFIL), created in 1978, were stationed in Lebanon to confirm Israel's "withdrawal from Lebanon" and to "restore international peace and security" when they died. At the time, Secretary-General Annan charged Israel with "the apparently deliberate targeting" of a "United Nations observer post."[22]

There has been discouragingly little reflection on or discussion of these tragedies. As I contemplate these events, it appears that Secretary-General Annan has spoken with more alarm about the use of force by U.S. and coalition forces in Iraq than about these instances of mass murder elsewhere in the world. Little wonder that the United Nations has shown no great vision or inclination to creating an organized and effective response to end these horrors and injustices and to ensure that they cannot recur.

Any death or suffering at the hands of a barbaric tyrant or terrorist is a senseless tragedy. That would include the bombing of the UN head-quarters in Baghdad in the summer of 2003, which killed seventeen peo-ple and injured at least one hundred. Among the dead was Sergio Vieira de Mello, an admired and veteran UN official. That bombing was one more heartbreaking reminder to all of us that the UN is as vulnerable to terrorism and tyranny as its member states.

Kofi Annan's departure from the United Nations in December 2006 marks the end of forty years of service, including a ten-year term as the secretary-general. Behind him he leaves a mixed legacy. There are those who recall his tenure with admiration, bestowing honors and awards on him for his efforts to reform the UN, combat AIDS, negotiate peace, and defend human rights. Others see Annan's legacy in the scandals of his time in office, in the UN's failures of will in Rwanda and Srebrenica, for example, and in the ongoing crises in Darfur and Iraq. They will ques-tion the bitter irony of his actions and words, wondering how and why he deemed himself the arbiter of morality and the rule of law in the face of the grave failures under his leadership.

Annan's legacy is damaged in particular by the UN's oil-for-food program, under which—as investigators have determined—"Saddam Hussein raked in $1.7 billion in kickbacks from participating companies and $11 billion in oil-smuggling profits."[23] The corruption of the oil-for-food program, and of many other professionals and programs that came in contact with it, badly tarnished the once-honorable global endeavor, whose noble purpose had been to help the Iraqi people and to keep the peace.

In addition to these more tangible achievements and failures during his tenure, I believe history will show that Secretary-General Annan sought to use his tenure to expand the independent authority of the UN, and his own powers, at the expense of sovereign rights and the rule of law.

In 1945 the UN Charter was crafted, in part, "to save succeeding gen-erations from the scourge of war; to practice tolerance and live together in peace with one another and unite our strength to maintain interna-tional peace and security."[24] The secretary-general who replaces Annan, South Korea's Ban Ki-Mun, admitted in October 2006 that he faces "dif-ficult foreign policy challenges." But he expressed a determination to "rebuild trust among all stakeholders," because "the political will of member states cannot be forged in an atmosphere of distrust." Ban acknowledged the "current divisiveness" at the United Nations as "worrisome," and vowed to "stay the course with ongoing reform . . . so that we may build the twenty-first-century Secretariat for a twenty-first-century organization."[25]

There has been much debate over the relevance of the United Nations and its future place in our global community. By its very nature, the UN—in a way unlike any other institution—will doubtless continue to challenge member states, the press, and others to remain reform minded. Despite the UN's inconsistent record of effectiveness, the American commitment to the UN should, and I hope will, remain unwavering. We should look beyond the UN's deficiencies and focus instead on both the sound principles of law and the hope, however qualified, for greater world stability that is expressed in the UN Charter.

However, it is also wise to approach the United Nations with a clear sense of its limitations in terms of keeping peace and building nations. The UN Security Council, after all, is like a committee, the UN Secretariat like a bureaucracy. And they operate like committees and bureaucracies—only more so. Action is by consensus. Consensus is hard to build, and sometimes watered down by compromise. Consensus is particularly hard to build within the construct of the UN. The Security Council provides each of its members—the United States, the United Kingdom, China, France, and Russia—with veto power on substantive matters. A single veto can overrule all other members of the Security Council, including the votes of the ten nonpermanent members, who serve two-year terms. The veto was devised to account for differences of power and influence in the UN, but it also allows members of the Security Council to protect their sovereign interests. This body was never intended to convene to declare war, but instead to protect the rights of its member states to do so, in the interest of self-preservation and security.

As of this writing, the idea of expanding the Security Council is under reform review. The key issue is whether additional permanent members—with veto power—will be added. Assuming that the Security Council is not deprived of its power, the United States should remain actively engaged with the UN, because its Charter provides us, and the other member states, built-in protections. Yet the question of how to re-store confidence in the UN persists, an unexpected outcome of conflict in the Gulf—and in other conflicts around the world where the UN has sought to keep peace, to build nations, and to expand beyond its legitimate bounds. Only the future officers and functionaries of the UN can repair what damage has been done. They can, and must, start by organizing themselves with transparency and functioning with integrity.

As America's Founding Fathers understood, it is not easy to structure a single government to provide both popular governance and accountability. Democratic government requires that representatives of the people make major decisions and that these representatives be elected by (and remain accountable to) some large portion of the electorate. Foreign policy also requires that those representatives make policy about remote, unfamiliar subjects. The development of the "experience, instruction, habit, and all the homely species of practical wisdom that are required to make sensible rules about everyday life are required for those who would govern and those who would advise those who would govern" of which Alexis de Tocqueville wrote would not come easy to those who endeavor to be governed democratically. As Tocqueville observed in his master-piece, *Democracy in America*:

> I do not hesitate to state that it is especially in the conduct of foreign re-lations that democracy appears to be decidedly inferior to other gov-ernments. Experience, instruction, habit, and the science of petty occurrences that is called good sense directs the ordinary course of so-ciety in the domestic affairs of a country. But it is not adequate for for-eign affairs . . . concrete personal experience is less relevant in making judgments about foreign affairs than it is in domestic affairs.

It requires patience, perspective, and knowledge—not just of one's own country and its culture, but also of all the countries with which we interact—to ensure that our foreign policy keeps pace with the changing realities of every era and generation. Globalization and technology have confronted us with new and unprecedented issues, with allies and adver-saries who can have instantaneous impact inside our borders no matter how remote they may be geographically. Tocqueville's observations remind us of the importance of vigilance, of the need to balance our desire to promote democracy against the new vulnerabilities of our national security in this new world.

CLASH OF CIVILIZATIONS

Most serious students of foreign policy in our time are familiar with Samuel Huntington's fascinating book, *The Clash of Civilizations and the Remaking of the World Order*. I have been an admirer of Professor Huntington all my professional life, but upon its publication in 1996 I was critical of the book's first chapter and doubted whether such a clash of civilizations was at hand. But it quickly became clear to me that I was wrong. When *The Clash of Civilizations* was published, we were already on the razor's edge. September 11, 2001, brought that home to America, and to me.

That lesson eluded Secretary-General Annan, who even a month before his term expired was still resisting the notion that there exists any such clash of civilizations. "We must start by reaffirming—and demonstrating—that the problem is not the Koran, nor the Torah or the Bible," he declared, and called for increased opportunities for young people as a credible alternative to hate and extremism. "We must give them a real chance to join in improving the world order," Annan continued, "so they will no longer feel the urge to smash it."[26] But the secretary-general failed to explain how one negotiates with groups whose intent is to smash the world and who cannot be dissuaded by invitations to enfranchisement. Islamic fundamentalism is an ideology of expansionist tyranny, propelled by an unrelenting will to dominate other nations, cultures, and religions. Just as during the cold war, when exporting Communism was the most powerful intellectual paradigm, exporting a radicalized version of Islam—under the cover of its religious status—has become today's most powerful paradigm.

Today we confront this clash around the world, and it is our most urgent task. In Afghanistan, Iraq, and elsewhere we deal with people who have no experience of democracy and who have no institutions capable of sustaining democracy if left without support. The people of Afghanistan were overrun by the Taliban, in a cloud of chaos and violence, in the power vacuum left after the Soviets withdrew in defeat. The people of Iraq, who lived under Saddam Hussein's oppression and terror for forty years, have developed habits, values, and a way of life very different from anything we know. We have our lessons—and decisions—before us.

Today, we confront more dangers in the world than at any time in our history. We know now that our policies must be made within the context of an expanding globe, and with an expanding awareness that democracy and the rule of law have little meaning to many nations with which we must pursue constructive relations. We can never predict all the consequences of our actions, but we surely can proceed with greater wisdom if we can adjust our worldview based on understanding our mistakes.

As the Balkan wars illuminated important differences in the political sensibilities and reflexes of the Old Continent and the United States, the Iraq war has illuminated differences with our relationship within the world community itself. One stark disparity that crystallized during the 2003 Iraq invasion was that the coalition of former U.S. allies (including France, Germany, Greece, Norway, and Canada) formed during the first Gulf war could no longer be relied upon to stand with the United States in the second war. Likewise, countries in the region (such as Egypt, Kuwait, Morocco, Qatar, Saudi Arabia, Syria, and the United Arab Emirates) that had stood with the United States in 1991 declined their support in 2003. With the end of the cold war, however, new alliances were offered in 2003, with nations such as Albania, Azerbaijan, Japan, Latvia, Lithuania, Macedonia, Romania, Slovakia, and Uzbekistan.[27]

After the current Bush administration's stunning and swift success during the initial invasion of Iraq, dramatized by Saddam Hussein's prompt overthrow, it is clear that some of the lessons learned from the first Gulf war were missed. Former secretary of defense Donald Rumsfeld, the architect of the Iraq war, has been quoted as saying, "We are going to go in, overthrow Saddam, get out. That's it."[28] But what happened in Afghanistan after the Soviets withdrew could not be allowed to occur again when the U.S. attacked after September 11, and neither could it be allowed to happen in Iraq. Iraq presented a very different set of circumstances from Afghanistan, however. These are things we ought to have known and taken into account when weighing our decision to invade in 2003.

Iraq lacked practically all the requirements for a democratic government: rule of law, an elite with a shared commitment to democratic procedures, a sense of citizenship, and habits of trust and cooperation. The administration's failure involved several issues, but the core concern is that they did not seem to have methodically completed the due diligence required for reasoned policy-making because they failed to address the aftermath of the invasion. This, of course, is reflected by the violence, sectarian unrest, ethnic vengeance and bloodshed we see in Iraq today.

The key to putting Iraq on the path of democracy today is to help to establish law and order. This policy is already part of the Bush administration plan, but as of this writing their strategy remains unclear. However, history offers hope for Iraq's future. Battles in other countries that had seemed unwinnable have come to peace—and victory. The conflict in the former Yugoslavia was fueled by what once appeared to be endless hatred and ethnic divides, but that nation is now on a slow road to democracy. The centuries-old conflict in Northern Ireland has drawn toward a peaceful close, as have former rivals in South Africa who abandoned apartheid's animosity and violence in favor of building a multiethnic democracy, stability, and peace.

Today, we battle multiple threats on several fronts, obliging us to develop foreign policy strategies to deal simultaneously with the terrorists and with their state sponsors. As we continue to fight in Iraq and Afghanistan, and grapple with such other hot spots as Iran and North Korea, it is natural for Americans to seek examples to guide our actions in this new world order, with its astonishing potential for mass casualties and its proliferation of outlaw nation-states. Many of the forces we confront today are more barbaric than modern, in their goals and means alike.

When I am asked for guidance as we move forward into this new and uniquely dangerous century, I am drawn to the prescient principles of the Reagan Doctrine.

Like most American "doctrines," the Reagan Doctrine emerged in response to circumstances. Above all, it was concerned with the moral legitimacy of U.S. support—including military support—for insurgencies under certain circumstances: where there are indigenous opponents to a government that is maintained by force, rather than popular consent. The Reagan Doctrine addressed such questions as: Is it morally and legally acceptable for the United States to support indigenous armed movements against such governments? Does such support constitute unjustified and illegal interference in their internal affairs?

The Reagan Doctrine expressed *solidarity* with the struggle for self-government as against one-party dictatorship. It did not require offering armed resistance, but it did permit such measures. It did not address the question of U.S. military involvement or involvement of U.S. forces in any particular contest. It was a broader doctrine that postulated the moral legitimacy of American military aid under certain circumstances and offered moral guidelines for offering such aid. Policy under the Reagan Doctrine was established *by prudential determination of the national interest in particular context.* It denied that assisting in the overthrow of an existing government is always wrong. Rather, it highlighted the need to weigh the legitimacy of such acts within their political and moral context: the nature of the government, the role of a foreign force, and the existence of resistance.[29] Moreover, even if such an act were justified, the Reagan Doctrine did not dictate that such action was always wise; rather, it counseled that the long-term costs and benefits of such action should be carefully weighed before taking any steps. Because once we intervene in a given situation, we are accountable for its outcome.

When I served as head of the U.S. delegation to the United Nations Human Rights Commission in 2003, I was reminded that the commission, like the United Nations itself, is a mélange of every region, every people, and every culture—and sometimes of radically contradictory values. As a result of this diversity, some members of the Human Rights Commission can barely understand other members. In my experience, in fact, it seemed that many of the members had no interest in understanding one another. It was apparent that not all nations were primarily concerned with ensuring the freedom of their people or were eager to devise constructive solutions to problems. Rather, many of them wanted to impose their own will on other regions, nations, and lives. The United States and the Western Group seek constructive solutions to problems. At least we think we do. The resulting clashes at the UNHRC were, indeed, often clashes between vastly different civilizations.

However, it is important to note that the principles of the Human Rights Commission were designed to protect the governed, even though many of the members of the commission do not. To quote the United Nations' Universal Declaration of Human Rights, adopted in 1948:

> Whereas disregard and contempt for human rights have resulted in barbarous acts which have outraged the conscience of mankind. . . .

> Whereas it is essential, if man is not to be compelled to have recourse, as a last resort, to rebellion against tyranny and oppression, that human rights should be protected by the rule of law.

The rule of law is the foundation of our constitutional government— that is, of government based on consent. Government based on consent is government by the governed. That is a crucial distinction between democracy and all other forms of government. On that distinction, too, the UN was grounded from its inception. Democratic principles can influence the dialogue among nations as long as the UN remains true to its Charter. There is hope in that, as well.

Today, as we confront these new challenges, Americans must acknowledge that not all governments we deal with are governed based on consent, even as we know that the historic role of our own government has been to promote democracy. Here, again, cultures clash, but the lessons of the post–cold war years must guide our future course if we are to remain a free and stable nation in the coming century. One such lesson is that the rule of law and the sovereign rights of nations are cornerstones for world order. Another is that national security must not take a backseat to our desire to promote democracy to cultures unable or unwilling to accept our overture. A third such lesson is that sanctions and other means may achieve ends more slowly than military force, but preserving our military to wage war only when our national security is at risk is perhaps the best way to keep peace.

The past fifteen years have seen violence perpetuated by coercive elements whose goal is not peace. Throughout history, such groups have sought to impose their will on others. Where these forces have surfaced in the modern world, there has seldom been a peace to keep. Yet we have dispersed peacekeepers to help alleviate the situations, and these peacekeeping operations have gradually expanded to include a variety of activities. They monitored human rights practices, oversaw elections, maintained cease-fires, separated adversaries, delivered humanitarian aid, and repatriated refugees, to name a few. Their duties have become so diverse that the concept of peacekeeping today might be taken to refer to almost any activity, in any region, at any time.

Yet, as these peacekeeping efforts expanded in scope and across the globe, a critical element seems to have been overlooked: too often the peacekeepers' initial mission was subsumed by a multitude of other tasks, which often lacked objective and cohesion. Sometimes the peacekeeping forces themselves have been left defenseless, and even in need of rescue themselves, as their environment devolved further into warfare. Eventually, the new realities of peacekeeping included kidnapping, starvation, and death—all the result of military forces being deployed not to wage war but to "*keep* peace" in a region where there was no peace from the start.

It is not surprising that the will of the nations providing such peacekeepers has sometimes wavered, and that these nations or their military leaders have sometimes proved unwilling to risk lives in such circumstances. In the 1990s, by a twist of policy nuance, we began to deploy peacekeepers into these challenging new circumstances, assigning them to build nations as a way of ensuring lasting peace. Because the ideal of human rights is at the core of our view of law and culture, we took upon ourselves the task of building democratic nations where there were no institutions to bolster the societies against the chaos inherent in democratic rule—or, more so, in the march toward democracy. Nation building became a military activity rather than one reserved for more gradual influences from market and cultural forces, and so the arrival of democracy in some nations came at the end of a sword, not the end of a cultural and political evolution instigated by the governed themselves.

In the process, our own military has too often been turned away from its main purpose—to wage war to keep peace, a peace in which more transformational forces would be freed to resurrect societies from the ruins of decay and war. In practice, the practice of nation building further distracted the UN from its own mission and opened the door for an expansionist UN and Secretariat. Our future foreign policy need not tempt the UN to challenge our sovereign rights to self-defense or to seek legitimacy in anything other than the will of its collective of nations. Neither our foreign policy nor our international role requires the use of our remarkable military simply because they are capable of enduring a crucible without an objective. Our military needs and deserves to remain dedicated to its true and singular role: to preserve peace and security by waging and winning war.

Our Declaration of Independence expresses a dream and a doctrine of government by consent. An important part of our history has been devoted to making a reality of this dream for all Americans. As the world has shrunk, we have, rightly or wrongly, sought to share the dream beyond our borders. We need to take stock and revisit this impulse, however well intentioned it may be.

For the United States, the enjoyment and protection of the rights stipulated in our Declaration of Independence and institutionalized in our Constitution lie at the heart of our identity as a nation. The struggle to ensure that those rights are enjoyed by each and every one of our citizens—a struggle that is still in progress—has been the engine of our history and our development as a nation. Of course, it is reasonable that those of us who enjoy the benefits of freedom are motivated to remember the millions who do not, the millions who are vulnerable to coercive forces of domination and injustice.

Yet it is a different matter entirely to commit military resources to keep peace in such areas, where often no peace can be kept, or to build nations in our own image before they are ready for our freedoms—or even want them. The military need not do the work of sanctions and diplomacy. As we carry on in this new century, we would do well to remember the importance of balancing the twin goals of our foreign policy: preserving national security and promoting democratic principles. And we must remember that historic conflicts between enemies can be won on moral force, without firing a single bullet or missile; that cultural, market, political, and perhaps religious forces can be far more transformative in areas of the world where chaos and violence reign; and that America can contribute to the building of nations by any and all of these means—while preserving our military and reserving our sovereign right to wage war to maintain true peace.

POSTSCRIPT
Allan Gerson

Reflecting on her work at the U.S. Mission to the United Nations from 1981 to 1985, Jeane Kirkpatrick later published the most definitive articulation of her views in an anthology of essays published by the Council on Foreign Relations in 1989: *Right v. Might: International Law and the Use of Force*. I was her counsel at the UN post during these years, and I served as her coauthor on the essay entitled, "The Reagan Doctrine, Human Rights, and International Law."

That article demonstrated her innate affinity to the ideas of John Stuart Mill, whom she quoted liberally to the effect that the first principle of law is necessarily the equal application of the law, or reciprocity, and that unilateral compliance was unacceptable. Secondly, she drew her cue from Mill in distinguishing intervention that is rightful from that which is not, while recognizing that what is right may not always be prudent.

The Reagan Doctrine, she wrote, "is, as we understand it, above all concerned with the moral legitimacy of U.S. support—including military support—for insurgencies under certain circumstances." She never saw a legitimate basis for the direct use of U.S. force in support of democracy. Rather, the use of force was legitimate when it was wielded to support indigenous insurgents in opposition to a government maintained by force and not democratic consent. As she put it, "The Reagan Administration did not create these resistance movements"; nor did it initiate the policy of providing support. This was in contrast to what the Kennedy, Johnson, Nixon, and Carter administrations had done, respectively, in

Vietnam and Cambodia; Southeast Asia and Angola; and Afghanistan. In short, the Reagan Doctrine permitted assistance in self-defense, and as a means of counterintervention where an enemy like the Soviet Union was using military support to keep in power governments that had no popular consent. As such, the Reagan Doctrine was one that expressed solidarity with democracy, but was prudent and conservative in the use of military force to support democracy. It was a doctrine of counterintervention. As she put it, "In Kant's view, intervention to bring down despotic governments was to be encouraged. The Reagan Doctrine does not go this far, but it has the same philosophical underpinnings."

Today, the future of U.S. foreign policy is essentially a debate between when and in what sequence security and promotion of democracy can occur. For Jeane Kirkpatrick, the choice was always clear. The sequence, if force was to be used at all, was first to gain security, and only then to encourage freedom. If there could be no security, there could be no freedom. By that token, if Iraq could not be secure, there could be no thought of promotion of democracy. And Iraq could only be made secure if U.S. force was established on the basis of legitimacy.

It is sad that the Bush administration, in its conduct of the Iraq war, did little to pursue the same standards of informed analysis that Jeane Kirkpatrick had tried to foster. The failure to do so has made matters worse, not better. Kirkpatrick understood the limitations of humanitarianism. She believed deeply in what she had learned as a student at Columbia in the 1950s, where she was deeply affected by the disclosures of the horrors of the Holocaust. She saw in what had occurred vindication of Sigmund Freud's thesis in *Civilization and its Discontents*, that we are born not as angels, but more like animals, and that the task of civilization is to civilize and to defend against barbarism. She applauded Ronald Reagan for declaring to the world that the American people had the necessary energy and conviction to defend itself, as well as a deep commitment to peace and democracy. But she believed most of all in being careful, very careful, before ever descending down the path of direct use of military might. As a nation, and as individuals, we have, she believed, a civic duty to articulate that direct resort to force should always be undertaken as a last resort, in a manner compatible with our nation's ideals and our understanding of international law.

APPENDIX

SUMMARY OF DAYTON PEACE AGREEMENT

The Dayton talks culminated in the initialing of a General Framework Agreement for Peace in Bosnia and Herzegovina. The parties were the Republic of Bosnia-Herzegovina, the Republic of Croatia, and the Federal Republic of Yugoslavia. The Contact Group of Nations (the United States, the United Kingdom, France, Russia, and Germany) and the European Union special negotiator witnessed the agreement.

GENERAL FRAMEWORK AGREEMENT

- Bosnia and Herzegovina, the Federal Republic of Yugoslavia, and Croatia agree to respect the sovereignty of one another and settle future disputes by peaceful means.

- The parties agree to fully respect the commitments made in the four annexes of the agreement and to respect the human rights of the habitants in the region and the rights of refugees and displaced persons.

- The parties agree to implement the peace settlement and cooperate with the entities authorized by the United Nations to carry out this task. They also agree to collaborate in the prosecution of war crimes and other violations of international humanitarian law.

ANNEX 1-A: MILITARY ASPECTS

- The parties commit themselves to re-create as quickly as possible normal conditions of life in Bosnia and Herzegovina and to respect the cease-fire that began on October 5, 1995.

- Foreign forces that are present in Bosnia will leave its territory within thirty days.

- The parties invite into Bosnia and Herzegovina a multinational military implementation force (IFOR) that will be under the authority of the United Nations (UN) and under the command of the North Atlantic Treaty Organization (NATO). The tasks of this force will include monitoring and helping ensure compliance with the military aspects of this agreement. To achieve this, IFOR will have unimpeded freedom of movement by land and control over airspace; if necessary, it will have the right to use violence.

ANNEX 1-B: REGIONAL STABILIZATION

- The Republic of Bosnia and Herzegovina, the Federation, and the Bosnian Serb Republic will begin negotiation within the next seven days with the objective of agreeing on measures that will build confidence within forty-five days. These may include the exchange of information between the parties and the implementation of restrictions for carrying out military exercises. These conversations will take place under the auspices of the Organization for Security and Cooperation in Europe (OSCE).

- The parties, as well as Croatia and the Federal Republic of Yugoslavia, must begin negotiations within thirty days to agree on limits on the holding of military armament, including combat vehicles, aircraft, helicopters, and tanks.

- If the parties do not establish limits in these categories within the next thirty days, the agreement provides numerical limits with which all parties will comply.

- The OSCE will conduct negotiations with the parties to establish a regional balance in the former Yugoslavia.

ANNEX 2: INTER-ENTITY BOUNDARY

- An inter-entity boundary line between the Federation and the Bosnian Serb Republic will be established.

- Sarajevo will be reunified within the Federation and will be open to all people of the country.

- Gorazde will be linked to the Federation by a land corridor composed of two roads.

ANNEX 3: ELECTIONS

- The parties agree to create the conditions necessary for organizing free and fair elections; these include ensuring the right of citizens to vote in secret without being intimidated, freedom of expression in the press, and freedom of association.

- Elections will be conducted within six to nine months for the presidency and House of Representatives of Bosnia and Herzegovina, the House of Representatives of the Federation, the National Assembly and presidency of the Bosnian Serb Republic, and, where possible, for local offices. International organizations will supervise the electoral process.

- Refugees and all those who were displaced by the conflict will have the right to vote in their original place of residence.

ANNEX 4: CONSTITUTION

- Bosnia and Herzegovina will adopt a new constitution that protects the human rights and freedom of movement of its citizens throughout the territory. The country will consist in two entities: the Federation and the Bosnian Serb Republic.

- The central government will include a president, two legislative chambers, and a constitutional court. Direct elections should be conducted to elect the president and one of the legislative chambers. The central government will be responsible for monetary policy, law enforcement, and foreign policy.

ANNEX 5: ARBITRATION

- The Federation and the Bosnian Serb Republic agree to resolve disputes between them through a system of arbitration and to establish the necessary mechanisms for arbitration to take place.

ANNEX 6: HUMAN RIGHTS

- A commission on human rights will be established with the objective of protecting the fundamental freedoms of all persons in Bosnia and Herzegovina. The commission will be composed of a human rights ombudsman who will be in charge of investigating human rights violations and a human rights chamber that will issue decisions.

- UN human rights agencies, the OSCE, and the International Criminal Tribunal for the Former Yugoslavia will monitor the situation regarding human rights violations.

ANNEX 7: REFUGEES AND DISPLACED PERSONS

- The parties agree to allow refugees and displaced persons to return safely to their homes and to permit the free movement of persons throughout the country.

- A Commission for Displaced Persons and Refugees will be created with the objective of returning property or giving just compensation for its loss.

ANNEX 11: INTERNATIONAL POLICE TASK FORCE

- The parties request the establishment by the UN of an international police task force (IPTF) with the objective of providing advice and training to the members of local law enforcement organizations. A commissioner appointed by the UN will be in charge of the IPTF.

NOTES

INTRODUCTION

1. Mikhail Gorbachev, National Museum of American History Archives, January 1992, http:/www.americanhistory.si.edu/subs/history/timeline/end@2000, The National Museum of American History.

2. President Ronald Reagan, Address to members of the British Parliament, June 8, 1982.

3. President George W. Bush, President Discusses War on Terror at National Endowment for Democracy, October 6, 2005.

1. IRAQ INVADES KUWAIT

1. David McCullough, *Truman* (New York: Simon and Schuster, 1992), 776–7.

2. In Secretary of State James Baker's words, "It took some arm-twisting to convince the Latins to denounce Noriega by name for stealing the election. The old doctrine of non-interventionism and fear of U.S. power still paralyzed the organization." James A. Baker III, *The Politics of Diplomacy Revolution, War and Peace, 1989–1992*, with Thomas M. DeFrank (New York: G.P. Putnam's Sons, 1995), 183.

3. National Security Directive 26, "U.S. Policy Toward the Persian Gulf," October 2, 1989, http://www.bushlibrary.tamu.edu/research/nsd/NSD/NSD%2026.0001.pdf.

4. Ibid. NSD-26 reaffirmed our strategic interests in the region and, with caveats conveying our concerns, generally confirmed the previous policy of engaging Iraq: "Normal relations between the United States and Iraq would serve our longer term interests and promote stability in both the Gulf and the Middle East. The United States government should propose economic and political incentives for Iraq to moderate its behavior and to increase our influence with Iraq."

5. Don Oberdorfer, "Missed Signals in the Middle East," *Washington Post Magazine*, March 17, 1991, W19.

6. The transcript of the July 25, 1990 meeting between April Glaspie originally appeared in an Arabic-language version leaked from Iraqi sources to ABC News, who translated it and then passed it on to the *New York Times*. The translated version appeared as "Excerpts from Iraqi Document on Meeting With U.S. Envoy," *New York Times*, September 23, 1990, page 19. In 1998 the U.S. government declassifed the document, which appears on the Margaret Thatcher Foundation website, as "Gulf War: US Embassy Baghdad to Washington, 25 July 1990," margaretthatcher.org.

7. Charter of the United Nations, Article 51.

8. George Bush, "Remarks and an Exchange with Reporters on the Iraqi Invasion of Kuwait," *Weekly Compilation of Presidential Documents* 26, August 5, 1990, 1207.

9. Entire text of UN Security Council Resolution 661:

The Security Council,

Reaffirming its Resolution 660 (1990) of 2 August 1990,

Deeply concerned that that resolution has not been implemented and that the invasion by Iraq of Kuwait continues with further loss of human life and material destruction,

Determined to bring the invasion and occupation of Kuwait by Iraq to an end and to restore the sovereignty, independence and territorial integrity of Kuwait,

Noting that the legitimate Government of Kuwait has expressed its readiness to comply with resolution 660 (1990),

Mindful of its responsibilities under the Charter of the United Nations for the maintenance of international peace and security,

Affirming the inherent right of individual or collective self-defence, in response to the armed attack by Iraq against Kuwait, in accordance with Article 51 of the Charter,

Acting under Chapter VII of the Charter of the United Nations,

1. *Determines* that Iraq so far has failed to comply with paragraph 2 of Resolution 660 (1990) and has usurped the authority of the legitimate Government of Kuwait;

2. *Decides*, as a consequence, to take the following measures to secure compliance of Iraq with paragraph 2 of Resolution 660 (1990) and to restore the authority of the legitimate Government of Kuwait;

3. *Decides* that all States shall prevent:

    (a) The import into their territories of all commodities and products originating in Iraq or Kuwait exported there from after the date of the present resolution;

(b) Any activities by their nationals or in their territories which would promote or are calculated to promote the export or trans-shipment of any commodities or products from Iraq or Kuwait; and any dealings by their nationals or their flag vessels or in their territories in any commodities or products originating in Iraq or Kuwait and exported there from after the date of the present resolution, including in particular any transfer of funds to Iraq or Kuwait for the purposes of such activities or dealings;

(c) The sale or supply by their nationals or from their territories or using their flag vessels of any commodities or products, including weapons or any other military equipment, whether or not originating in their territories but not including supplies intended strictly for medical purposes, and, in humanitarian circumstances, foodstuffs, to any person or body in Iraq or Kuwait or to any person or body for the purposes of any business carried on in or operated from Iraq or Kuwait, and any activities by their nationals or in their territories which promote or are calculated to promote such sale or supply of such commodities or products;

4. *Decides* that all States shall not make available to the Government of Iraq or to any commercial, industrial, or public utility undertaking in Iraq or Kuwait, any funds or any other financial or economic resources and shall prevent their nationals and any persons within their territories from removing from their territories or otherwise making available to that Government or to any such undertaking any such funds or resources and from remitting any other funds to persons or bodies within Iraq or Kuwait, except payments exclusively for strictly medical or humanitarian purposes and, in humanitarian circumstances, food stuffs;

5. *Calls upon* all States, including States non-members of the United Nations, to act strictly in accordance with the provisions of the present resolution notwithstanding any contract entered into or license granted before the date of the present resolution;

6. *Decides* to establish, in accordance with Rule 28 of the provisional rules of procedure of the Security Council, a Committee of the Security Council consisting of all the members of the Council, to undertake the following tasks and to report on its work to the Council with its observations and recommendations:

(a) To examine the reports on the progress of the implementation of the present resolution which will be submitted by the Secretary-General;

(b) To seek from all States further information regarding the action

taken by them concerning the effective implementation of the provisions laid down in the present resolution;

7. *Calls upon* all States to co-operate fully with the Committee in the fulfillment of its task, including supplying such information as may be
sought by the Committee in pursuance of the present resolution;

8. *Requests* the Secretary-General to provide all necessary assistance to the
Committee and to make the necessary arrangements in the Secretariat
for the purpose;

9. *Decides* that, notwithstanding paragraphs 4 through 8 above, nothing in
the present resolution shall prohibit assistance to the legitimate Government of Kuwait, and *calls upon* all States;

   (a) To take appropriate measures to protect assets of the legitimate
Government of Kuwait and its agencies;

   (b) Not to recognize any regime set up by the occupying Power;

10. *Requests* the Secretary-General to report to the Council on the progress
of the implementation of the present resolution, the first report to be
submitted within thirty days;

11. *Decides* to keep this item on its agenda and to continue its efforts to put
an early end to the invasion by Iraq.

10. Baker, *The Politics of Diplomacy*, 277. Baker's interpretation is curious, as
the last five presidents were Ronald Reagan, Jimmy Carter, Gerald Ford, Richard
Nixon, and Lyndon Baines Johnson. Only Carter's presidency really fits Baker's description.

11. Ibid., 277.

12. Michael R. Gordon and Bernard E. Trainor, *The General's War* (Boston:
Little Brown, 1995), 49.

13. George Bush, "Annual State of the Union Address," *Weekly Compilation of
Presidential Documents* 27, January 29, 1991, 90. In his State of the Union Address in
1991, President Bush said,

We will succeed in the Gulf. And when we do, the world community will have
sent an enduring warning to any dictator or despot, present and future, who
contemplates outlaw aggression. The world can therefore seize this opportunity to fulfill the long-held promise of a new world order—where brutality will
go unrewarded, and aggression will meet collective resistance.

14. William Schneider, "Consensus Holds, But for How Long?" *National Journal* 22, no. 35 (September 1, 1990): 2102. Schneider points out that the American
public generally believed that protecting our oil supply was the most important reason for our actions in the Gulf. He writes:

[in mid-August 1990] the Gallup Organization Inc. asked Americans "Why do you think we are involved in the Iraqi situation and why are our troops in Saudi Arabia?" The leading answer (49%) was that we were there to protect our oil supplies. By comparison, 35% gave internationalist reasons that we were there to defend other countries or stop Iraqi aggression. When the CBS News-New York Times poll asked people to choose between two objectives, 46% said America was sending troops mainly because the price of oil would increase too much if Iraq controlled the oil fields, and 30% said America was there mainly to help its friends.

15. Robert Thompson, "Japanese Companies in Saudi Oil Link Study," *Financial Times*, May 29, 1991; Steven Weisman, "Japan, Courting Israel, Joins Move to Scrap UN's Stand on Zionism," *New York Times*, December 13, 1991.

16. George Bush, "On Deployment of U.S. Troops to Persian Gulf," *Weekly Compilation of Presidential Documents* 26, August 8, 1990, 1216. See also William Schneider, "Consensus Holds, But for How Long?" 2101. In an address to the nation concerning the Gulf crisis, President Bush stated,

> We succeeded in the struggle for freedom in Europe because we and our allies remained stalwart. Keeping the peace in the Middle East will require no less. We're beginning a new era. This new era can be full of promise, an age of freedom, a time of peace for all peoples. But if history teaches us anything, it is that we must resist aggression or it will destroy our freedoms. Appeasement does not work. As the case in the 1930s we see in Saddam Hussein an aggressive dictator threatening his neighbors. . . . *This is not an American problem or a European problem or a Middle East problem, it is the world's problem.* [Emphasis added.]

17. These six points are paraphrased from remarks by Secretary of Defense Caspar W. Weinberger to the National Press Club. See Caspar W. Weinberger, "The Uses of Military Power," Speech before the National Press Club, November 28, 1984. See John T. Correll, "The Use of Force," *Air Force Magazine*, December 1999, Vol. 82, No. 12. http://www.afa.org/magazine/Dec1999/1299force.asp.

18. Edward Lucas, "Storm Clouds Gather Over White House," *Independent*, May 3, 1991, 10.

19. Robert M. Kimmitt, undersecretary of state for political affairs, citing President Bush in "Economic and National Security," *U.S. Department of State Dispatch*, June 3, 1991.

20. Baker, The Politics of Diplomacy, 304.

21. Ibid., 278.

22. Ibid., 279.

23. Ibid., 307.

24. Ibid., 304.

25. See Steven Hurst, "The Foreign Policy of the Bush Administration," in Search of a New World Order (New York: Cassell, 1999), 4.

> This ability to use the UN as a vehicle for U.S. policy was vital to Bush both domestically and internationally. In the latter context it gave legitimacy to the U.S. goal of expelling Iraq from Kuwait. It also ensured that Iraq would be isolated diplomatically and, thanks to UN resolutions imposing sanctions, economically.... Domestically, the securing of UN backing was a vital step towards securing congressional and public support for the major U.S. role in the Gulf being contemplated.

26. George Bush, "Remarks of President George Bush upon Presenting the Presidential Medal of Freedom to Margaret Thatcher," Weekly Compilation of Presidential Documents 27, March 7, 1991, 264.

27. Baker, The Politics of Diplomacy, 279.

28. Margaret Thatcher, The Downing Street Years (New York: HarperCollins Publisher, 1993), 821.

29. Ibid.

30. Ibid.

31. Ibid., 167.

32. Other policymakers also speculated openly about the price of the coalition. For example, during a November 28, 1990, hearing of the Armed Services Committee, Senator William Cohen (R-ME) noted: "According to this morning's news, China apparently is using the threat of a veto or an abstention from voting to purchase goodwill, mainly an adjustment of their trade status. This is extortion by another name, in my opinion." Senator William Cohen, Hearing of the Senate Armed Services Committee, "Persian Gulf Crisis," November 28, 1990.

33. See Weinberger Doctrine, principles 2 and 5.

34. Jeane Kirkpatrick, "Though Our Allies May Waiver . . . ," Washington Post, January 7, 1991.

35. Baker, The Politics of Diplomacy, 314.

36. Note: There are no democratic governments in the Gulf; but there is a great deal of oil.

37. Dan Goodgame and Michael Duffy, "Read My Hips," Time, October 22, 1990, 26–27.

38. On December 31, 1991, when the Soviet Union dissolved into fifteen separate countries, Russia, the largest one of these, was the Soviet Union's successor in the UN Security Council and other international organizations.

39. President George Bush, "Address to the 46th Session of the United Nations General Assembly," Weekly Compilation of Presidential Documents 27, September 23, 1991, 1324.

40. Gerald Butt, "Iraq Showing No Signs of Relaxing Grip on Kuwait," *Daily Telegraph*, August 11, 1990.

41. Aleksandr Yakovlev, August 20, 1990, on Moscow television.

42. Senator Sam Nunn, 102nd Congress (January 11, 1991), *Congressional Record* 137, pts.6-8:S190.

43. Senator George Mitchell, 102nd Congress (January 10, 1991), *Congressional Record* 137, pts. 6-8:S102.

44. Senator Claiborne Pell, Ibid., S125.

45. Vice President Dan Quayle, "America's Objectives in the Persian Gulf," Address at Seton Hall University, South Orange, New Jersey, November 29, 1990; U.S. Department of State Dispatch, December 10, 1990, 310.

46. Richard M. Nixon, "Why," *New York Times*, January 6, 1991. Also, Jeane Kirkpatrick, "Turn Saddam Back," *Washington Post*, December 17, 1990.

47. Nixon, "Why."

48. "Speech by Saddam Hussein on the 70th Anniversary of the Establishment of the Iraqi Army," Baghdad Domestic Service, in Arabic, 0805 GMT, January 6, 1991. Also, Ofra Bengio, *Saddam Speaks on the Gulf Crisis* (Tel Aviv: Tel Aviv University, 1992), 156, 159.

49. *Al Qadisiya*, January 5, 1991.

50. "Text of Appeal by the General Secretariat of the Popular Islamic Conference Organization," Baghdad, Republic of Iraq Radio, 1734 GMT, January 18, 1991.

51. Jeane Kirkpatrick, "Fantasies That Doomed Saddam," *Washington Post*, January 20, 1991.

52. Ibid. The Koran passage is from *The Book of the Pilgrimage*, verse 39.

53. Ibid.

54. Judith Miller, citing Iraqi information minister Latif Jassem, "Iraq's Seesaw Diplomacy: Threats and Entreaties," *New York Times*, October 17, 1990. Also, Judith Miller, "Mideast Tensions: King Hussein on Kuwait and Dashed Hope," *New York Times*, October 16, 1990; Jeane Kirkpatrick, "The Threat Must Be Real," *Washington Post*, October 22, 1990.

55. Jim Wolf, "U.S. Rejects Any Deal Letting Iraq Hang on to Part of Kuwait," *Reuters*, PM Cycle, October 17, 1990.

56. Carl von Clausewitz, *On War* (Middlesex, England: Penguin Books, 1968), 111. Also, Kirkpatrick, "The Threat Must Be Real."

57. von Clausewitz, *On War*, 104. Also, Jeane Kirkpatrick, "Will We Liberate Kuwait?" *Washington Post*, November 12, 1990.

58. Eliot Cohen, "Iraq: Why and How the U.S. Should Strike," *Commentary* 90, no. 5 (November 1990): 27.

59. George H. W. Bush, "Why We Are in the Gulf," *Newsweek*, November 26, 1990, 28.

60. George Bush, " State of the Union Address," *Weekly Compilation of Presidential Documents* 27, January 29, 1991, 90.

61. Hoffman, David, "Baker Wants UN to Approve Force; More Tank Divisions to Be Deployed," *Washington Post*, November 8, 1990.

62. George Bush, "Address to the Nation Announcing Allied Military Action in the Persian Gulf," *Weekly Compilation of Presidential Documents* 27, January 16, 1991, 50.

63. Baker, *Politics of Diplomacy*, 278–281, 305.

64. Daniel O. Graham, "Q: Should the U.S. Build a Space-Based Missile Defense? Yes: Only a Space-Borne System Can Counter Missile Threats," *Insight on the News*, September 11, 1995, p. 18. Also, "Patriot Missiles Ineffective During Gulf War, Analysis Claims," *United Press International*, January 28, 1993.

65. Alan Riding, "Confrontation in the Gulf," *New York Times*, January 1, 1991.

66. Senator Sam Nunn, Hearings of the Senate Armed Services Committee, "Persian Gulf Crisis," 27–28 November 1990.

67. George Bush, "Address to the Nation Announcing Allied Military Action in the Persian Gulf," 50.

68. Ibid.

69. Ibid.

70. Donald Kagan, "The General's War: Book Review," *Commentary* 99, no. 6 (June 1995): 41, citing Laurence Freedman and Efram Karsh, *The Gulf Conflict, 1990–1991: Diplomacy and the New World Order* (Princeton, NJ: Princeton University Press, 1993).

71. Charles Lane, "The Legend of Colin Powell," *New Republic*, April 17, 1995, 70.

72. George Bush, "Kuwait Is Liberated," *Weekly Compilation of Presidential Documents* 27, February 27, 1991, 224c.

73. Elie Kedourie, "Iraq: The Mystery of American Policy," *Commentary* 91, no. 6 (June 1991): 15–19, 17.

74. John Kelly, assistant secretary of state, Near East and South Asian affairs, Hearing of the Europe and Middle East Subcommittee of the House Foreign Affairs Committee, June 26, 1991. Kelly also stated,

> The United Nations, in all of the resolutions it passed with regard to Iraq and the debate and vote in the American Congress on authorizing the use of force against Iraq, set as their goals the liberation of Kuwait, the restoration of the legal government of Kuwait, and a return to peace and stability in the area. Neither in the UN debates nor in other enunciations of military objectives was the conquest of Baghdad or of the removal of Saddam Hussein enunciated as an objective.

75. *McNeil/Lehrer NewsHour* (March 27, 1991). Schwarzkopf continued: "And the President, you know, made the decision that we should stop at a given time, at a given place. That did leave some escape routes open for them to get back out and I think that it was a very humane decision and a very courageous decision on his part also."

76. Ibid.

77. Ibid.

78. George Bush and Brent Scowcroft, *A World Transformed* (Knopf: New York, 1998), 487.

79. George Bush, "A Question of Policy." *Weekly Compilation of Presidential Documents* 27, March 27, 1991, 369.

80. Bush and Scowcroft, *A World Transformed*, 482.

81. Ibid., 483.

82. Ibid., 486–487.

83. Ibid., 488–490.

84. Col. David H. Hackworth, "The Snake That Slithered Off for Another Day," *Palm Beach Post*, September 15, 1996.

85. Bush and Scowcroft, *A World Transformed*, 487.

86. Max Van der Stoel, special rapporteur of the Commission on Human Rights, "Report on the Situation of Human Rights in Iraq," prepared in accordance with Commission Resolution 1991/74, February 18, 1992, paragraph 100, p. 28.

87. Ibid., paragraphs 100, 104, pp. 28–29.

88. Ibid., paragraph 152, p. 64.

89. UN Security Council Resolution 688 (S/Res/688), April 5, 1991.

90. See Charter of the United Nations, Article 2(4).

91. UN Security Council Resolution 688 "On Repression of Iraqi Civilians"

The Security Council,

Mindful of its duties and its responsibilities under the Charter of the United Nations for the maintenance of international peace and security,

Recalling Article 2, paragraph 7, of the Charter of the United Nations,

Gravely concerned by the repression of the Iraqi civilian population in many parts of Iraq, including most recently in Kurdish populated areas which led to a massive flow of refugees toward and across international frontiers and to cross border incursions, which threaten international peace and security in the region,

Deeply disturbed by the magnitude of the human suffering involved,

Taking note of the letters sent by the representatives of Turkey and France to the United Nations dated 2 April 1991 and 4 April 1991, respectively (S/22435 and S/22442),

Taking note also of the letters sent by the Permanent Representative of the

Islamic Republic of Iran to the United Nations dated 3 and 4 April 1991, respectively (S/22436 and S/22447),

Reaffirming the commitment of all Member States to the sovereignty, territorial integrity and political independence of Iraq and of all States in the area,

Bearing in mind the Secretary-General's report of 20 March 1991 (S/22366),

Condemns the repression of the Iraqi civilian population in many parts of Iraq, including most recently in Kurdish populated areas, the consequences of which threaten international peace and security in the region;

Demands that Iraq, as a contribution to removing the threat to international peace and security in the region, immediately end this repression and expresses the hope in the same context that an open dialogue will take place to ensure that the human and political rights of all Iraqi citizens are respected;

Insists that Iraq allow immediate access by international humanitarian organizations to all those in need of assistance in all parts of Iraq and to make available all necessary facilities for their operations;

Requests the Secretary-General to pursue his humanitarian efforts in Iraq and to report forthwith, if appropriate on the basis of a further mission to the region, on the plight of the Iraqi civilian population, and in particular the Kurdish population, suffering from the repression in all its forms inflicted by the Iraqi authorities;

Requests further the Secretary-General to use all the resources at his disposal, including those of the relevant United Nations agencies, to address urgently the critical needs of the refugees and displaced Iraqi population;

Appeals to all Member States and to all humanitarian organizations to contribute to these humanitarian relief efforts

Demands that Iraq cooperate with the Secretary-General to these ends;

Decides to remain seized of the matter.

92. William Drozdiak, "Europeans to Press Bush to Back Enclave Plan, EC Responds to Outrage Over Kurds' Plight," *Washington Post*, April 11, 1991.

93. George Bush, "Remarks at Maxwell AFB War College, Montgomery, Alabamba," *Weekly Compilation of Presidential Documents* 27, April 13, 1991, 431.

94. George Bush, "Address to the United Nations General Assembly in New York City," *Weekly Compilation of Presidential Documents* 28, September 21, 1992, 1697.

95. Bush and Scowcroft, *A World Transformed*, 491–492.

96. UN Department of Public Information, *The Blue Helmets, A Review of United Nations' Peacekeeping*, 18 UN Doc. DPI/1065, Sales No. E.90 I.18 (1990), Part I, pp. 5–7.

97. UN Security Council Resolution 688 (S/Res/688), April 5, 1991.

98. UN Security Council Resolution 794 (S/Res/794), December 3, 1992.

99. UN Security Council Resolution 687, (S/Res/687), April 3, 1991.

100. Thomas Pickering, United Nations press release, "Security Council Says Iraq's Unconditional Agreement to Implement Obligations of Resolutions Essential Precondition for Reconsideration of Sanctions," *Federal News Service*, February 19, 1992. Also, John M. Goshko, "UN Charges Iraq with Violations: New Confrontation Over Arms Looms," *Washington Post*, March 12, 1992.

101. Rowan Scarborough and Frank J. Murray, "Iraq Backs Down, Allows Inspections; Bush Keeps Pressure High," *Washington Times*, July 27, 1992.

102. See Article III, "Treaty on the Non-Proliferation of Nuclear Weapons," International Atomic Energy Agency Information Circular. April 22, 1970. http://www.iaea.org/Publications/Documents/Infcircs/Others/infcirc140.pdf. Also, "Statue of the International Atomic Energy Agency (IAEA)," July 29, 1957, Article XII, Agency Safeguards, which states:

> A. With respect to any Agency Project, or other arrangement where the Agency is requested by the parties concerned to apply safeguards, the Agency shall have the following rights and responsibilities to the extent relevant to the project of arrangement: . . . 6. To send into the territory of the recipient state or States inspectors, designated by the Agency *after consultation with the State or States concerned . . . and with any other conditions prescribed in the agreement between the Agency and the State or States concerned*. [Emphasis added.]

Iraq has consistently violated the terms of Section A6 of Article XII, which states, "Inspectors designated by the Agency shall be accompanied by representatives of the authorities of the State concerned, if that State so requests, *provided that the inspectors shall not thereby be delayed or otherwise impeded in the exercise of their functions*." [Emphasis added.] In Resolution 687 on April 3, 1991, the Security Council charged the IAEA with identifying and destroying Iraq's nuclear weapons capabilities.

103. See "IAEA Statute," Article IV, Membership and Article VI, Board of Governors.

104. George Bush, "Annual State of the Union Address," *Weekly Compilation of Presidential Documents* 27, January 29, 1991, 90.

105. Lawrence Martin, "Peacekeeping as a Growth Industry," *National Interest*, no. 32 (Summer 1993).

2. SAVING SOMALIA

1. Barry M. Blechman and Tamara Cofmen Wittes, *Defining Moment: The Threat and Use of Force in American Foreign Policy Since 1989*, Committee on International Conflict Resolution, Occasional Paper No 1. (Washington, DC: National Research Council, 1998).

2. Joel S. Migdal described the characteristics of such "states" in his book *Strong Societies and Weak States: State Society Relations and State Capabilities in the Third World* (Princeton, NJ: Princeton University Press, 1988). Migdal's model and description resemble the pattern of political development offered in Gabriel Almond and G. Bingham Powell, *Comparative Politics, a Developmental Approach* (Boston: Little Brown, 1966), and other studies in that series. However, Migdal focused sharply on state/society boundaries, interactions, and development. In fact, such states were less "failed" than weak.

3. Gerald B. Helman and Steven R. Ratner, "Saving Failed States," *Foreign Policy*, no. 89 (Winter 1992): 3–20.

4. Defining an internal breakdown as a "threat to international peace and security" is how the Security Council claims jurisdiction over a situation such as that in Somalia. Charter of the United Nations, Chapter VII, Article 39.

5. "Interim Leader OK's Caretaker Cabinet," *Los Angeles Times*, February 3, 1991.

6. That Egypt had a long-standing close relationship with Said's government made Boutros-Ghali suspect in the eyes of Said's enemies, especially Farah Aideed.

7. Anton Ferreira, "More Than 1000 Somalis Die Every Day—Red Cross," *Reuters Library Report*, October 5, 1992. Also, William Claiborne and Keith B. Richburg, "U.S. Envoy Arranges Talks Between Top Somali Warlords," *Washington Post*, December 11, 1992. The United Nations estimated that "out of a total population of 8 million, approximately 4.5 million Somalis required urgent external assistance. Of those some 1.5 million people were at immediate risk of starvation, including 1 million children." Boutros Boutros-Ghali, *The United Nations and Somalia*, 1992–1996 (New York: United Nations Department of Public Information, 1996), 5.

8. Mohamed Sahnoun, *Somalia: The Missed Opportunities* (Washington, DC: United States Institute of Peace Press, 1994), 18.

9. Ibid., 28–29.

10. UN Security Council Resolution 751 (1992), April 24, 1992.

11. The violence in Mogadishu in July 1992 was judged too great to permit Senator Nancy Kassebaum from visiting the city. Jane Perlez, "UN Observer Unit to Go to Somalia," *New York Times*, July 20, 1992.

12. Boutros Boutros-Ghali, "Report of the secretary-general on the situation in Somalia, proposing the deployment of four additional security units, each with 750 troops, in Bossacco, Berbera, Kismayo, and the Southwest," S/24480, August 24, 1992, and addendum, S/24480/Add.1, August 28, 1992, in Boutros-Ghali, *The United Nations and Somalia*, 187.

13. Sahnoun, *Somalia*, 27.

14. Robert Kaplan, "Continental Drift," *The New Republic*, December 28, 1992 (v. 207, n. 27), p. 28.

15. Sahnoun, *Somalia*, 37–40.

16. Ibid., 28.

17. *Ibid.*, and Thomas W. Lippman, "UN Chief Faulted in Somalia Mess," *Washington Post*, August 29, 1994.

18. Mohamed Sahnoun, "This Way Out (of Somalia)," *Pittsburgh Post-Gazette*, November 7, 1993.

19. Michael Maren, *The Road to Hell* (New York: The Free Press, 1997), 214–15.

20. "The Somalia intervention was a unique geopolitical event . . . In sum, Bush's intervention in Somalia contained the seeds of a new doctrine: that Americans would fight for human and moral values, in contrast to the cold war, when it was willing to fight only for its strategic interests." In William G. Hyland, *Clinton's World* (Westport, CT: Praeger Publishers, 1999), 54.

21. UN Security Council Resolution 678 (1990), November 29, 1990.

22. Somalia was not the first instance of the deliberate use of hunger and famine as a weapon of war in Africa. Mengistui Haile Mariam had produced mass famine in Ethiopia in the decade before famine developed in Somalia, and many thousands of Ethiopians died. One important difference was the absence of foreign troops in Ethiopia. Another difference was in the lesser control exercised by any one faction leader in Somalia.

23. Iraq's invasion of Kuwait had created an international problem, not an internal problem. Both Iraq and Kuwait were recognized as wholly independent members of the United Nations.

24. Charter of the United Nations, Chapter VII, Article 39, et seq.

25. UN Security Council Resolution 688 (1991), April 5, 1991, on Repression of Iraqi Civilians, "The Security Council . . . Gravely concerned by the repression of the Iraqi civilian population in many parts of Iraq, including most recently in Kurdish populated areas which led to a massive flow of refugees toward and across international frontiers and to cross-border incursions, which threaten international peace and security in the region."

26. Not only was there a humanitarian catastrophe, but the Bush administration believed that Somalia "was a problem it was able to solve, indeed one that only the U.S. was capable of solving, and at relatively little cost. . . . The differences between Bosnia and Somalia were thus quite clear and rest fundamentally in the perception of the Bush administration, and particularly of the Pentagon, that the mission in Somalia was achievable." In Steven Hurst, *The Foreign Policy of the Bush Administration: In Search of a New World Order* (New York: Cassell, 1999), 220.

27. John M. Goshko, "UN Orders U.S.-Led Force Into Somalia," *Washington Post*, December 4, 1992.

28. Eagleburger quoted: "This is a tragedy of massive proportions . . . and, underline this, one that we could do something about." In Patrick Glynn, "The 'Doable' War: Somalia v. Bosnia. Now," *New Republic*, August 16, 1993, 15.

29. George Bush, "Address to the 46th Session of the United Nations General Assembly in New York City," *Weekly Compilation of Presidential Documents* 27, September 23, 1991, 1324.

30. Alan Elsner, "Baker Attacked in Congress for UN Peacekeeping Expenditures," Reuters Library Report, March 3, 1992.

31. Secretary of State James A. Baker, III, "Hearing of the Commerce, Justice, State and Judiciary Subcommittee of the House Appropriations Committee, State Department Fiscal Year 1993 Funding Proposal," March 3, 1992.

32. Ibid.

33. Boutros Boutros-Ghali, *An Agenda for Peace, Preventative Diplomacy, Peacemaking and Peacekeeping* (New York: United Nations, 1992).

34. The reasoning resembled that in the somewhat similar decision that was made concerning Iraq's repression of Kurds and Shiites after the Gulf War (Res. 688), where massive human rights violations by the government of Iraq were said to constitute a serious threat to international peace and security.

35. Boutros Boutros-Ghali, "Empowering the United Nations," *Foreign Affairs* 71, no. 5 (December 1992): 89.

36. Ibid., 91. He commented, "It is difficult to avoid wondering whether the conditions exist for successful peacekeeping in what was Yugoslavia." In fact, the necessary conditions did not exist in either the former Yugoslavia or Somalia.

37. Boutros-Ghali, *An Agenda for Peace*, 13–38, 41–45.

38. Letter dated November 29, 1992, from the secretary-general to the president of the Security Council presenting five options for the Security Council's consideration, S/24868, November 30, 1992, in Boutros-Ghali, *The United Nations and Somalia*, 209–212.

39. Ibid., 212. In explaining his idea of how the fifth option would operate and his clear preference for that course of action, Boutros-Ghali writes, "The focus of the Council's immediate action should be to create conditions in which relief supplies can be delivered to those in need. Experience has shown that this cannot be achieved by a United Nations operation based on the accepted principles of peacekeeping. There is now no alternative but to resort to Chapter VII of the Charter . . . If forceful action is taken, it should preferably be under United Nations Command and Control. If this is not feasible, an alternative would be an operation undertaken by Member States acting with the authorization of the Security Council." In Boutros-Ghali's mind, U.S. troops undertaking the Somali operation under the command and control of a competent U.S. military and its democratically elected leaders was less "preferable" than the submission of U.S. troops to the personal command and control of Secretary-General Boutros-Ghali.

40. The commander in chief of the U.S. Central Command (also the commander of Operation Restore Hope) argued in an article published the following year that "disarmament was . . . neither realistically achievable nor a prerequisite for

the core mission of providing a secure environment for relief operations." See Joseph P. Hoar, "A CINC's Perspective," *Joint Force Quarterly*, no. 2 (Autumn 1993): 58.

41. John Bolton, "Wrong Turn in Somalia," *Foreign Affairs* 73, no. 1 (January/February 1994) (New York: Council on Foreign Relations): 61.

42. On these early understandings and misunderstandings, see especially Robert Oakley, "An Envoy's Perspective," *Joint Force Quarterly*, no. 2 (Autumn 1993): 46. Also John L. Hirsch and Robert B. Oakley, *Somalia and Operation Restore Hope: Reflections on Peacemaking and Peacekeeping* (Washington, DC: United States Institute of Peace, 1995). They stated that "the top UN officials rejected the idea that the U.S. initiative should eventually become a UN peacekeeping operation" (13), but George Bush had said, "once we have created that secure environment, we will withdraw our troops, handing the security mission back to a regular UN peacekeeping force" (14).

43. Rowan Scarborough, "Somalia Dangers Weighed," *Washington Times*, December 2, 1992.

44. "Letter dated 29 November 1992 from the Secretary-General to the President of the Security Council presenting five options for the Security Council's consideration," S/24868, November 30, 1992, in Boutros-Ghali, *The United Nations and Somalia*, 209.

45. UN Security Council Resolution 794, S/Res/794 (1992), December 3, 1992, paragraph 10.

46. Written statement of President-elect Bill Clinton, issued December 3, 1992.

47. George Bush, "The People of Somalia . . . the Children . . . Need Our Help," *Weekly Compilation of Presidential Documents* 28, December 4, 1992, 2329.

48. George Bush, "Letter to Congressional Leaders on the Situation in Somalia," *Weekly Compilation of Presidential Documents* 28, December 10, 1992, 2338.

49. "Congressional Leaders Offer Comments on Somalia Plan," CNN News, Live Report, Transcript #175-3, December 4, 1992.

50. Peter Applebome, "Mission to Somalia," *New York Times*, December 13, 1992.

51. Boutros Boutros-Ghali, "Statement made by the Secretary-General to the people of Somalia on United Nations action on security, humanitarian relief, and political reconciliation in Somalia," UN Press Release SG/SM4874, December 8, 1992, in Boutros-Ghali, *The United Nations and Somalia*, 218.

52. Boutros Boutros-Ghali, "Letter dated 8 December 1992 from the Secretary-General to President Bush of the United States discussing the establishment of a secure environment in Somalia and the need for continuous consultations," in Boutros-Ghali, *The United Nations and Somalia*, 216.

53. Paul Lewis, "Mission to Somalia," *New York Times*, December 13, 1992.

54. George Bush, "Address to the Nation on the Situation in Somalia," *Weekly Compilation of Presidential Documents* 28, December 4, 1992, 2329.

55. Barton Gellman, "United States and United Nations Differ over Best Way to Silence Somalia's Many Guns," *Washington Post*, December 23, 1992.

56. Eagleburger quoted: "This is a tragedy of massive proportions . . . and one that we could do something about," in Glynn, "The 'Doable' War," 15.

57. In his address to the nation as the operation in Somalia was beginning, President Bush said, "To the people of Somalia I promise this: We do not plan to dictate political outcomes. We respect your sovereignty and independence. Based on my conversations with other coalition leaders, I can state with confidence: We come to your country for one reason only, to enable the starving to be fed." In Michael MacKinnon, *The Evolution of U.S. Peacekeeping Policy Under Clinton: A Fairweather Friend* (London: Frank Cass Publishers, 2000), 17. Also, an account of the limits the Bush administration placed on American involvement in Somalia, including a treatment of its haggling with the secretary-general over those limits, is found in Bolton, "Wrong Turn in Somalia," 56–66.

58. On American public opinion regarding intervention in Somalia, see Steven Kull and Clay Ramsay, "U.S. Public Attitudes on Involvement in Somalia" (College Park, MD: University of Maryland Program on International Policy Attitudes, October 26, 1993).

59. Alison Mitchell, "Legislator Faults UN Over Somalia," *New York Times*, January 11, 1993.

60. Jennifer Parmelee, "UN Hosts Meeting of Somalia Factions," *Washington Post*, January 5, 1993.

61. Keith B. Richburg, "U.S. Envoy in Somalia Viewed As Linchpin of Reconciliation; In Absence of Leaders, Many Groups Seek Oakley's Counsel," *Washington Post*, February 2, 1993.

62. Jonathan T. Howe, "The United States and United Nations in Somalia: The Limits of Involvement," *Washington Quarterly* 18, no. 3 (Summer 1995): 47–62, 53.

63. Stuart Auerbach, "It Just Seems Our Job Is Done: Tired Marines, Eager to Return Home, Await UN Takeover in Somalia," *Washington Post*, March 1, 1993.

64. Stuart Auerbach, "Oakley Calls Mission in Somalia a Success; U.S. Envoy Laments Leaving Before Marines," *Washington Post*, March 2, 1993.

65. Ibid.

66. Stuart Auerbach, "UN Assailed for Delay in Takeover in Somalia; Envoy Seeks to Clear Way to Pare U.S. Role," *Washington Post*, February 21, 1993.

67. Julia Preston, "Shift to UN Targeted for May 1," *Washington Post*, March 4, 1993.

68. Boutros Boutros-Ghali, "Further Report of the Secretary-General submitted in pursuance of paragraphs 18 and 19 of Resolution 794 (1992)," in Boutros-Ghali, *The United Nations and Somalia*, 251.

69. Elizabeth Drew, *On the Edge: The Clinton Presidency* (New York: Simon and Schuster, 1994), 317–320.

70. Boutros Boutros-Ghali, "Further Report of the Secretary-General submitted in pursuance of paragraphs 18 and 19 of Resolution 794 (1992), proposing that the mandate of UNOSOM II cover the whole country and include enforcement powers under chapter of the Charter," S/25354, March 3, 1993, and addenda S/25354/Add.1, March 11, 1993, and S/25354/Add.2, March 22, 1993, paragraph 100, in Boutros-Ghali, *The United Nations and Somalia*, 255.

71. Ibid., 254.

72. See specifically paragraph A4 of UN Security Council Resolution 814, S/Res/814 (1993), March 26, 1993.

73. Michael Maren, *The Road to Hell*, 217–218.

74. In March 1993, a leadership transition occurred from UNITAF to UNOSOM II, but the public was not made aware of the shift in strategic direction because the operation in Somalia had dropped from the major news headlines. The polls showed that fewer people were responding, which meant that fewer people were following the events in Somalia. A report by the Aspen Institute stated that the number of people responding to the events in Somalia had dropped from 52 percent in January 1999 to 16 percent in June 2000. See Mackinnon, *U.S. Peacekeeping Policy Under Clinton*, 77.

75. Stuart Auerbach, "Gunmen in Mogadishu Battle Marines, Allies for Second Day," *Washington Post*, February 26, 1993.

76. Daniel Williams and John Lancaster, "Somali Violence May Delay U.S. Withdrawal," *Washington Post*, February 26, 1993.

77. UN Security Council Resolution 814, paragraphs 7 and 14.

78. Ibid., paragraph 14, 4(a)-(g). In her statement on the vote, Albright said, "Yet, we are certain of this: each element of the program for Somalia is necessary to its overall success and that country's recuperation. Through his Special Representative the Secretary-General must oversee the continued ceasefire, disarmament, and maintenance of security. . . ." "Statement by Ambassador Madeleine K. Albright, United States Permanent Representative to the United Nations, in the Security Council, in explanation of the vote on the situation in Somalia," USUN press release 37-(93), March 26, 1993.

79. "Further Report of the Secretary-General submitted in pursuance of paragraphs 18 and 19 of Resolution 794 (1992), proposing that the mandate of UNOSOM II cover the whole country and include enforcement powers under Chapter VII of the Charter," S/25354, March 3, 1993, and addenda S/25354/Add.1, 11 March 1993, and S/25354/Add. 2, March 22, 1993, paragraph 92, in Boutros-Ghali, *The United Nations and Somalia*, 254.

80. Altogether, the Security Council charged Boutros-Ghali and Howe with "rehabilitating [the] political institutions and economy" of Somalia. UN Security Council Resolution 814, paragraph 4, March 26, 1993. One case study of Somalia concludes that "in Somalia, largely as a result of inattention at political levels, the

United States allowed its military forces to transform their original humanitarian mission into a coercive activity intended to enforce a peaceful settlement of the underlying political conflict by disarming factions in a particular section of the country." Blechman and Wittes, *Defining Moment*, 14.

81. Paul Lewis, "UN Is Developing Control Center to Coordinate Growing Peacekeeping," *New York Times*, March 28, 1993.

82. In a speech before the Los Angeles World Affairs Council, Clinton said, "We will stand up for our interests, but we will share burdens, where possible, through multilateral efforts to secure the peace, such as NATO, and a new voluntary UN rapid deployment force. In Bosnia, Somalia, Cambodia, and other torn areas of the world, multilateral action holds promise as never before, and the UN deserves full and appropriate contributions from all the major powers." In "The 1992 Campaign; Excerpts from Clinton's Speech on Foreign Policy Leadership," *New York Times*, August 14, 1992.

83. William Cran, writer, producer, and director, "Ambush in Mogadishu, " *Frontline*, Public Broadcasting System, Transcript #1704, September 28, 1998, 11.

84. Ambassador Robert Oakley, in Cran, "Ambush in Mogadishu," 11.

85. Bolton, "Wrong Turn in Somalia," 56.

86. "Statement by Ambassador Madeleine Albright" (press release).

87. See "Addis Ababa Agreement concluded at the first session of the Conference on National Reconciliation in Somalia, March 27, 1993," in Boutros-Ghali, *The United Nations and Somalia*, 264.

88. Drew, *On the Edge*, 319.

89. Morton Halperin and David J. Scheffer, *Self-Determination in the New World Order* (Washington, DC: Carnegie Endowment for International Peace,1992).

90. A copy of this memo was passed to me confidentially.

91. Ibid.

92. Ashton B. Carter, William J. Perry, and John D. Steinbruner, "A New Concept of Collective Security," A Report for the Brookings Institution. 1992 (Washington, DC).

93. Senator Richard Lugar, in Cran, "Ambush in Mogadishu," 11.

94. Blechman and Wittes, *Defining Moment*. See also Halperin and Scheffer, *Self-Determination*.

95. Letter from President Clinton to congressional leaders, July 1, 1993.

96. "Executive summary of the report prepared by Professor Tom Farer of American University, Washington, D.C., on the June 5, 1993, attack on the United Nations forces in Somalia," S/26351, August 24, 1993, in Boutros-Ghali, *The United Nations and Somalia* 296–300, paragraph 15, 299. Mohammed Sahnoun was especially critical of the decision to target Aideed and of the "investigation" that preceded it. Farer's emphasis on Aideed's military forces was particularly significant in light of the Ranger disaster in Mogadishu.

97. Letter from President Clinton to congressional leaders, July 1, 1993.

98. General Anthony Zinni (director of operations for UNITAF, November 1992 to May 1993. After October 1993, assistant to the special envoy in Somalia, Robert Oakley, in negotiations with Aideed for a truce and the release of U.S. Ranger Michael Durant), interview in Cran, "Ambush in Mogadishu," 7.

99. Robert Oakley, in Cran, "Ambush in Mogadishu," 4, 15.

100. United Nations Operation in Somalia II, *United Nations*, last update, March 21, 1997

101. Ibid., 13.

102. "Executive summary of the report prepared by Professor Tom Farer of American University, Washington, D.C., on the June 5, 1993, attack on the United Nations forces in Somalia," S/26351, August 24, 1993, in Boutros-Ghali, *The United Nations and Somalia* 296–300, paragraph 15, 299. Mohammed Sahnoun was especially critical of the decision to target Aideed and of the "investigation" that preceded it. Farer's emphasis on Aideed's military forces was particularly significant in light of the Ranger disaster in Mogadishu.

103. Robert Oakley, interview in Cran, "Ambush in Mogadishu," 8–9.

104. General Anthony Zinni (director of operations for UNITAF, November 1992 to May 1993. After October 1993, assistant to the special envoy in Somalia, Robert Oakley, in negotiations with Aideed for a truce and the release of U.S. Ranger Michael Durant), interview in Cran, "Ambush in Mogadishu," 7.

105. "Letter dated 27 August from Ambassador Madeleine Albright, President of the UN Security Council to the Secretary-General concerning the Council's intention to study the recommendations in Professor Farer's report (Document 62) on the reestablishment of Somali Police Forces," S/26375, August 29, 1993, in Boutros-Ghali, *The United Nations and Somalia*, 300–301. UN Security Council Resolution 865, S/Res/865(1993), September 22, 1993.

106. Captain Haad (sector commander for General Aideed's militia), in Cran, "Ambush in Mogadishu," 14–15.

107. Specialist Jason Moore, U.S. Army Ranger (1992–95), in Cran, "Ambush in Mogadishu," 15.

108. General Thomas Montgomery (ret.), deputy UN commander (1993–94), in Cran, "Ambush in Mogadishu," 6.

109. Ibid., 6.

110. Drew, *On the Edge*, 321.

111. CIA special agent Gene Cullen, in Cran, "Ambush in Mogadishu," 16.

112. Ibid., 17.

113. Patrick J. Sloyan, "A Look at . . . the Somalia Endgame; How the Warlord Outwitted Clinton's Spooks," *Washington Post*, April 3, 1994.

114. Cran, "Ambush in Mogadishu," 17.

115. General Thomas Montgomery (ret.), in Cran, "Ambush in Mogadishu," 17.

116. A dramatic and carefully researched account of the firefight in Mogadishu was written by Mark Bowden: *Black Hawk Down: A Story of Modern War* (New York: Atlantic Monthly Press, 1999).

117. William Garrison, commander, Joint Special Operations Command, Hearing of the Senate Armed Services Committee, "U.S. Military Operations in Somalia," May 12, 1994.

118. When the U.S. Marines returned to Somalia to help evacuate UN forces, they were armed with nonlethal weapons.

119. Jim O'Connell, "Members Call for More Consultation on Somalia After This Week's Big Battle," *Roll Call*, October 7, 1993.

120. Cran, "Ambush in Mogadishu," 29–30.

121. Senator Sam Nunn, Hearing of the Senate Armed Services Committee, "U.S. Military Operations in Somalia," May 12, 1994.

122. Ibid.

123. Patrick J. Sloyan, "Full of Tears and Grief; For Elite Commandos, Operation Ended in Disaster," *Newsday*, December 7, 1993, 6.

124. Sahnoun, *New Perspectives*; Boutros-Ghali, *The United Nations and Somalia*.

125. Senator Strom Thurmond, Hearing of the Senate Armed Services Committee, "U.S. Military Operations in Somalia," May 12, 1994.

126. General Thomas Montgomery, former deputy commander of UN operations in Somalia, Hearing of the Senate Armed Services Committee, "U.S. Military Operations in Somalia," May 12, 1994.

127. Madeleine Albright, Hearing of the Senate Foreign Relations Committee, October 20, 1993.

128. Ibid.

129. Peter Tarnoff, Hearing of the Senate Foreign Relations Committee, "U.S. Policy in Somalia," October 19, 1993.

130. Howe, "The United States and United Nations in Somalia," 47–62.

131. Michael Mandelbaum, "Foreign Policy As Social Work," *Foreign Affairs* 75, no. 1 (January/February, 1996): 16.

132. Madeleine Albright, in Charles Bierbauer, "Interview with Ambassador Madeleine Albright and Congressman Dave McCurdy," *Newsmaker Saturday*, CNN transcripts, #172, June 12, 1993.

133. William Jefferson Clinton, "Press Conference by the President," Washington, DC, October 14, 1993.

134. General Thomas Montgomery, former deputy commander of UN operations in Somalia, Hearing of the Senate Armed Services Committee, "U.S. Military Operations in Somalia," May 12, 1994.

135. See Thomas Henriksen, *Clinton's Foreign Policy in Somalia, Bosnia, Haiti,*

and North Korea (Stanford: Hoover Institution, 1996), 11. "The introduction of elite U.S. combat forces, the militarization of the humanitarian mission, and the change in scope of American efforts were all undertaken without an adequate explanation to the public or to Congress. Thus, when disaster struck later in the Mogadishu streets, the American people were unprepared."

136. Thomas Montgomery, interview in Cran, "Ambush in Mogadishu," 1.

137. Ibid.

138. Kenneth Allard, *Somalia Operations: Lessons Learned* (Washington, DC: National Defense University Press, 1995). See also Colonel Kenneth Allard (ret.), interview in Cran, "Ambush in Mogadishu," 5.

139. Walter B. Slocombe, nominee to be undersecretary of defense for policy, Hearing of the Senate Armed Services Committee, "Defense Nominations," August 10, 1994.

140. Boutros Boutros-Ghali, "Further report of the Secretary-General on UNOSOM, submitted in Pursuance of Paragraph 14 of Resolution 897 (1994), with annex containing the text of the declaration issued by Somali political leaders in Nairobi on 24 March 1994," S/1994/614, May 24, 1994, paragraph 22, in Boutros-Ghali, *The United Nations and Somalia*, 357.

141. Paul Lewis, "Report Faults Commanders of UN Forces in Somalia," *New York Times*, May 20, 1994.

142. Barbara Crossette, "UN Falters in Post–Cold War Peacekeeping, But Sees Role as Essential," *New York Times*, December 5, 1994.

143. Richard Haass, *Intervention: The Use of American Military Force in the Post–Cold War World* (Washington, DC: The Brookings Institution, 1999), 155.

144. Bowden, *Black Hawk Down*, 413.

145. Ibid., 335–336.

146. See testimony of General Thomas Montgomery, former deputy commander of UN operations in Somalia, and General William Garrison, commander, Joint Special Operations Command, Senate Armed Services Committee Hearing, "U.S. Military Operations Committee in Somalia," May 12, 1994.

147. John H. Cushman, Jr., "The Somalia Mission: Forces; How Powerful U.S. Units Will Work," *New York Times*, October 8, 1993.

148. Harold E. Bullock, *Peace by Committee, Command and Control Issues in Multinational Peace Enforcement Operations* (Maxwell AFB, AL.: Air University Press, 1995).

149. Howe, "The United States and United Nations in Somalia," 53.

150. Ibid.

151. Bowden, *Black Hawk Down*, 335–336.

152. General Joseph P. Hoar, Somalia-An Envoy's Perspective, *Joint Forces Quarterly* (Autumn 1993).

153. In a formal statement on October 7, President Clinton stated "We have obligations elsewhere . . . [it is not America's job to] rebuild Somalia society." See Hyland, *Clinton's World*, 58.

154. Keith B. Richburg, citing the U.S. ambassador to Somalia, Daniel Simpson, "Somalia Slips Back to Bloodshed; Anarchy, Death Toll Grow as UN Mission Winds Down," *Washington Post*, September 4, 1994.

155. William J. Clinton, Remarks to the 49th Session of the United Nations General Assembly, September 26, 1994.

3. HAITIANS' RIGHT TO DEMOCRACY?

1. Gerald B. Helman and Steven R. Ratner, "Saving Failed States," *Foreign Policy* 89 (Winter 1992–1993): 3.

2. Madeleine K. Albright, "The Testing of American Foreign Policy," *Foreign Affairs* 77, no. 6 (November/December 1998): 51.

3. Ibid., 52.

4. OAS-AG/RES. 1080 (XXI-0/91), June 5, 1991.

5. "Independent Haiti," Library of Congress Country Studies, December 1989, http://lcweb2.loc.gov/frd/cs/httoc.html.

6. World Bank, *Haiti: The Challenges of Poverty Reduction*, Report 17242-HA, Vol. 1 (August 1998): 14.

7. François "Papa Doc" Duvalier was Haiti's president from 1957 until 1971; he was succeeded by his son, Jean-Claude "Baby Doc" Duvalier, who was president from 1971 until he was ousted in 1986.

8. Aristide won about 67 percent of the vote; his closest competitor, Marc Bazin, in a field of about a dozen candidates, received only 14 percent. About 50 percent of registered voters turned out for the election. See Economist Intelligence Unit, *Dominican Republic, Haiti, Puerto Rico: Country Profile, 1991–92* (London: Business International Limited, 1991), 32; Henry F. Carey, "Electoral Observation and Democratization in Haiti," quoted in Kevin J. Middlebrook, ed., *Electoral Observation and Democratic Transitions in Latin America* (San Diego: University of California, 1998), 147–148.

9. World Bank, Haiti: The Challenges of Poverty Reduction, 14.

10. Ibid.

11. Maureen Taft-Morales, "Haiti: Issues for Congress," *CRS Issue Brief for Congress* (Congressional Research Service: Library of Congress), August 1, 2001, 8; International Monetary Fund, *Haiti: Selected Issues*, Report 01/04, (January 2001): 51.

12. IMF. Haiti: Selected Issues, 51.

13. EIU Country Profile 2000: Dominican Republic, Haiti, Puerto Rico (London: Economist Intelligence Unit, 2000), 42.

14. Taft-Morales, "Haiti: Issues for Congress," 8.

15. Ibid.

16. World Bank, Haiti: The Challenges of Poverty Reduction, 21.

17. Ibid.

18. Accounts in U.S. media indicate that several influential Americans with ties to the Clinton administration's policies in Haiti have reaped profits from financial ties established when that administration was making policy. Telephone service is one example. See "Haitian Connections: How Clinton's Cronies Cashed in on Foreign Policy," *Wall Street Journal*, May 29, 2001.

19. Editorial Desk, "Haiti's Disappearing Democracy," *New York Times*, November 28, 2000; House Committee on International Relations, "Gilman, Helms, and Goss Issue Statement on Haitian Election," 106th Congress, December 8, 2000; Rep. Porter Goss originally sponsored a bill "condemning the irregular interruption of the democratic political institutional process in Haiti" on March 8, 1999. http://thomas.loc.gov/cgi=bin/bdquery/D?d106:7:./temp/~bdUEGI:@@@ L&summ2=m&.

20. World Bank, Haiti: The Challenges of Poverty Reduction, 14.

21. In classified congressional briefings in October 1993, the CIA provided an analysis and psychological profile of Aristide that suggested he was mentally unstable, violent, and used medication to treat depression and severe mood swings. In December 1993, *Miami Herald* reporter Christopher Marquis wrote an article claiming to disprove allegations that Aristide had received psychiatric treatment in Canada and concluding that the CIA report was false. See Steven A. Holmes, "Administration Is Fighting Itself on Haiti Policy," *New York Times*, October 23, 1993; Christopher Marquis, "CIA Report on Aristide Was False; He Did Not Undergo Psychiatric Treatment," *Miami Herald*, December 2, 1993; Mark Danner, "The Fall of the Prophet," *New York Review of Books*, December 2, 1993 (see also Danner, "Haiti on the Verge, *New York Review of Books*, November 4, 1993, and "The Prophet," *New York Review of Books*, November 18, 1993); David Malone, *Decision-Making in the UN Security Council: The Case of Haiti, 1990–1997* (Oxford: Clarendon Press, 1998), 98.

22. Danner, "Haiti on the Verge," *New York Review of Books*, November 4, 1993.

23. Lally Weymouth, "Haiti's Suspect Savior: Why President Aristide's Return from Exile May Not Be Good News," *Washington Post*, January 24, 1993; see also Malone, *Decision-Making in the UN Security Council*, 61; "The Clever Soldiers of Haiti," *The Economist*, October 12, 1991.

24. For one account of this episode, see Danner, "The Fall of the Prophet."

25. Charles Lane, with Peter Katel, Tim Padgett, Ann McDaniel, and Daniel Glick, "Haiti: Why the Coup Matters," *Newsweek*, October 14, 1991.

26. Lee Hockstader, "Critics of Exiled Haitian President Focus on His Concept of Justice," *Washington Post*, October 7, 1991; Mark Falcoff, "What 'Operation Restore Democracy' Restored," *Commentary* 101, no. 5 (May 1996): 45.

27. Henry F. Carey, "Electoral Observation and Democratization in Haiti," 148. In a speech to his Lavalas followers, Aristide said, "Alone, we are weak; Together, we are strong; Together, we are the flood; The flood of poor peasants and poor soldiers, The flood of the poor jobless multitudes . . . the flood of all our poor friends. . . . Let the Flood descend!" (Danner, "The Fall of the Prophet," 44). Danner wrote, "Lavalas is a Creole word rich in connotations; [it] evokes not only 'flood,' as it is usually translated but its near cognate, 'avalanche'; for poor Haitians, the word evokes the image of the sweeping rains that spawn the torrents that course through the enormous slums . . . an image that transformed the poor millions of Haiti into a surging wave that could not be forestalled, a revolution that was unstoppable and inevitable" (Danner, "The Fall of the Prophet," 46).

28. Georges A. Fauriol, "The Military and Politics in Haiti," in Georges A. Fauriol, ed., *Haitian Frustrations: Dilemmas for U.S. Policy* (Washington, DC: CSIS, 1995), 22–23.

29. Ibid.

30. Thomas L. Friedman, "U.S. Suspends Assistance to Haiti and Refuses to Recognize Junta," *New York Times*, October 2, 1991.

31. SG/SM/4627 HI/4 of October 1, 1991; quoted in Malone, *Decision-Making in the UN Security Council*, 63.

32. Malone, *Decision-Making in the UN Security Council*, 63, 209.

33. "Remarks by United States Secretary of State James Baker before the Organization of American States Meeting on the Situation in Haiti, Washington, DC," *Federal News Service*, October 2, 1991.

34. OAS-MRE/RES 1/91.

35. Executive Order 12775, "Prohibiting Certain Transactions with Respect to Haiti," *Public Papers of the Presidents*, October 4, 1991 (published in the *Federal Register*, October 7, 1991).

36. "The President's News Conference," *Public Papers of the Presidents*, October 4, 1991.

37. James A. Baker III, *The Politics of Diplomacy: Revolution, War and Peace, 1989–1992* (New York: G. P. Putnam's Sons, 1995), 602.

38. See Malone, *Decision-Making in the UN Security Council*, 65.

39. Ibid, 259.

40. OAS-MRE/RES 2/91.

41. MRE/RES 1/91 and MRE/RES 1/92; A/RES/46/7 of October 11, 1991.

42. Lee Hockstader, "Haitian Premier: If Aristide Returns, He's a Dead Duck; Embargo Will Mean Civil War," *Washington Post*, October 25, 1991; Howard W. French, "U.S. Will Impose a Trade Ban on Haiti," *New York Times*, October 30, 1991.

43. John M. Goshko, "Bush Strengthens Curb on Haiti Trade," *Washington Post*, October 30, 1991.

44. Lee Hockstader, "Embargo of Haiti at Issue," *Washington Post*, November 27, 1991; Canute James, "Executives Urge End to Haiti Trade Ban," *Journal of Commerce*, December 6, 1991; Carla Anne Robbins, Robin Knight, and Peter Green, "A Diplomatic Stalemate Leaves the Haitian Poor Adrift," *U.S. News & World Report*, December 9, 1991.

45. Howard W. French, "Sanctions Said to Fuel Haitian Exodus by Sea," *New York Times*, November 23, 1991.

46. Ibid.

47. Jeane Kirkpatrick, "Hurting Haiti's Poorest," *Washington Post*, December 9, 1991, editorial; page A21.

48. Associated Press, May 30, 1992.

49. OAS-MRE/RES 3/92. The same resolution, of May 17, 1992, recommended that members deny visa privileges and freeze the assets of the Haitian military commanders in power.

50. "Statement by President George Bush on the Haitian Trade Embargo," *Federal News Service*, May 28, 1992.

51. Lori Fisler Damrosch, ed., *Enforcing Restraint: Collective Intervention in Internal Conflicts* (New York: Council on Foreign Relations Press, 1993), 17.

52. Damrosch, *Enforcing Restraint*, 375.

53. Elizabeth D. Gibbons, *Sanctions in Haiti: Human Rights and Democracy under Assault* (Washington, DC: Center for Strategic and International Studies/ Praeger, 1999), 110–111.

54. David Binder, "Clinton Urges Haitian Leader to Appoint a New Premier," *New York Times*, July 23, 1993.

55. Elizabeth Drew estimated the number of attachés at between forty and sixty. See Elizabeth Drew, *On the Edge: The Clinton Presidency* (New York: Touchstone, 1995), 333.

56. Dan Balz and Richard Morin, "Public Losing Confidence in Clinton Foreign Policy," *Washington Post*, May 17, 1994.

57. Adapted from Jeane Kirkpatrick, "Invade Haiti?" *Baltimore Sun*, May 24, 1994.

58. The Friends of the Secretary-General for Haiti, a group of countries charged with special responsibility for shaping Security Council policy toward Haiti, included Canada, France, the United States, and Venezuela, and later Argentina (from 1994) and Chile (from 1996).

59. See Gibbons, *Sanctions in Haiti*, 11.

60. William Schneider, "The Carterization of Bill Clinton," *National Journal*, October 1, 1994.

61. President Clinton's Address to the Nation on the Situation in Haiti, *Federal News Service*, September 15, 1994.

62. Poll results cited in "This Week with David Brinkley," *ABC News*, September 18, 1994; "Majority of U.S. Voters Opposed to Invasion of Haiti," *ABC News: World News Sunday*, September 18, 1994.

63. After U.S. troops landed in Haiti, a poll taken on September 19 showed 52 percent opposed (Michael R. Kagay, "Occupation Lifts Clinton's Standing in Poll, But Many Americans Are Skeptical," *New York Times*, September 21, 1994).

64. The forces included approximately 1,300 troops to guard vital installations, 600 police monitors, and 100 police trainers (Steven Greenhouse, "Showdown in Haiti," *New York Times*, September 19, 1994).

65. Greenhouse, "Showdown in Haiti."

66. Thomas M. Franck, "The Emerging Right to Democratic Governance," *American Journal of International Law* 86 (1992): 46.

67. Ibid.

68. Adapted from Jeane Kirkpatrick, "Is Democracy an Entitlement?" *Washington Post*, September 12, 1994.

69. Helman and Ratner, "Saving Failed States," 10.

70. Gareth Evans and Mohamed Sahnoun, "The Responsibility to Protect," *Foreign Affairs* (November/December 2002): 99–110.

71. Helman and Ratner, "Saving Failed States," 3.

72. Ibid., 12.

73. Jeane Kirkpatrick, "Why It's Smart to Bet on a Haiti Invasion," *Los Angeles Times*, August 21, 1994.

74. The agreement was also called the Carter-Cédras Accord.

75. Adapted from Jeane Kirkpatrick, "The Theory and Practice of Clintonism," in Jeane Kirkpatrick, Jacqueline Tillman, et al., *Security and Insecurity: A Critique of Clinton Policy at Mid-Term* (Washington, DC: Empower America, 1994).

76. Greenhouse, "Showdown in Haiti."

77. Adapted from Jeane Kirkpatrick, "Peacekeeping: Go Easy," *New York Post*, May 9, 1995.

78. Kenneth T. Walsh, Bruce B. Auster, and Tim Zimmerman, "Good Cops, Bad Cops," *U.S. News & World Report*, September 26, 1994.

79. See Kenneth J. Cooper and Helen Dewar, "Congress Urges 'Prompt' Troop Withdrawal from Haiti," *Washington Post*, October 7, 1994.

80. Adapted from Jeane Kirkpatrick, "The Theory and Practice of Clintonism."

81. Adapted from Jeane Kirkpatrick, "Clinton at Mid-Term," in Jeane Kirkpatrick, Jacqueline Tillman, et al., *Security and Insecurity*.

82. S/1995/305, "Report of the Secretary-General on the United Nations Mission in Haiti," April 13, 1995, documents the composition of UNMIH troops and UNMIH civilian police.

83. Tara Sonenshine, "What's Going Right in Haiti?" *Washington Times*, March 7, 1995.

84. See Larry Rohter, "February 19–25: Taking Charge; Haiti Cripples Its Army and Schedules Its Elections," *New York Times*, February 26, 1995.

85. "The United Nations and the Situation in Haiti" (New York: United Nations Department of Public Information, March 1995), prepared by the Department of Public Information, United Nations—as of September 1996. http://www.un.org/Depts/DPKO/Missions/unmih_b.htm; David Malone, *Decision-Making in the UN Security Council*, 122–124.

86. See Charles Lane, "Island of Disenchantment," *New Republic*, September 29, 1997. A U.S. embassy cable of September 1, 1995, stated that the FBI investigation of the Bertin assassination was at a standstill owing to lack of cooperation from the government of Haiti and that persons close to the president who were thought to be implicated in execution-style killings continued to hold their positions.

87. Douglas Farah, " 'From Death to Life:' One Year After President's Return, Haiti Fitfully Democratizes," *Washington Post*, September 30, 1995.

88. Robert A. Pastor, "Mission to Haiti #3, Elections for Parliament and Municipalities, June 23–26, 1995" (Atlanta: Carter Center, July 17, 1995).

89. Robert A. Pastor, "A Popular Democratic Revolution in a Predemocratic Society: The Case of Haiti," in Robert I. Rotberg, ed., *Haiti Renewed: Political and Economic Prospects* (Washington, DC: Brookings Institution Press, 1997), 131.

90. Ron Howell, "Haiti Votes, in Peaceful Confusion," *Newsday*, June 26, 1995; International Republican Institute (IRI) News Release, June 24, 1995. See also IRI News Release of June 26 and "Haiti Election Alert," June 27, 1995.

91. Pastor, "A Popular Democratic Revolution in a Predemocratic Society," 131.

92. Ibid., 127.

93. Ibid., 131.

94. Ibid., 127.

95. "No Major Irregularities in Haiti Vote," *United Press International*, June 25, 1995.

96. "Half of Electorate Turns Out for Poll, Marked by Irregularities," *Agence France Presse*, June 26, 1995.

97. State Department briefing, *Federal News Service*, June 27, 1995.

98. Anita Snow, "U.S. Official: Elections Were Least Violent in Haiti's History," *Associated Press*, June 28, 1995.

99. United Nations Security Council, "Report of the Secretary-General on the United Nations Mission in Haiti," S/1995/614, July 24, 1995.

100. Dominique Levanti, "International Observers Hail Fairness of Haitian Vote," *Agence France Presse*, June 28, 1995.

101. See OAS, "Final Report of the OAS Electoral Observation Mission to the Legislative and Municipal Elections in Haiti," CP/doc.2703/96, February 20, 1996, and corr. 1, March 5, 1996; Pastor, "Mission to Haiti #3"; United Nations Security Council, "Report of the Secretary-General on the United Nations Mission in Haiti."

102. Préval had "such a close relationship with President Jean-Bertrand Aristide that supporters call[ed] them 'the twins,' despite their sharp differences in temperament and upbringing." (Douglas Farah, "Aristide's 'Twin' on Center Stage," *Washington Post*, December 17, 1995.)

103. "Gilman Charges Cover-up as White House Claims Executive Privilege over Documents on Political Killings," press release issued September 25, 1996.

104. Haiti: "Steps Forward, Steps Back: Human Rights 10 Years after the Coup," September 27, 2001, http://www.web.amnesty.org/library/index/engAMR360102001?OpenDocument.

105. OAS, "Final Report: The Election Observation Mission for the Legislative, Municipal, and Local Elections in Haiti, February to July 2000" (OAS, Unit for the Promotion of Democracy, 2001), Electoral Observations in the Americas Series, No. 28, OEA/Ser.D/XX, SG/UPD/II.28, December 13, 2000.

106. OAS, "The OAS Electoral Observation Mission in Haiti: Chief of Mission Report to the OAS Permanent Council," July 13, 2000, http://www.haitipolicy.org/archives/Archives/June-August2000/oas9.htm. See also OAS, "Final Report . . . February to July 2000."

107. OAS, "Electoral Observation Mission in Haiti: OAS Will Not Observe Haiti's July 9 Second Round Elections," Press Release, July 7, 2000, http:/www.upd.oas.org/EOM/Haiti/haitiobservation200010.htm.

108. OAS, "Chief of Mission Report."

109. Yves Colon, "U.S. Warns against Travel to Haiti before Elections," *Miami Herald*, November 19, 2000.

110. Edward Cody, "Divided and Desperate Haiti Braces for Aristide's Return," *Washington Post*, February 1, 2001; Yves Colon, "Aristide Urged to Condemn Violence," *Miami Herald*, January 12, 2001.

111. Colon, "Aristide Urged to Condemn Violence."

112. The United States did not send an official delegation to the inauguration; however, the State Department said that "in support of national interests in Haiti and in view of our historic ties to the Haitian people," U.S. ambassador to Haiti Brian Dean Curran represented Washington (Eric Green, "U.S. Urges Aristide, Opposition to Address Haiti's Difficulties," *Washington File*, U.S. Department of State, February 7, 2001.)

113. "Annan Recommends Winding Up UN Mission to Haiti," *Agence France Presse*, November 28, 2000.

114. Ibid.

115. "Report of the Secretary General on the OAS Mission and of the Joint OAS/CARICOM Mission to Haiti," June 3, 2001, www.oas.org/Assembly2001/documentsE/AG264.htm.

116. "United States Urges Calm Among Political Parties in Haiti," *Washington File*, U.S. Department of State, March 12, 2001.

117. AG/RES. 1831 (XXXI-O/01), June 5, 2001.

118. Quoted in "New Elections OK'd in Haiti," *Miami Herald*, July 18, 2001.

119. "OAS Secretary-General Condemns Violence, Calls for the Continuation of Political Dialogue in Haiti," OAS, E-165/01, July 31, 2001.

120. Michael Norton, "Talks Collapse in Haiti," *Associated Press*, October 14, 2001.

121. Michael Norton, "Police Defeat Coup Attempt in Haiti," *Miami Herald*, December 18, 2001.

122. Nancy San Martin, "OAS Report Says Attack in Haiti on Palace Was Not Coup Attempt," *Miami Herald*, July 2, 2002; OAS, "Report of the Commission of Inquiry into the Events of December 17, 2001, in Haiti," OEA/Ser. G, CP/INF 4702/02, July 1, 2001.

123. Michael Norton, "Top OAS Official Leaves Haiti Empty-Handed after Vain Attempt to Jump Start Talks to Break Political Impasse," *Associated Press*, July 10, 2002.

124. Ibid.

125. OAS, CP/RES. 822 (1331/02), September 4, 2002.

126. Tim Johnson, "Haiti on Road to Ruin, OAS Leader Says," *Miami Herald*, November 7, 2002; Marika Lynch, "Critics Doubt Haiti Vow to Disarm Political Gangs," *Miami Herald*, November 19, 2002.

127. Johnson, "Haiti on Road to Ruin."

128. For example, "Florida Voyage Spurs Charges," *Washington Post*, October 31, 2002; Charles Babin and Jennifer Maloney, "Sharpton Urges Release of Haitians," *Miami Herald*, November 3, 2002; Andrew Elliott and Larry Lebowitz, "Haitians a Threat, INS Says," *Washington Post*, November 7, 2002; Raymond A. Joseph, "Aristide's Refugee Politics," *Wall Street Journal*, October 31, 2002; Sabra Ayres, "U.S. Policy Favors Cuban Refugees," *Washington Post*, November 14, 2002.

129. Adapted from Jeane Kirkpatrick, "A Pro-Democracy Policy," *Washington Post*, July 13, 1992.

130. Jeane Kirkpatrick, "The Theory and Practice of Clintonism." in Jeane Kirkpatrick, Jacqueline Tillman, et al., *Security and Insecurity*.

131. Anthony P. Maingot, *Current History* (February 1994).

4. THE BALKAN WARS: MAKING WAR TO KEEP THE PEACE

1. James Gow, Triumph of the Lack of Will: International Diplomacy and the Yugoslav War (New York: Columbia University Press, 1997), 328.

2. A good account of early Serb efforts at ethnic cleansing in Kosovo is the monograph "Dismissals and Ethnic Cleansing in Kosovo," published in October 1992 by the International Confederation of Free Trade Unions (ICFTU), Rue Montagnc aux-Herles-Potageres 37–41 B 1000, Brussels.

3. Janez Drnovsek, *Escape from Hell*, trans. Greg Davies, Robert Metcalfe, and Toby Robertson (Martigny, France: Editions Latour, 1996), 167. Drnovsek saw special irony in the Bush administration's reflexive effort to encourage the preservation of the state.

4. Ibid.

5. Ibid., 171.

6. James A. Baker III with Thomas M. DeFrank, *The Politics of Diplomacy Revolution, War and Peace, 1989–1992* (New York: G. P. Putnam's Sons, 1995), 636.

7. Ibid.

8. Laura Silber and Allan Little, *Yugoslavia: Death of a Nation* (New York: Penguin USA, 1997), 191. First published in the United States by TV Books in 1996.

9. *The President's Fiscal Year 1990 Budget Request for Eastern Europe*, March 7, 1990 (Washington, DC: U.S. Government Printing Office, 1990), 38–40.

10. Eagleburger testimony in *President's Fiscal Year 1991 Budget Request for Eastern Europe: Hearing Before the Committee on Foreign Affairs, House of Representatives*, March 7, 1990 (Washington, DC: U.S. Government Printing Office, 1990), 40.

11. Baker, The Politics of Diplomacy, 480.

12. Ibid., 481.

13. "The U.S. Commitment to Reform," U.S. Department of State, August 12, 1991, 596–98, 527.

14. Colin L. Powell, "Why Generals Get Nervous," *New York Times*, October 8, 1992.

15. Silber and Little, *Yugoslavia*, 201.

16. Noel Malcolm, *Bosnia: A Short History* (New York: New York University Press, 1996), 225.

17. Ibid., 226.

18. UN Security Council Resolution 713, September 25, 1991.

19. UN Security Council Resolution 1021, November 22, 1995.

20. Cyrus Vance's opinion stated that the arms embargo continued in force and applied to all areas of Yugoslavia and all Yugoslav republics. The Security Council then adopted Resolution 727 (January 8, 1992), which affirmed that the arms embargo should apply in accordance with the report.

21. Ibid., vii.

22. Misha Glenny, *The Fall of Yugoslavia*, 3rd ed., rev. (New York: Penguin Books, 1996), 136.

23. Roy Gutman, *A Witness to Genocide.* (New York: Macmillan Publishing Company, 1993). ix.

24. Ibid., xi.

25. Warren Zimmerman, *Origins of a Catastrophe: Yugoslavia and Its Destroyers* (New York: Random House and Times Books, 1999), 194.

26. Tadeusz Mazowiecki, special rapporteur for the UN Commission on Human Rights, wrote eighteen detailed accounts of the human rights situation in the former Yugoslavia between July 1992 and his resignation in August 1995, after the Srebrenica massacre. His reports provide detailed accounts of most of the violations of human rights, including the systematic rape of women and girls of all ages in many Bosnian towns and villages, and the number of pregnancies and abortions resulting from these rapes. His accounts were submitted to the Commission on Human Rights, a subdivision of the Economic and Social Council.

27. Former Yugoslavia, UNPROFOR, United Nations peace-keeping profile, last update August 31, 1996.

28. Samantha Power, Breakdown in the Balkans: A Chronicle of Events, January, 1989 to May, 1993 (Washington, D.C.: Carnegie Endowment for International Peace, 1993).

29. Silber and Little, *Yugoslavia*, 232–42.

30. Gutman, *A Witness to Genocide.*

31. John F. Burns, "Mitterand Flies into Sarajevo: Shells Temper 'Message of Hope,' " *New York Times*, June 29, 1992.

32. Chapter VII of the UN Charter authorizes the use of force. A Security Council resolution adopted under Chapter VII provides such authorization.

33. Seth Faison, "UN Chief Mired in Dispute With Security Council," *New York Times*, July 24, 1992, p. A3.

34. See Jeane Kirkpatrick's testimony before the Senate Foreign Relations Committee. *American Policy in Bosnia: Hearing Before the Subcommittee on European Affairs of the Committee on Foreign Relations, United States Senate*, February 18, 1993 (Washington: U.S. Government Printing Office, 1993), p. 16. Levy has written about this incident in his untranslated *Le Lys et la Cendre* (Lilies and Ashes) (Paris: Grasset, 1996).

35. Gutman, *A Witness to Genocide*, xvii.

36. Ibid.

37. William Drozdiak, "NATO Sets Plans to Enforce Ban on Serb Flights; U.S., Allies Are Said to Consider Other Balkan Military Options," *Washington Post*, December 16, 1992.

38. The Associated Press obtained a copy of the September 1 orders issued to Serbian police officers and authorities, governing permissible conduct in areas they occupied. The Center for Security Policy issued a Decision Brief (Document No.

92-D 104, September 1, 1992) that reproduced the orders, headlined "Everything but the Yellow Star: Milošević's Script for a Serbian Holocaust."

39. Clarence K. Streit, "32 Nations Gather to Help Refugees," *New York Times*, July 6, 1938.

40. Ibid.

41. Baker, *The Politics of* Diplomacy, 649.

42. Baker testimony in *The START Treaty: Hearings Before the Committee on Foreign Relations, United States Senate*, June 23, 1992 (Washington, DC: U.S. Government Printing Office, 1992), 6.

43. "Developments in Yugoslavia and Europe, August 1992," August 4, 1992 (Washington, DC: U.S. Government Printing Office, 1992), 1993–98.

44. The three Foreign Service officers who resigned in protest went to work on behalf of a U.S. policy they could support. They played a major role in the organization and management of the Balkan Action Group, a bipartisan group of people who shared their views.

45. The London Conference, August 26–28, 1992, under the leadership of Lord Peter Carrington. Department of State Dispatch, September, 1992, Supplement 7.

46. Lawrence Eagleburger, "Intervention at the London Conference on the Former Yugoslavia" (speech given on August 26, 1992, Department of State Dispatch, August 31, 1992.

47. On July 27, 1995, Mazowiecki resigned in protest against UN hypocrisy. (See *Washington Post*, July 28, 1995.)

48. Brendan Simms, *Unfinest Hour: Britain and the Destruction of Bosnia*. (London: Allen Lane, Penguin Press, 2001), 68, 115–119, 129.

49. Don Oberdorfer, "Bush, Major Seek Action on Serbia; UN Enforcement of Flight Ban Urged; START Progress Seen," *Washington Post*, December 22, 1992.

50. UN Security Council Resolutions.

51. John M. Goshko, "Bush Threatens Military Force if Serbs Attack Ethnic Albanians," *Washington Post*, December 29, 1992. See also Simms, *Unfinest Hour*, 56.

52. Simms, *Unfinest Hour*.

53. Ibid., 349.

54. McNeil/Lehrer NewsHour, August 29, 1992.

55. David Owen, *Balkan Odyssey* (New York: Harcourt Brace & Company, 1995). Dedicated to "Cyrus Vance and Thorvald Stoltenberg and all those from the UN and the EU who worked with us in the International Conference on the Former Yugoslavia." Owen reviews the history of differences between the United States and the United Kingdom concerning Yugoslavia, 8–13.

56. Peter Maass, "Warfare, Genocide Reemerge in Face of Bosnian Peace Plan," *Washington Post*, February 9, 1993.

57. Christine Spolar and Julia Preston, "Rivals Debate U.S. Role in Balkan Crisis, Call for Support of UN Plan," *Washington Post*, February 5, 1993.

58. Robert Mauthner, "Bosnia Negotiators Count Considerable Achievements: Robert Mauthner Talks to Lord Owen about the Peace Process," *Financial Times* (London), February 1, 1993..

59. Owen, *Balkan Odyssey*, 15. Owen had some curious views on how to improve the situation on the ground. He wrote to John Major about the urgency of action, saying of that Lord Carrington's efforts were likely to fail, and adding "Only the United Nations has both the experience and the authority to impose international order."

60. Elaine Sciolino, "U.S. Faces a Delicate Task in Intervening in Negotiations on Bosnia," *New York Times*, February 12, 1993.

61. Op-ed, "Clinton in Effect Supports Current Bosnia Peace Plan," *New York Times*, February 11, 1993.

62. Colin Powell, "U.S. Forces: Challenges Ahead." *Foreign Affairs* 71, (Winter 1992): 32–45.

63. "Rules of Engagement," Force Commander's Policy Directive Number 13 (FCPD 13) issued March 24, 1992, and revised July 19, 1993, by the UN Department of Peace-Keeping Operations.

64. Peter Maass, "Top Official Assassinated in Bosnia," *Washington Post*, January 9, 1993.

65. "Rules of Engagement" (FCPD 13). Rule 16(b) states, "When it becomes necessary to open fire, force is to be used only until the aggressor has stopped firing."

66. Don Oberdorfer, "U.S. Intervention in Bosnia Urged," *Washington Post*, January 14, 1993.

67. Ibid.

68. Peter Maass, "A Cry for Help from a Frozen Hell," *Washington Post*, January 13, 1993.

69. Ibid.

70. Ibid.

71. John F. Burns, "UN to Ask NATO to Airdrop Supplies for Bosnians," *New York Times*, January 12, 1993.

72. Ibid.

73. Jonathan C. Randal, "Preserving the Fruits of Ethnic Cleansing," *Washington Post*, February 11, 1993.

74. John M. Goshko, "U.S. Options Dwindling on Bosnia; Voicing Frustration, Clinton Vows to Press Serbs Via Belgrade," *Washington Post*, April 7, 1993.

75. Ibid.

76. John M. Goshko, "U.S. Takes More Active Balkans Role" *Washington Post*, February 10, 1993.

77. UN Security Council Resolution 819.

78. Tadeusz Mazowiecki, special rapporteur for the UN Commission on Human Rights, report of April 1996, 21.

79. In 1992, Boutros-Ghali openly complained to the Security Council in the context that the West was more interested in Bosnia than the catastrophe in Somalia.

80. Stephen Bates, "Thatcher Attacks Inaction Over Serbian Aggression," *The Guardian*, April 14, 1993; Patricia Wynn Davies and Annika Savill, "Thatcher Demands the Arming of Bosnia," *The Independent*, April 14, 1993.

81. Thatcher also expressed her views about Bosnia in an interview with Harry Smith on *CBS This Morning*, April 14, 1993.

82. Visits to the U.S. by Foreign Heads of States and Governments 1993, U.S. Department of State. President Alia Izetbegovic went to Washington to attend the Muslim-Croat federation agreement on March 17–19, 1994 and September 25, 1994 to the UN General Assembly in New York to discuss the Bosnia conflict with President Clinton.

83. Warren Christopher on *Nightline*, June 7, 1993.

84. Daniel Williams and John M. Goshko, "Reduced U.S. World Role Outlined but Soon Altered," *Washington Post*, May 26, 1993.

85. Ann Devroy, "Nailing Down a Bosnia Policy; Top Clinton Aides, After Finger-Pointing, Meet to Seek United Front," *Washington Post*, June 7, 1995.

86. John F. Burns, "Attacks on Bosnian Muslims Are Intensifying After Pause," *New York Times*, July 4, 1993.

87. Warren Christopher at Hearing of the House Foreign Affairs Committee. Subject: Foreign Aid Budget for FY94, May 19, 1993. Text available from LexisNexis Congressional. Bethesda, MD: Congressional Information Service.

88. Ibid., May 18, 1993.

89. Burns, "Attacks on Bosnian Muslims Are Intensifying."

90. Ibid.

91. Jonathan C. Randal, "UN, Croatia in Dispute," *Washington Post*, February 5, 1993.

92. Uli Schmetzer, "How West Let Croatia Sneak Arms," *Chicago Tribune*, August 20, 1995.

93. Annika Savill, "Mitterrand Sends in Troops to Outflank Kohl," *Independent* (London), June 24, 1993.

94. Chuck Sudetic, "Serbs Block UN Relief Convoy to Besieged Town in Bosnia," *New York Times*, March 16, 1994.

95. Daniel Williams, "U.S. Backs UN Plan for Bosnia Airstrikes," *Washington Post*, February 8, 1994.

96. Bill Gertz, "White House Retreats on Idea of UN Army," *Washington Times*, March 8, 1994.

97. Walter B. Slocombe, Testimony to the U.S. House of Representatives Com-

mittee on Foreign Affairs, Subcommittee on Europe and the Middle East (February 2, 1994), 8–9.

98. Report of the Secretary-General Pursuant to General Assembly Resolution 53/35 (1998), *The Fall of Srebrenica*, A/54/549, November 15, 1999, p. 6.

99. Ibid.

100. 1994: Market Massacre in Sarajevo, *BBC News*, February 5, 1994.

101. Joshua Muravchik, "Yellow Rose: The UN Bosnia Commander Should Quit," *New Republic*, December 5, 1994;Jennings, 3.

102. ABC, *Peter Jennings Reporting*, "The Peacekeepers: How the UN Failed in Bosnia," April 24, 1995.

103. UN Security Council Resolution 836 (1993).

104. Paul Lewis, "Conflict in the Balkans; UN About to Step Up Action on Serbia," *New York Times*, December 18, 1992, 14.

105. "Le conflit en Bosnie-Herzegovine: M. Boutros-Ghali autorise les responsables de la FORPRONU a recourir a l'aviation." *Le Monde*. January 31, 1994.

106. Rick Atkinson, "Sarajevo's Shell-Shattered Market Was 'Like a Butcher Shop,' " *Washington Post*, February 8, 1994.

107. Zlatko Dizdarevic, *Sarajevo: A War Journal* (New York: Fromm International, 1993), 174.

108. Oaul Lewis, "Terror in Sarajevo: UN Seeks Power for Bosnia Strikes," *New York Times*, February 7, 1994.

109. For a good account of this incident and comments, see Michael R. Gordon, "NATO Craft Down Four Serb Warplanes Attacking Bosnia," *New York Times*, March 1, 1994.

110. Ibid.

111. The description of the replacement was provided to me by a top UNPROFOR commander. The description of the treatment of UN forces in the Banja Luka area is from John Pomfret, "UN Wants to Withdraw from Bijac: Forced Withdrawal Would Be First of War," *Washington Post*, December 7, 1994. Pomfret described Serb mistreatment of UN forces after two NATO strikes in late November, including locking Canadian troops in jail cells and refusing requests for food, resupply, and medical treatment. Complaints to the Bosnian Serbs about such treatment were met with the reply that those troops would be better treated if UN air strikes on Serb bases ended.

112. For an account of Scott O'Grady's ordeal, see Daniel Williams, " 'I'm Ready to Get the Hell Out of Here,' " *Washington Post*, June 9, 1995.

113. Jim Hoagland, "Bosnia: The UN's Moral Rot," *Washington Post*, June 9, 1995.

114. Jeane J. Kirkpatrick, "Round and Round They Go, But Still No Help for Bosnia," *New York Post*, Worldview, February 14, 1994.

115. Ashton P. Carter, William J. Perry, and John D. Steinbruner, *A New Concept of Cooperative Security* (Washington, DC: Brookings, 1992), 25.

116. Madeleine Albright, "Enforcing International Law," address to Bureau of International Organization Affairs, Philadelphia Bar Association, June 15, 1995.

117. "Peace Operations: Heavy Use of Key Capabilities May Affect Response to Regional Conflicts" (Washington, DC: General Accounting Office, 1995).

118. George J. Church, "Pity the Peacekeepers: The Serbs Respond to Nato Bombings by Chaining Hostages Near Potential Targets and the Stalemate Resume," *Time Magazine*, June 5, 1995.

119. "Dutch Government Quits over Srebrenica," *BBC News*, April 16, 2002.

120. John Tagliabue. "Former Bosnian Serb Officer Admits Guilt in "95 Massacre." *New York Times*, May 7, 2003; Marlise Simons, "Prosecutors Say Document Links Milošević to Genocide," *New York Times*, June 20, 2003.

121. This account is based on that of Roy Gutman in *Newsday*, May 26, 1994.

122. Michael Dobbs and Christine Spolar, "Anybody Who Moved or Screamed Was Killed; Thousands Massacred on Bosnia Trek in July," *Washington Post*, October 26, 1995.

123. Report of the Secretary-General, *The Fall of Srebrenica*, paragraphs 235–238.

124. Richard Holbrooke, *To End a War* (New York: Random House, 1998), 65.

125. Holbrooke, *To End a War*, 65.

126. Ibid.

127. Michael Rose, *Fighting for Peace* (London: Harvill Press, 1998), 143.

128. On the tension between the United Kingdom and the United States, see Simms, *Unfinest Hour*.

129. UN High Commissioner for Refugees, *The State of the World's Refugees 1997: A Humanitarian Agenda*, January 1997, http://www.reliefweb.int/rw/lib.nsf/db900SID/LGEL-5SAH4M?Open Document.

130. Admiral Leighton, U.S. Department of Defense, Defense Department Briefing, February 1, 1996. See also *Federal News Service*, February 2, 1996.

131. For an informed discussion of the unwritten commitments, see Paul R. Williams, "Promise Them Anything," *Weekly Standard*, December 18, 1995.

132. Sheri Fink, "Truth in the Balkans," *Washington Post*, November 12, 2003.

133. Claire Trean, "Massacres de Srebrenica: les députés concluent à un échec de la France; Dans son rapport rendu public, Jeudi 29 Novembre, la mission d'information parlementaire estime que Paris partage avec ses partenaires la responsabilité de l'abandon, en juillet 1995, de l'enclave musulmane de Bosnie orientale où sept mille personnes furent tuées par les forces serbes," *Le Monde*, November 30, 2001.

134. There are many accounts of the horrors at Srebrenica. Of special note are David Rohde, *Endgame: The Betrayal and Fall of Srebrenica* (New York: Farrar, Strauss Giroux, 1997); Wayne Bert, *The Reluctant Superpower: United States Policy in*

Bosnia, 1991–1995 (New York, St. Martin's Press, 1997); and James Gow, *Triumph of the Lack of Will: International Diplomacy and the Yugoslav War* (New York: Columbia University Press, 1997).

135. William Shawcross, *Deliver Us from Evil: Peacekeepers, Warlords, and a World of Endless Conflict* (New York: Simon & Schuster, 2000), 154.

136. Jeane J. Kirkpatrick, "A Man of Courage Leaves in Disgust," *Washington Post,* August 14, 1995.

137. William Shawcross, *Deliver Us from Evil: Peacekeepers, Warlords and the World of Endless Conflicts* (New York: Simon & Schuster, 2000.

5. KOSOVO

1. Warren Zimmerman, "The Last Ambassador, A Memoir of the Collapse of Yugoslavia" *Foreign Affairs* 74, no. 2 (March/April 1995): 3.

2. Alain Pellet, "The Opinion of the Badinter Arbitration Committee: A Second Breath for the Self-Determination of Peoples," *European Journal of International Law* 13, no. 1 (2002), http://www.ejil.org/journal/vol3/no1/art12html.

3. "Communiqué in the Yugoslav Crisis," U.S. Department of State Dispatch, August 1, 1992.

4. Kosovo Report, 58.

5. John M. Goshko, "Bush Threatens Military Force if Serbs Attack Ethnic Albanians," *Washington Post,* December 29, 1992.

6. "Conflict in the Balkans; Christopher's Remarks on Balkans: 'Crucial Test,' " Reuters, in the *New York Times,* February 11, 1993.

7. Pierre Sane, secretary-general of Amnesty International, wrote, "It can be argued that the chronic neglect of the warning in these reports and the almost complete absence of redress for all Kosovo's people has been one of the chief catalysts for the current conflict." *Kosovo: A Decade of Unheeded Warnings, Amnesty International's Concerns in Kosovo,* vol. 1, *May 1989 to March 1999,* Al Index: EUR 70/39/ 1999 and EUR 70/40/99, May 1999.

8. Noel Malcolm, *Kosovo: A Short History* (London: Papermac, 1998), 350.

9. Sabrina Ramet, *Balkan Babel: The Disintegration of Yugoslavia from the Death of Tito to the War for Kosovo,* 3rd ed. (Boulder, CO: Westview Press, 1999), 51.

10. Not until 1996 did the Kosovo Liberation Army (KLA), which had been based in Switzerland, claim responsibility for any violent attacks in Kosovo. However, according to the Independent International Commission on Kosovo, the KLA's first violent action was the killing of a Serb policeman in 1995. See the commissions report, *Kosovo Report: Conflict, International Response, Lessons Learned* (Oxford, England: Oxford University Press, 2000), 51.

11. Ramet, *Balkan Babel,* 309.

12. Ibid.

13. UN Security Council Resolution 1146, (S/RES1146, January 28, 1998).

14. Amnesty International *Report on Yugoslavia, 1998*, covering the period from January to June 1998, http://web.amnesty.org/library/Index/ENGEURO 10021998?open&of=ENG-EST.

15. Ibid.

16. Tyler Marshall, U.S. Tells Serbia Kosovo Violence Risks Reprisals, *Los Angeles Times*, Part A; page 1; March 8, 1998. *Federal Republic of Yugoslavia: Amnesty International's Current Recommendations Concerning the Crisis in Kosovo Province.*

17. Secretary of State Madeleine K. Albright and Italian Foreign Minister Lamberto Dini. Press Briefing at the Ministry of Foreign Affairs, Rome, Italy. Released by the Office of the Spokesman, U.S. Department of State, March 7, 1998.

18. The U.S. troops stationed in Macedonia were part of the UN Preventive Deployment Force (UNPREDEP).

19. Wesley K. Clark, *Waging Modern War* (New York: Public Affairs, 2001), 108.

20. Madeleine Albright, *Madame Secretary* (New York: Miramax, 2003), 406.

21. Ibid., 406.

22. Clark, *Waging Modern War*, 109.

23. Secretary of State Madeleine Albright, Remarks on Kosovo at the Contact Group Ministerial, London, UK, March 10, 1998.

24. The Contact Group, consisting of Germany, France, Italy, Russia, the U.K.,and the U.S., was initially established in 1992 as a diplomatic representative of the international community to deal with events in Bosnia.

25. Steven Erlanger, "Yugoslavs Try to Outwit Albright Over Sanctions," *New York Times*, March 23, 1998.

26. News Summary, Associated Press, in the *New York Times*, April 1, 1998.

27. Ibid. Chinese representative Shed Guofeng is being quoted.

28. "Report on the Visit of Ambassador Sheffer to the Border Between the Former Republic of Macedonia and Kosovo" (Washington, DC: U.S. Department of State, April 7, 1999).

29. Milošević held a national referendum on whether to accept international mediation in the Kosovo crisis. Supposedly, the referendum had a 74 percent turnout and 95 percent of Serb voters rejected mediation. However, there were reports of fraud, and many ethnic Albanians boycotted the vote. See "Yugoslavia Warns of War," Associated Press, April 24, 1998.

30. Thomas W. Lippman, "U.S. Pressures Allies on Yugoslavia," *New York Times*, April 25, 1998.

31. The meeting took place on April 29, 1998. "Contact Group Imposes Embargo on Yugoslavia over Kosovo," Agence France Presse, May 9, 1998.

32. Ibid.

33. Steven Erlanger, "Clinton Meets Delegation from Kosovo Seeking Talks," *New York Times*, May 30, 1998.

34. Independent International Commission on Kosovo, *Kosovo Report*, 72.

35. Clark, *Waging Modern War*, 112.

36. William Drozdiak and John F. Harris, "U.S. Backs Europe on Serbia; EU Imposes Sanctions as NATO Options," *Washington Post*, May 9, 1998.

37. "Blair Spells Out Tough Line on Kosovo," Agence France Presse, June 6, 1998.

38. Ibid.

39. World News Digest, *Facts on File*, June 4, 1998.

40. Capt. Michael Doubleday, a Pentagon spokesman, said, "[E]very time there is a significant troop development, we are very concerned about questions like the desired end and what is the command and control structure"; Elizabeth Becker, "Pentagon Sees Risk in Going into Kosovo," *New York Times*, February 11, 1999.

41. Ibid.

42. William Drozdiak, "NATO Plans Exercises to Pressure Milošević," *Washington Post*, June 12, 1998.

43. David Hoffman, "Milošević to Meet Kosovo Moderate; Serb Visits Yeltsin, Yields on Talks," *Washington Post*, June 17, 1998.

44. Clark, *Waging Modern War*, 120.

45. Craig Whitney, "France Urges Allies to Define Plan for Autonomy for Kosovo," *New York Times*, June 25, 1998.

46. Craig Whitney, "Western Officials Say Accord on Kosovo Seems Uncertain," *New York Times*, July 1, 1998.

47. John M. Goshko, "Security Council Urges Negotiations over Kosovo," *Washington Post*, August 12, 1998.

48. This complaint was made by the German government. See Roger Cohen, "Kosovo Crisis Strains Relations Between the U.S. and Europe," *New York Times*, November 10, 1998. The Cohen testimony can be found in *U.S. Policy and NATO Military Operations in Kosovo: Hearings Before the Committee on Armed Services, United States Senate*, July 20, 1999 (Washington: U.S. Govenment Printing Office, 2000), pp. 241–251.

49. Ibid.

50. UN Security Council Resolution 1199 (S/RES/1199, September 23, 1998).

51. Barbara Crossette, "Kosovo Rebels' Political Chief Calls for UN Representation," *New York Times*, September 18, 1999, p. A3.

52. Human Rights Watch, "A Week of Terror in Drenica: Humanitarian Law Violations in Kosovo, 1999," http://www.hrw.org/reports/1999/kosovo.

53. UN Security Council Resolution 1203 (S/RES/1203, October 24, 1998).

54. UN Security Council Resolution 1207 (S/RES/1207, November 17, 1998).

55. Jeffrey Smith and William Drozdiak, "Serbs' Offensive Was Meticulously Planned," *Washington Post*, April 11, 1999.

56. Ibid.

57. Independent International Commission on Kosovo, *Kosovo Report*, 80.

58. Smith and Drozdaik, "Serbs' Offensive."

59. Thomas W. Lippmann, "Deadline Set for Kosovo Accord; Threat of NATO Forces Underlies Feb. 19 Limit on Peace Talks," *Washington Post*, January 30, 1999.

60. Editorial, "The Logic of Kosovo," *New York Times*, May 2, 1999.

61. Human Rights Watch, "Under Orders: War Crimes in Kosovo: March to June 1999: An Overview," http://www.hrw.org/reports/2001/kosovo.

62. Human Rights Watch, "War Crimes in Kosovo: Executive Summary," http://www.hrw.org/reports/2001/kosovo/undword.htm.

63. *Erasing History: Ethnic Cleansing in Kosovo* (Washington, DC: U.S. Department of State, May 1999).

64. Independent International Commission on Kosovo, *Kosovo Report*, 90.

65. Smith and Drozdiak, "Serbs' Offensive."

66. "Serbs Stage Rock Concert Against NATO Strikes," Deutsche Presse-Agentur, April 4, 1999.

67. Hilary Mackenzie and Mike Trickey, "Serb Soldiers Deserting: U.S.: Antiwar Demonstrations Break Out in Third Yugoslavian City," *Montreal Gazette*, May 20, 1999.

68. Eve-Ann Prentice and Michael Binyon, "Anti-War Mob Lynches Serbian Mayor," *The Times* (London) May 20, 1999.

69. Ibid.

70. Jane Perlez, "Crisis in the Balkans: Overview; NATO Approves Naval Embargo Oil Going to Serbs," *New York Times*, April 24, 1999.

71. Daniel Williams, "Serbia Yields to NATO Terms," *Washington Post*, June 4, 1999.

72. Blaine Harden, "Crisis in the Balkans: Doing the Deal," *New York Times*, June 6, 1999.

73. Ibid.

74. "Military Technical Agreement Between NATO and Yugoslavia," Reuters, June 9, 1999.

75. Michael Dobbs, "Milošević Claims Victory, Lauds 'Best Army in the World' " *Washington Post*, June 11, 1999.

76. John Ward Anderson, "Refugees Are Willing to Delay Trip Home; Many Waiting for NATO's Assurance of Safety," *Washington Post*, June 11, 1999.

77. U.S. participation in Operation Allied Force, as a percentage of NATO total:

Sorties 60 %

Support (all types) 71 %

Electronic warfare 90 %

Strike 53 %
Intelligence/reconnaissance 90 %

Ordnance
Precision guided munitions 80 %
Cruise missiles 95 %

See statement of Secretary of defense William S. Cohen to the Senate Armed Services Committee, July 20, 1999. http://armed-services.senate.gov/statemnt/1999/990720 wc.pdf.

78. *Ethnic Cleansing in Kosovo: An Accounting*, Executive Summary (Washington, DC: U.S. State Department, December 1999).

79. Ibid.

80. Ivo Daalder, "U.S. diplomacy Before the Kosovo War," Senate Committee on Foreign Relations, Subcommittee on European Affairs, September 28, 1999.

81. One day after Milošević was handed over to the International War Crimes Tribunal, "[i]nternational governments and organizations meeting in Brussels . . . pledged almost $1.3bn (£924m) in aid and loans to help rebuild the Yugoslav economy" (*BBC News*, "Yugoslavia Wins $1.3bn Aid Pledges," June 29, 2001).

82. Serbia and Montenegro agreed to remain in the union for a period of three years, after which each would have the right to separate (International News, "New Serbia and Montenegro to Last at Least Three Years," Agence France Presse, March 14, 2002).

83. The Accord on Principles in Relations Between Serbia and Montenegro was signed on March 14, 2002. See also Perparim Isufi, "Kosovars Jittery over Final Status," Institute for War and Peace Reporting, November 14, 2002.

84. Ibid.

85. In Operation Joint Guardian, the United Kingdom contributed twelve thousand troops with headquarters in Pristina. Germany deployed eighty-five hundred troops in the southwest part of Kosovo, with the headquarters in Prizren. France deployed seven thousand troops in the northwestern sector, with headquarters in Kosovska Mitrovica. Italy contributed two thousand troops in the western part of Kosovo, with headquarters in Pec. The United States provided a force of approximately seven thousand in the southeastern part of Kosovo, as of June 11, 1999. http://www.globalsecurity.org/military/ops/joint_guardian.htm.

86. The Russians tried to fly in additional forces through Romanian, Hungarian, and Bulgarian airspace, and they attempted to negotiate the status of their forces in Kosovo after their presence was already established, http://www.kosovo.mod.uk/jointguardian.htm. See also, MOD briefing; Russian Foreign Minister Ivanov

claimed it was a mistake when Russian troops entered Pristina, June 12, 1999, http://www.kosovo.mod.uk/brief120699.htm.

87. Initial reports suggested Russia would contribute nine hundred troops. http://www.kosovo.mod.uk/brief160699.htm.

88. Judy Dempsey and Irena Guzelova, "Serb Attitude to Kosovo Election Seen as Test of Progress," *Financial Times* (London), November 17, 2001.

89. Ibid.

90. "The Kosovo Election Process."

91. Barbara Crossette, "Kosovo Rebels' Political Chief Calls for UN Representation," *New York Times*, Section A; page 3, September 18, 1999.

92. Ivo H. Daalder and Michael E. O'Hanlon, *Winning Ugly: NATO's War to Save Kosovo*, (Washington, DC: Brookings, 2001), 91.

93. Ibid.

6. CONCLUSION: AFGHANISTAN AND IRAQ

1. John Stuart Mill, *Dissertations and Discussions: Political, Philosophical and Historical*, vol. 3 (London, 1875), 176. Reprinted from Mill, "A few Words on Non-Intervention," *Fraser's Magazine* (December 1859).

2. President George W. Bush, *Presidential Address to the Nation*, October 7, 2001.

3. Kofi Annan, Transcript of Press Conference by Secretary-General Kofi Annan at Palais Des Nations, Geneva, *Press Release SG/SM?8011/Rev.1*, November 1, 2001.

4. President Hamid Karzai, News Conference: President Bush and President Karzai, *CQ Transcripts Wire*, September 26, 2006.

5. In 2006, the UN Human Rights Commission was dismantled and renamed the UN Human Rights Council.

6. Allan Gerson, "Why We're There, *Washington Times*, April 30, 2003.

7. Condoleezza Rice, Condoleezza Rice Discusses President's National Security Strategy, *Office of the Press Secretary*, Waldorf Astoria Hotel, New York, New York, October 1, 2002.

8. Ibid.

9. Kofi Annan, interview by Owen Bennett-Jones, *BBC News*, September 16, 2004.

10. Colin Powell (Address to the United Nations, February 3, 2003).

11. Ibid.

12. Ibid.

13. Ibid.

14. Kofi Annan, UN Press Briefing by the Office of the Spokesman for the Secretary-General, March 10, 2003.

15. Fred Eckhard, Highlights of the Noon Briefing, by the Office of the Spokesman for the Secretary General, September 16, 2004.

16. Kofi Annan, The Secretary-General Address to the General Assembly, September 23, 2003.

17. United Nations Charter, Chapter XV, Article 99.

18. Fred Eckhard, Highlights of the Noon Briefing by Fred Eckhard, Spokesman for the Secretary-General of the United Nations, September 16, 2004.

19. John Danforth, Remarks by Ambassador John C. Danforth, US Representative to the United Nations, on Sudan and Iraq, at the Security Council Stakeout, September 16, 2004.

20. Phillip Gourevitch, "The Optimist," *The New Yorker*, February 24, 2003.

21. Humanitarian and security situations in western Sudan reach new lows, UN agency says, *UN News Service*, December 5, 2003.

22. Kofi Annan, Statement from the Secretary-General, July 25, 2006.

23. Colum Lynch, "Oil-for-Food Panel Rebukes Annan, Cites Corruption," *Washington Post*, September 8, 2005.

24. Preamble to the Charter of the United Nations, *United Nations Charter*, San Francisco, 1945.

25. Ban Ki-Mun, Transcript of Press Conference by Secretary-General-Designate Ban Ki-Mun, October 14, 2006.

26. Kofi Annan, "Annan refutes notion of 'clash of civilizations,' points to youth as key to end mistrust," *UN News Service*, November 13, 2006.

27. U.S. State Department, March 2003.

28. John Barry and Michael Hirsh, "A Warrior Lays Down His Arms," *Newsweek*, November 20, 2006.

29. Jeane J. Kirkpatrick and Allan Gerson, "Right v. Might: International Law and the Use of Force," *Council of Foreign Relations*, 1989, 20–24.

ACKNOWLEDGMENTS

I would like to thank my husband, Evron, who was my intellectual companion, my best friend, my greatest counsel—and my first literary agent. To our children, Douglas, John, and Stuart, I thank you for your inspiration and for the joy you have given me in my life's most important work, as your mother.

Kate Campaigne has been a tireless support and good friend throughout my completion of this work. With tenacity, Teresa Hartnett has been my trusted ambassador to book publishing, and my friend. Cal Morgan, my indefatigable editor, has energized my work with his appreciation, sharp eye, and keen mind: I thank him for his commitment.

I especially thank my students for the privilege of teaching them and for the pleasure of learning from them. I encourage you to champion freedom by treasuring democracy. Never waver in your commitment to speak truth.

INDEX